Collective Dynamics of Nonlinear and Disordered Systems

G. Radons · W. Just · P. Häussler (*Eds.*)

Collective Dynamics of Nonlinear and Disordered Systems

With 153 Figures

Springer

Günter Radons
Institut für Physik, Theoretische Physik I
Technische Universität Chemnitz
Reichenhainer Str. 70
09126 Chemnitz, Germany
e-mail: radons@physik.tu-chemnitz.de

Wolfram Just
School of Mathematical Science
Queen Mary, University of London
Mile End Road
London E1 4NS, UK
e-mail: w.just@qmul.ac.uk

Peter Häussler
Institut für Physik
Technische Universität Chemnitz
Reichenhainer Str. 70
09126 Chemnitz, Germany
e-mail: haeussler@physik.tu-chemnitz.de

ISBN 978-3-642-43820-2 Springer Berlin Heidelberg New York

Springer is a part of Springer Science+Business Media

springeronline.com

Softcover re-print of the Hardcover 1st edition 2005

Typesetting: Authors, editors and DA-TEX, Leipzig
Production: LE-TEX Jelonek, Schmidt & Vöckler GbR, Leipzig
Cover production: *design & production, Heidelberg*

Printed on acid-free paper 55/3144/YL - 5 4 3 2 1 0

Preface

Dynamical aspects of nonlinear systems, and in particular dynamical behaviour far from equilibrium, is one of the few areas in theoretical physics which has been very active during the last decade. Whereas the early approaches in Nonlinear Dynamics have dealt with systems of few degrees of freedom the concepts developed in this context have been, meanwhile, applied fruitfully for the investigation of collective behaviour in many particle physics. As a very prominent and important physical example we consider the glass transition and low temperature properties in general. Here those mechanisms which generate complex aperiodic or disordered structures and the interaction between static disorder and dynamical processes are at the centre of interest. In that context the concepts of Nonlinear Dynamics have proven to be of considerable relevance in condensed matter physics.

So far such aspects have not been covered in a consistent way by the existing literature, in particular as far as graduate textbooks are concerned. With the present monograph we intend to fill such a gap. Each chapter is written by an expert in his field. But in contrast to the original literature and conference proceedings the material is presented in such a way that it is accessible for graduate students. With the present monograph we intend that the state of the current research at the borderline between Nonlinear Dynamics and disordered systems proliferates among different disciplines of natural science in order to emphasise the general mechanism for the generation of complex structures.

From the general point of view the whole subject can be grouped in three parts which show, however, numerous links between the different topics. All aspects are illustrated by experimental applications, although our special focus is on theoretical concepts.

I Pattern Formation and Growth Phenomena
Growth processes are typically related with pattern formation, instabilities, and complex dependence on external parameters. These phenomena are endemic for nonequilibirum pattern formation, e.g. the kinetics of ordering processes, the formation of amorphous or liquid phases, crystal growth, electrochemical deposition, erosion, or general front dynamics and surface growth processes. From the theoretical point of view reaction–diffusion systems, nonlinear transport equations, or stochastic field the-

ories are prominent models describing the afore mentioned physical phenomena. On a mesoscopic scale these models permit the identification of the elementary building blocks of complex patterns as well as the dynamical interaction of these entities. The resulting dynamical morphologies of surfaces and fronts cover ordered, fractal, and disordered structures. General order and selection principles are of particular interest, as they are condensed e.g. in group theoretical considerations and universal scaling laws.

II Dynamics of Disordered Systems
The generation of disordered structures as it can be found in physics e.g. in supercooled liquids and glasses constitutes a quite prominent example for pattern formation. The complex dynamics of these systems is governed by a whole spectrum of different time scales as well as by the occurrence of spatially confined ordered structures. Whereas the first feature may be responsible for self generated structural disorder as visible e.g. in the glass transition, the second feature can generate spherically periodic local order. From the theoretical perspective such features can be characterised by quantities borrowed from the physics of static disorder as well as by quantifiers developed in the context of Nonlinear Dynamics. From such a point of view the properties of glasses and other aperiodic systems constitute a bridge between complex dynamical behaviour and the dynamics of disordered systems.

III Complex and Chaotic Behaviour
For the investigation of complex chaotic behaviour the notions and concepts of Nonlinear Dynamics have been adapted recently to the peculiarities of systems with a large number of degrees of freedom, for instance in terms of Lyapunov and dimension spectra. Such approaches can be applied for the analysis of disordered systems. In particular, one gains a deeper understanding of the ergodic properties of the underlying dynamics. In that sense these concepts are the basis of numerous theoretical concepts. In particular, the characterisation of different time scales, of frozen degrees of freedom, and of aging processes becomes feasible. Thus such concepts are the centre of interest of a deeper microscopic understanding of equilibrium and nonequilibrium thermodynamics.

The present monograph summaries the scientific content of a WE Heraeus summer school held at Chemnitz University of Technology in August 2002. We owe the WE Heraeus foundation our deepest gratitude, as without the continuous and intense support from the foundation the whole project would not have been feasible. Whereas the scientific merits are due to our colleagues who have written the individual chapters, any remaining deficiencies must be attributed to the editors.

Chemnitz / London,
January 2004

Günter Radons
Peter Häussler
Wolfram Just

Contents

Part I

Pattern Formation and Growth Phenomena

Introduction to Part I

Pattern formation is a ubiquitous non-equilibrium phenomenon which on one hand is simply fascinating, but on the other hand has many practical implications. It is one of the most prominent nonlinear collective dynamical effects, and therefore entails a long scientific tradition. Macroscopic patterns arising in fluids as a consequence of hydrodynamic instabilities are known for at least one century and are connected with prominent names such as Rayleigh-Benard or Taylor-Couette. In the meantime the importance of the pattern formation concept has been recognized in very diverse fields such as biology, laser physics, chemistry, and many others.

Quite recently pattern formation on a much smaller scale attracted the interest of the scientific community. Due to the development of advanced microscopy techniques such as the scanning tunnelling microscope, patterns that arise on a submicron scale during the epitaxial growth of solid surfaces, became observable. Their origin and understanding is the topic of the first contribution by Joachim Krug. He presents a phenomenological continuum theory, which provides a unified framework for various pattern forming mechanisms for surface growth. They include the Villain instability, mound and meandering instabilities, or patterns arising due to electromigration. Replacing the adsorption of atoms or molecules by the "deposition of vacancies" leads from growth to erosion phenomena which are caused e.g. by the bombardment of surfaces with high energy ions, and which can also be explained by such theories. The theoretical aspects include the wavelength selection problem and nonlinear evolution scenarios such as coarsening and chaotic bubbling.

The second contribution by Eckehard Schöll also deals with a very recent development, the nonlinear dynamics and pattern forming mechanisms in nanostructured semiconductors. The self-organized formation of current filaments, stationary or moving fronts, and other complex spatio-temporal or chaotic behavior are elucidated. Thereby the role of a negative differential conductivity is emphasized. Totally new aspects arise with the possibility of applying time-delayed feedback control, which is introduced and explored in this article. This opens the possibility to stabilize patterns and motions which are usually unstable and therefore not observable. The consequences of such an approach are discussed for resonant tunnelling diodes and for semiconductor superlattices.

The previous two articles dealt with pattern formation in selected applications of recent interest. In both cases insights are gained mostly within the framework of continuum theories. The latter constitute the central topic also in the contribution of Rudolf Friedrich. Now, however, no special application is aimed at, but consequences of symmetries of the system at hand are elaborated. The group theoretical results obtained can be applied to macroscopic or microscopic phenomena as long as a continuum description holds. It provides a characterization of non-equilibrium phase transitions or bifurcations through the classification of the order parameter equations which determine the patterns being formed. Finally the role of Goldstone modes for higher instabilities is explained and the experimentally relevant implications of fluctuations for dissipative solitons, such as filaments, are clarified.

The last article in this part by Peter Häussler deals with structure formation on a microscopic level. It is concerned with self-organization principles which aim at explaining the short-range ordered structure in disordered solids such as alloys. The paper advocates spherical periodic order and resonant behavior between electronic and structural degrees of freedom as the organizing principle, which can explain many properties in different families of alloys. Consequences e.g. for the density of states, transport properties, and dynamic excitations are discussed. In many respects this contribution leads over to Part II, which is exclusively devoted to the dynamical properties of disordered systems.

Kinetic Pattern Formation at Solid Surfaces

Joachim Krug*

Fachbereich Physik, Universität Duisburg-Essen,
45117 Essen, Germany
krug@thp.uni-koeln.de

Die Form ist überall dieselbe, und muss es auch, denn ein *Geist ist, der sie denkt.*[2]
Johann Wilhelm Ritter

1 Introduction

The emergence of regular spatial patterns from inanimate natural processes has been a source of fascination and wonder since the beginnings of scientific exploration. A quantitative understanding of the mechanisms underlying pattern formation is a rather recent achievement, which has been possible through a strong concerted research effort in nonlinear physics and mathematics during the past two or three decades. This, by now classic, body of work [1] has been mostly concerned with pattern formation phenomena in macroscopic systems, such as hydrodynamic instabilities and chemical oscillations. More recently, patterns in granular materials such as sand ripples have also been addressed from related points of view [2, 3].

The focus of the present contribution is on submicron-scale patterns like mounds, ripples and step bunches, which form on solid surfaces under nonequilibrium conditions. This subject differs from macroscopic pattern formation in that (i) the visualisation of the patterns of interest requires advanced microscopy techniques, notably scanning probe microscopes, and (ii) the formation of the patterns often relies on specific atomic processes, which must be analyzed in detail to reach quantitative agreement between theory and experiment [4]. Nevertheless a phenomenological continuum approach to pattern formation at solid surfaces has been developed, to some extent in analogy to the established macroscopic theories of pattern formation. Such an approach has proven to be useful, because it allows for a compact, unified description of a variety of mechanisms, as well as for the efficient analytical or numerical modeling of global aspects of the surface morphology.

---

* Present address: Institut für Theoretische Physik, Universität zu Köln, Zülpicher Strasse 77, 50937 Köln, Germany

[2] *The shape is the same everywhere, as it has to be, because it is conceived by* one *mind.*

The purpose of this article is to provide an elementary introduction to the continuum approach. The article is based on lectures on pattern formation during epitaxial growth and erosion, which were delivered jointly with Thomas Michely. In addition to the topics covered in these lectures, phenomena related to surface electromigration and steering (the deflection of the trajectories of depositing atoms due to the attraction by the growing film surface) will be included here. The lectures also addressed atomistic aspects of growth and erosion, and the detailed relationship between the atomistic and continuum viewpoints. As extensive recent reviews on these issues are available elsewhere [4, 5, 6, 7], they will be treated only briefly here. We further emphasize that we are concerned only with *kinetically* (rather than energetically) driven morphological instabilities. This excludes the formation of hill-and-valley structures at thermodynamically unstable surfaces, as well as the broad class of patterns which form in heteroepitaxial growth due to the strain caused by the lattice mismatch between the substrate and the growing film [6, 7, 8].

The basic instability mechanisms are described in the next section. Section 3 discusses the emergence of the characteristic length scale of the patterns in the early time regime, while in Sect. 4 the nonlinear, late time evolution is discussed. Experimental examples are presented as appropriate to illustrate the theoretical concepts.

2 Instability Mechanisms

The basis of the continuum theory is the description of the surface morphology in terms of the height function h, which gives the surface position $z = h(\boldsymbol{r}, t)$ above a point $\boldsymbol{r} = (x, y)$ in the substrate plane at time t. The height function satisfies an evolution equation of the general form

$$\frac{\partial h}{\partial t} = \mathcal{F}(\nabla h, \nabla^2 h, ...) \; , \tag{1}$$

where the function $\mathcal{F}$ must be chosen appropriately to encode the processes that contribute to the morphological evolution. It depends on the derivatives of h but is independent of the height itself, because the evolution is invariant under constant shifts of the reference plane, $h \to h + C$. Moreover, the function $\mathcal{F}$ is constructed such that the evolution equation always admits solutions corresponding to a flat, featureless surface, either parallel to the reference plane, or, more generally, inclined relative to the reference plane with a tilt vector $\boldsymbol{m}$. In growth and erosion processes the mean height of the flat surface grows or recedes with some velocity v. The general flat solution of (1) is therefore of the form

$$h_0(\boldsymbol{r}, t) = \boldsymbol{m} \cdot \boldsymbol{r} + vt \; . \tag{2}$$

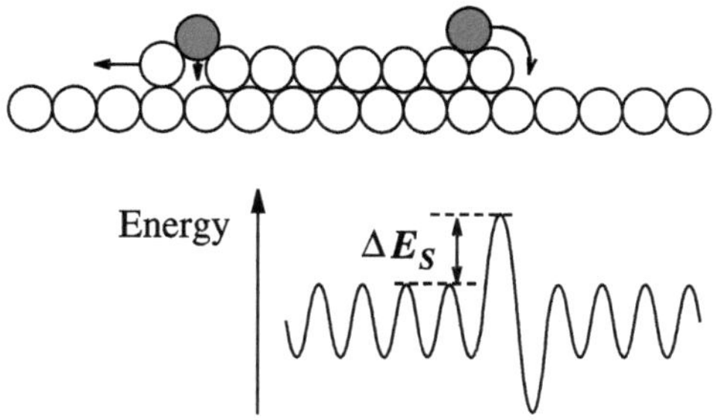

Fig. 1. The upper part of the figure shows the descent of adatoms (*shaded*) from an island. The adatom on the right descends through hopping, while the adatom on the left descends by exchanging its position with a step atom. The lower part of the figure illustrates the potential energy landscape experienced by the descending adatom on the right; the potential minima are aligned with the corresponding adsorption sites on the surface. The descent into the favorable position at the step edge is hindered by an additional step edge energy barrier ΔE_S. In the case of hopping, the existence of an additional barrier can be attributed to the poor coordination of the adatom in the transition state. The exchange process on the left allows the descending adatom to maintain the coordination to its neighbors, but it requires to move two atoms rather than just one

The emergence of a pattern on the surface is signalled by the instability of the flat solution (2). In the following subsections, the leading order contributions to $\mathcal{F}$ corresponding to various nonequilibrium processes will be derived, and the conditions for morphological instability will be explored.

2.1 Epitaxial Growth

In epitaxial growth from an atomic or molecular beam, atoms arrive at the surface at a rate set by the deposition flux F. They diffuse as *adatoms* over the atomically flat terraces and are incorporated into the crystal at steps of monoatomic height. As each step separates an upper and a lower atomic terrace, the incorporation of atoms can occur either from above or from below. Using field ion microscopy, Ehrlich and Hudda discovered in 1966 that diffusing adatoms tend to be reflected when they approach a descending step from above [9]. This implies that incorporation into steps is typically more facile from the lower terrace, and that *interlayer transport*, which involves step crossing, is reduced compared to the transport within one atomic layer. The corresponding potential energy landscape experienced by the adatom is illustrated in Fig. 1.

The Villain Instability

Jacques Villain first developed a continuum picture which shows that the step edge barrier illustrated in Fig. 1 generically implies a morphological instability

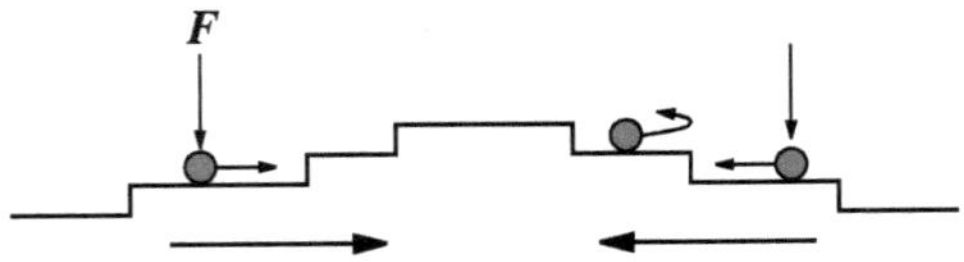

Fig. 2. Mechanism underlying the Villain instability (see text)

of the growing surface [10]. The essence of his argument is depicted in Fig. 2. It shows a small bump on the surface which may have been formed through a random fluctuation. The terraces on the slopes of the bump are bounded by one descending and one ascending step, and are called *vicinal* terraces. Because of the step edge barrier, the atoms deposited onto these terraces attach preferentially to the ascending step. The figure shows how this implies a net displacement of each adatom, between its point of impact and its point of incorporation, which is, on average, in the *uphill* direction. Within the continuum description, this implies a net mass current $\boldsymbol{j}(\nabla h)$, which is a function of the local slope ∇h. Assuming isotropy within the reference plane, the current can be written in the form

$$\boldsymbol{j}(\nabla h) = f(|\nabla h|^2)\,\nabla h\,. \tag{3}$$

The current is uphill when $f > 0$.

It is intuitively plausible that an uphill current will cause the bump to grow, and thus destabilize the surface. To analyze the instability in mathematical terms, we write down a surface evolution equation of the form (1). The mean film height grows at rate F, and the current $\boldsymbol{j}$ redistributes the deposited mass without changing the volume of the growing film. This implies a continuum equation of conservation type,

$$\frac{\partial h}{\partial t} = -\nabla \cdot \boldsymbol{j} + F\,, \tag{4}$$

which is obviously satisfied by the flat solution (2) with $v = F$.

For the time being, we specialize to a horizontal surface [$\boldsymbol{m} = 0$ in (2)]. To probe the stability of the flat state, we insert the ansatz $h(\boldsymbol{r}, t) = Ft + \varepsilon(\boldsymbol{r}, t)$ into (4) and expand to linear order in ε. This yields

$$\frac{\partial \varepsilon}{\partial t} = -f(0)\nabla^2 \varepsilon\,, \tag{5}$$

a diffusion equation with diffusion coefficient $-f(0)$. A linear partial differential equation like (5) is most conveniently solved by Fourier transformation. We make the ansatz for a perturbation of wavevector $\boldsymbol{q}$,

$$\varepsilon(\boldsymbol{r}, t) = \varepsilon_0 \mathrm{e}^{\mathrm{i}\boldsymbol{q}\cdot\boldsymbol{r} + \omega(\boldsymbol{q})t}\,, \tag{6}$$

and insert it into (5). We find that the growth rate of the perturbation is $\omega(\boldsymbol{q}) = f(0)|\boldsymbol{q}|^2$. Thus the flat state is unstable (the perturbation grows)

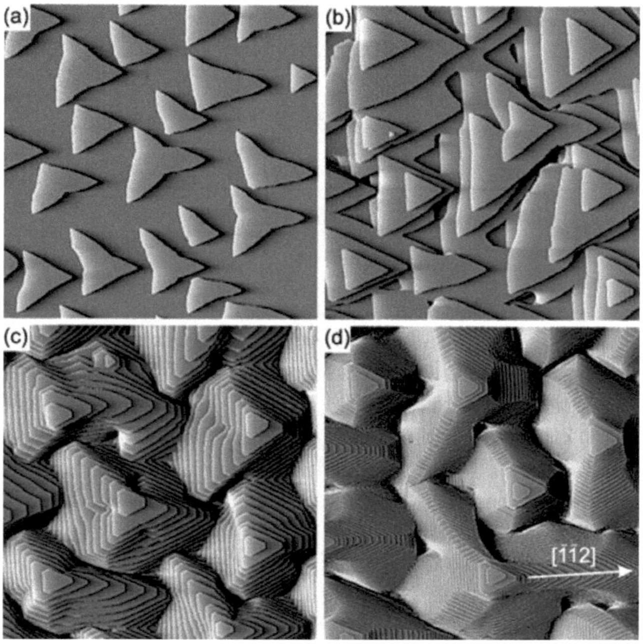

Fig. 3. Scanning tunnelling microscope images of a Pt(111) surface after deposition of (a) 0.35 monolayers (ML), (b) 3 ML, (c) 12 ML and (d) 90 ML of platinum [11]. The growth temperature was 440 K and the deposition flux $F = 7 \times 10^{-3}$ ML/s. The triangular shape of the islands and mounds reflects the threefold symmetry of the fcc(111) surface. The image size is 3450 Å× 3450 Å. Note that the different images do not show the same sample location. Courtesy of Thomas Michely

when $f(0) > 0$. Through the uphill current that it causes, the step edge barrier implies a *mound instability* of the growing surface. An experimental example of the mound morphology caused by this effect is shown in Fig. 3 (see [11] for details).

Vicinal Surfaces

It is straightforward to generalize the above stability analysis to surfaces of nonzero tilt [$\boldsymbol{m} \neq 0$ in (2)]. Without loss of generality, the tilt vector can be chosen along the x-axis, i.e. $\boldsymbol{m} = (m, 0)$ (see Fig. 4). We then find the following anisotropic generalization of (5):

$$\frac{\partial \varepsilon}{\partial t} = \nu_{\|} \frac{\partial^2 \varepsilon}{\partial x^2} + \nu_{\perp} \frac{\partial^2 \varepsilon}{\partial y^2} , \tag{7}$$

where the coefficients are given by [12, 13]

$$\nu_{\|} = -[f(m^2) + 2f'(m^2)m^2] = -\mathrm{d}j/\mathrm{d}m$$
$$\nu_{\perp} = -f(m^2) = -j(m)/m , \tag{8}$$

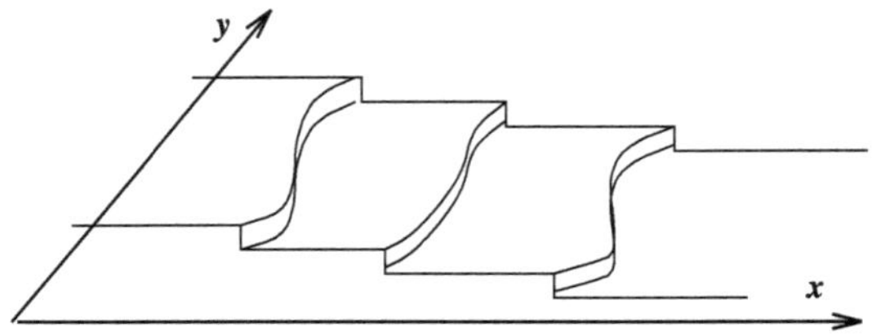

Fig. 4. Sketch of a vicinal surface

and $j(m) = mf(m^2)$ is the one-dimensional version of the current (3). The growth rate of a general perturbation (6) is now found to be $\omega(\boldsymbol{q}) = -\nu_\| q_x^2 - \nu_\perp q_y^2$. Thus the coefficients $\nu_\|$ and $\nu_\perp$ govern the stability of the surface against perturbations parallel to the tilt (with wavenumber q_x) and perpendicular to the tilt (with wavenumber q_y), respectively.

A perturbation parallel to the tilt implies that the initially uniform spacing between the surface steps becomes modulated. If such a perturbation is amplified (i.e., if $\nu_\| < 0$), the surface undergoes a *step bunching* instability, in which it breaks up into regions of high step density (the step bunches) separated by wide flat terraces. A perturbation perpendicular to the step implies that the individual steps become wavy, hence $\nu_\perp < 0$ implies that the surface undergoes a *step meandering* instability. Because steps cannot cross, the meander can be accommodated only if the steps are deformed in phase with each other.

To evaluate the coefficients $\nu_\|$ and $\nu_\perp$, we need to specify the function $f(\nabla h)$ in (3). We consider a *vicinal surface* of the staircase shape shown in Fig. 4, which is composed entirely of vicinal terraces; it consists of steps of a single sign, which run on average along the y–axis and form no closed loops (i.e., adatom or vacancy islands). Such a surface shape is maintained during growth, if each adatom deposited onto a terrace is able to reach one of the bordering steps before it encounters another adatom to form an island. This is called the *step flow* growth mode, and it requires that the spacing l between steps is small compared to the spacing l_D between islands formed on a completely flat surface (the island spacing l_D is evident in Fig. 3a).

It is easy to determine the surface current (3) for a surface growing in step flow mode. Let us assume for simplicity that the step edge barrier is very strong, so that all adatoms have to attach to the upper step of a terrace. Then each adatom travels a mean distance $l/2$ between its point of arrival and the step, and the magnitude of the current is $Fl/2$. Since the surface height gradient is $|\nabla h| = a/l$, where a denotes the thickness of an atomic layer, we have that $f(m^2) = (F/2)a/m^2$ and

$$j(m) = \frac{Fa}{2m} \,. \tag{9}$$

Comparison with (8) shows that $\nu_\| > 0$ and $\nu_\perp < 0$, hence the current stabilizes the surface against step bunching but makes it unstable with respect to

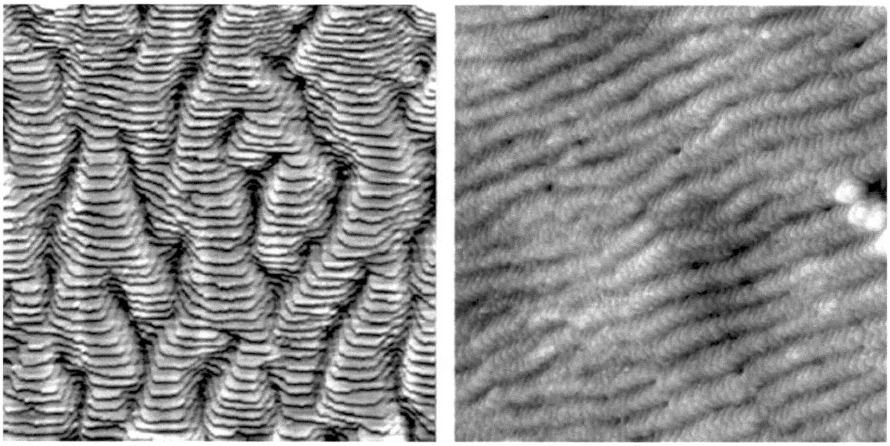

Fig. 5. Scanning tunnelling microscope images of step meandering on copper surfaces vicinal to Cu(100) [17]. Left panel: Cu(1,1,17) surface after deposition of 18 ML of Cu at 280 K with flux $F = 5{\times}10^{-3}$ ML/s. The image size is 1000 Å×1000 Å. Right panel: Cu(0,2,24) surface after deposition of 20 ML at 250 K with flux $F = 3 \times 10^{-3}$ ML/s; image size is 1300 Å × 1300 Å (from [17]). The mean step orientation is along the close–packed direction for the Cu(1,1,17) surface, and runs at 45° to the close–packed direction for the Cu(0,2,24) surface. Courtesy of Hans-Joachim Ernst

step meandering. While the first conclusion was reached already by Schwoebel and Shipsey [14] shortly after the discovery of the step edge barrier by Ehrlich and Hudda, the step meandering instability was predicted only much later, in 1990, by Bales and Zangwill [15]. Both these works relied on the model of step motion pioneered by Burton, Cabrera and Frank (BCF) [16], rather than on the continuum viewpoint presented here.

A detailed experimental study of growth-induced step meandering on copper surfaces vicinal to Cu(100) has been presented by Ernst and collaborators [17]; examples of the observed morphologies are shown in Fig. 5. The quantitative analysis of the patterns revealed, however, that the underlying mechanism in this case is *not* the one proposed by Bales and Zangwill. Instead, the meander is caused by the one-dimensional analogue of the Villain instability acting on the individual steps [18, 19, 20, 21]. The one-dimensional counterpart of the step edge barrier illustrated in Fig. 1 is an additional energy barrier which prevents atoms diffusing along a step edge from crossing kinks [22]. We will return to this issue below in Sect. 3.

2.2 Steering

In theories of epitaxial growth, it is usually assumed that the atoms arrive at the surface randomly and uniformly, without any dependence on the local surface morphology. The deposition flux could therefore be represented by a

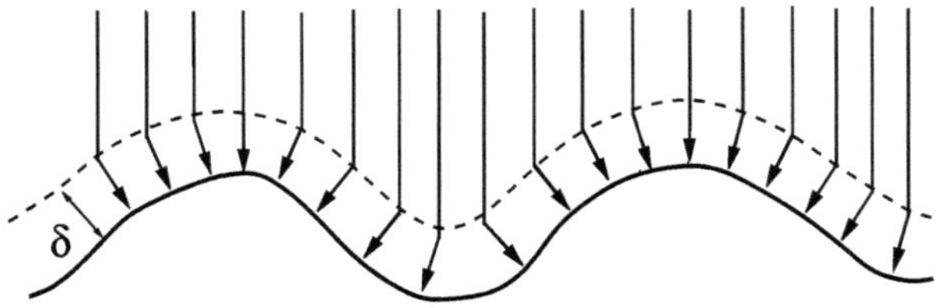

Fig. 6. Illustration of the destabilising effect of steering during growth. In this simplified picture, it is assumed that depositing atoms follow straight vertical trajectories until they are within a distance δ from the surface, at which point they are abruptly deflected in the direction of the surface normal. It is evident that this implies a larger flux to hilltops than to valleys

constant term F in (4). However, this assumption is not strictly true, because an atom approaching the surface feels the attractive force from the substrate atoms, which tends to deflect its trajectory away from a straight line. This effect is called *steering*, and it is well documented in molecular dynamics simulations [23, 24, 25]. The deflection becomes more pronounced the lower the kinetic energy of the incident atom, i.e., the lower the beam temperature. Experimentally, strong effects of steering have been observed in epitaxial growth of copper, when deposition occurs at near grazing incidence (the atom trajectories follow an angle of 80° from the surface normal) [26]. In this geometry steering implies an enhanced flux onto the top of islands near the ascending step, which leads to elongated island shapes and ripples. The same effect can be shown to induce step bunching on vicinal surfaces [25, 27].

Although steering affects the deposited atoms before they become adatoms, the consequences are rather similar to those of an uphill current along the surface, as it would be generated by a step edge barrier (Sect. 2.1) [25]. This idea was implemented already more than a decade ago in a simple one-dimensional growth model, where it leads to the formation of a morphology of well-separated columns [28].

In a continuum picture, in which morphological details such as steps and islands are ignored, the destabilising effect of steering can be represented as in Fig. 6. There it is assumed that the trajectories are deflected abruptly towards the surface normal as soon as the atoms come within a distance δ from the surface. The flux that reaches a portion of the surface is then the flux that is incident on the "virtual surface" indicated by the dashed line in Fig. 6, which is obtained by moving the original surface by a distance δ along its normal. For a surface morphology with a typical radius of curvature R, this implies an excess flux of order $F(\delta/R)$ at the hilltops and a corresponding reduction of the flux in the valleys. The flux becomes *curvature dependent* in a way that destabilises the flat surface.

For a one-dimensional geometry, the effect can be described by the continuum equation

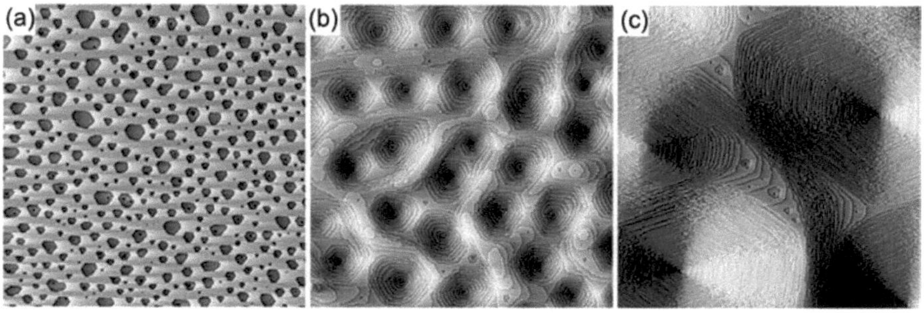

Fig. 7. Scanning tunnelling microscope images of pit formation on Pt(111) by erosion with 1 keV Xe^+ ions at 650 K. The amount of removed material is (a) 0.26 ML, (b) 6.2 ML and (c) 454 ML. The topograph size is 1600 Å× 1610 Å. Courtesy of Thomas Michely

$$\frac{\partial h}{\partial t} = F - F\delta \frac{1}{[1 + (\partial h/\partial x)^2]^{3/2}} \frac{\partial^2 h}{\partial x^2} = F(1 + \delta\kappa), \tag{10}$$

where κ is precisely the curvature of the curve $h(x,t)$. Equation (10) was derived in [29, 30] based on the picture illustrated in Fig. 6, however with a different interpretation: The scale δ was taken to be the size of the deposited particles, and the instability was related to the columnar microstructure of thin films grown at low temperatures. For small deviations $\varepsilon(x,t)$ from the flat surface, (10) reduces to $\partial\varepsilon/\partial t = -F\delta\partial^2\varepsilon/\partial x^2$, and similarly in two dimensions $\partial\varepsilon/\partial t = -F\delta\nabla^2\varepsilon$, which is the same as (5) in the case of the Villain instability. A continuum equation with such a destabilising term attributed to steering was proposed in [31] to describe the formation of mound-like structures in the growth of amorphous metal films.

2.3 Erosion by Ion Beams

It has been known for a long time that the erosion of a solid surface by ion bombardment gives rise to a variety of patterns such as pits and ripples [32, 33]. While much of the early results concerned amorphous solids, recent work has considered the erosion of single crystal metal surfaces [34, 35, 36, 37, 38, 39]. As an example, Fig. 7 shows the evolution of erosion pits during the bombardment of a Pt(111) surface with Xe^+ ions [34, 35]. The mechanism responsible for the formation of patterns like this will be described next, and we return to the case of amorphous surfaces in the second part of this subsection.

Erosion as Negative Growth

Evidently the shape of the pits seen in Fig. 7 is quite similar to an inverted version of the mounds in Fig. 3. Indeed, in the regime of interest here, the

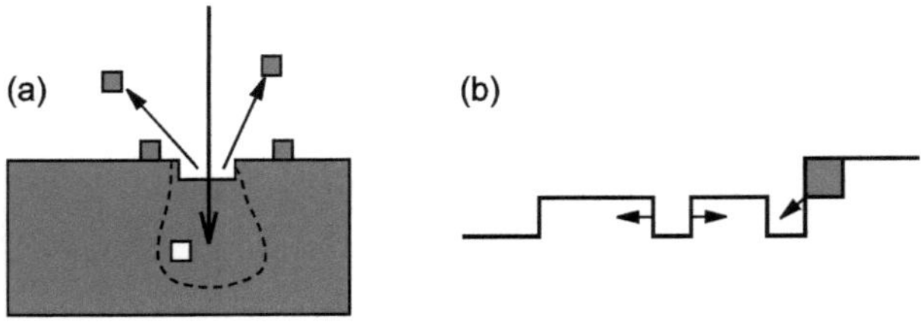

Fig. 8. Schematic of atomic processes in erosion. (**a**) An ion, indicated by the large arrow, penetrates into the solid and causes the material in the region enclosed by the dashed line to melt. After recrystallisation of the melted region, the net effect of the ion impact is the creation of a bulk vacancy, three surface vacancies and two adatoms. Two atoms are sputtered away. (**b**) Intralayer and interlayer transport of surface vacancies. The surface vacancy on the right passes to the next higher layer when the shaded atom jumps into it

erosion process can be described in close analogy to epitaxial growth. To understand why this is so, we need to consider in some detail the atomic processes involved (see Fig. 8); this discussion is valid at not too low temperatures, above 20% – 30% of the melting temperature [35]. The net effect of an ion impact on the surface morphology is the creation of a few surface vacancies and adatoms. As these are created in close spatial vicinity, the recombination of vacancies and adatoms takes place rapidly, leaving a few excess surface vacancies behind to account for those atoms which have been sputtered away[3]. On intermediate time scales, ion bombardment can therefore be viewed as a process in which vacancies are "deposited" onto the surface.

Once created, the surface vacancies diffuse over the terrace, as sketched in Fig. 8b. When a vacancy encounters a descending step edge, it simply disappears. In contrast, at an ascending step edge, it can pass to the higher layer only if a step atom jumps into it. This requires not only that the step atom moves to the lower layer, which may be impeded by a step edge barrier, but it requires, first, that the step atom detaches from its lateral neighbors. A surface vacancy is therefore subject to an additional step edge barrier even when no such barrier exists for adatom diffusion [39]. This implies that the Villain instability should be generically present in ion beam erosion, and many of the concepts developed in Sect. 2.1 carry over to the erosion of crystalline surfaces, however with the additional complication that one is dealing with currents of two diffusing species, adatoms and vacancies [35].

[3] It is also possible that the number of adatoms is larger than the number of surface vacancies, leading to net *growth* induced by ion bombardment, at the expense of the creation of bulk vacancies [40].

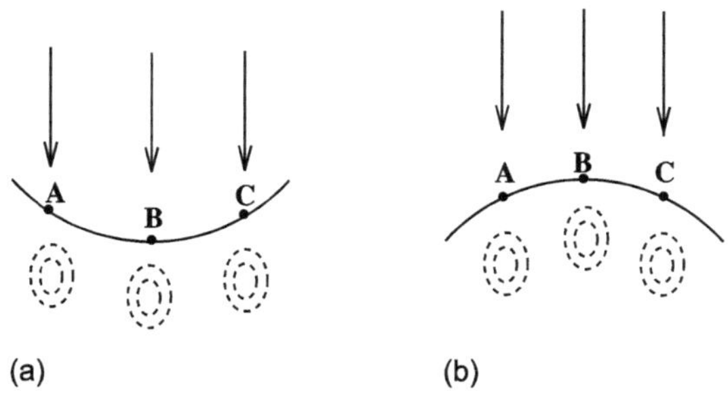

Fig. 9. Origin of the curvature dependence of the sputtering yield. Ions penetrating the surface at positions A, B and C deposit their energy as indicated by the dashed energy density contour lines. The energy deposited at B due to the ions incident at A and C is larger in (**a**) than in (**b**), because the centers of the corresponding ellipses are closer to the surface point B. Therefore the sputtering yield is larger in a valley (**a**) than on a hilltop (**b**)

The Bradley-Harper Instability

A different mechanism for pattern formation by ion bombardment was proposed by Bradley and Harper to account for the numerous observations of sputter-induced ripple formation on amorphous surfaces [41]. The mechanism is based on a curvature dependence of the sputtering yield Y, the number of sputtered atoms per incident ion, and is, in this sense, similar to the growth instability due to steering discussed in Sect. 2.2. Within Sigmund's theory of sputtering [42], it is assumed that the energy deposited by the penetrating ion is distributed according to a Gaussian with an elliptic shape, and the sputtering yield is proportional to the deposited power density at the surface. As shown in Fig. 9, this implies a larger sputtering yield in the valleys than on the hilltops. A similar mechanism has been proposed to act on a macroscopic scale in abrasive waterjet cutting [43].

Starting from Sigmund's theory, Bradley and Harper computed the morphology dependence of the sputtering yield to linear order in the surface modulation. The ion beam is taken to impinge on the surface at an angle θ from the normal. For $\theta > 0$ the horizontal projection of the ion beam singles out a direction within the surface plane, which implies in-plane anisotropy in the topography evolution, similar to the case of growth on vicinal surfaces discussed in Sect. 2.1. Choosing the beam direction along the x-axis, the leading terms of the evolution equation read [41]

$$\frac{\partial h}{\partial t} = -Y_0(\theta) - \frac{\mathrm{d}Y_0}{\mathrm{d}\theta}\frac{\partial h}{\partial x} + \nu_{\|}(\theta)\frac{\partial^2 h}{\partial x^2} + \nu_{\perp}(\theta)\frac{\partial^2 h}{\partial y^2}. \tag{11}$$

Here $Y_0(\theta)$ denotes the sputtering yield for a flat (unmodulated) surface. The second term on the right hand side describes the drift of surface features along the x-axis, while the stability of the flat surface is determined by the signs of the second derivative coefficients $\nu_{\|}$ and $\nu_{\perp}$ [compare to (7)]: An instability occurs when at least one of the two coefficients is negative, and the type of instability is determined by which of the negative coefficients is of larger absolute magnitude. Near normal incidence both coefficients are negative, as expected from Fig. 9, but near grazing incidence ($\theta \to 90°$) the instability along the beam direction is overcompensated by the fact that the (exposed) hilltops receive a larger ion flux than the valleys. Therefore $\nu_{\|} > 0$ and $\nu_{\perp} < 0$ for near-grazing incidence, which implies ripples running parallel to the beam direction. The ripple orientation rotates[4] at a critical angle θ_c, such that $\nu_{\|} < \nu_{\perp} < 0$ and the ripples run perpendicular to the beam direction for $\theta < \theta_c$. This scenario is well documented experimentally [32, 33, 44], and has also been observed in computer simulations [45].

2.4 Surface Electromigration

Surface electromigration is the biased motion of adatoms on the surface of a current–carrying solid. For metal surfaces, the dominant microscopic mechanism is the momentum transfer to the adatom caused by the scattering of conduction electrons (the *wind force*) [46, 47]; the direction of the electromigration force then coincides with the direction of electron flow. Much like a flow of air or water causes the formation of ripples on a sand surface, the mass flow induced by electromigration can give rise to patterns on the surface of a solid [48]. Such patterns were first observed in 1938 on the surfaces of burned-out tungsten filaments in incandescent lamps [49]. Recent experimental studies on the morphological effects of surface electromigration on silicon have uncovered a whole zoo of patterns with a complicated temperature dependence [50], but the microscopic coupling between the electric current and the adatom motion is not clearly understood for this system.

A continuum description of the current-induced destabilization of a solid surface in a one-dimensional geometry was presented by Frohberg and Adam in 1975 [51]. Here we briefly describe the two-dimensional theory of [48], emphasizing the similarity to the growth instabilities discussed in Sect. 2.1, and show how simple considerations explain some features of the experimentally observed patterns. In the absence of growth and evaporation, the evolution equation for the surface is given by the conservation law (4) with $F = 0$. In contrast to the situation in Sect. 2.1, where the current $\boldsymbol{j}$ in (4) was induced by the deposition flux in conjunction with the step edge barrier, here the mass current along the surface is simply driven by the local electric field $\boldsymbol{E}$. Hence we can write

$$\boldsymbol{j} = \sigma(\nabla h) e Z^* \boldsymbol{E} \ , \tag{12}$$

where σ is the *adatom mobility*, e is the unit charge and Z^* denotes the effective valence of the adatom, which contains the details of the microscopic

[4] In sputter erosion of crystalline surfaces, ripple rotation can also occur due to the crystalline anisotropy [37].

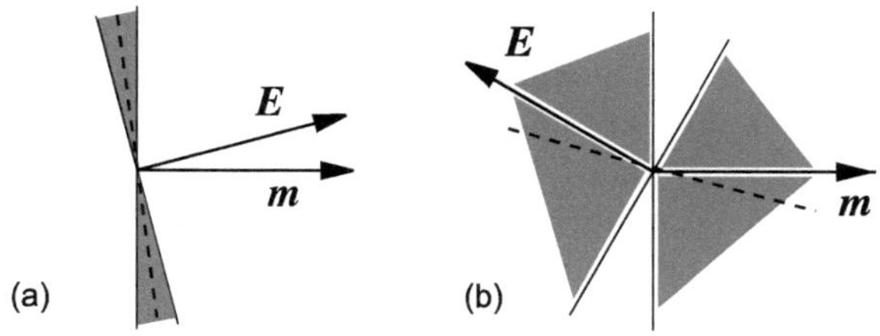

Fig. 10. (**a**), (**b**) Regions of unstable wavevectors (*shaded*) and most unstable direction (*dashed lines*) for the morphological instability caused by surface electromigration with different choices of the direction of the tilt vector $\boldsymbol{m}$ and the electric field vector $\boldsymbol{E}$

origin of the electromigration force. In the following we take $Z^* > 0$ without loss of generality.

The coupling between the electromigration current and the surface morphology arises because of the slope dependence of the mobility $\sigma(\nabla h)$. We expect that the mobility will be reduced in the presence of surface steps, which are able to trap the adatoms; the mobility should therefore be a decreasing function of $|\nabla h|$. Assuming isotropy in the plane, the mobility can then be expanded for small slopes in the form

$$\sigma(\nabla h) = \sigma_0 - \sigma_1 |\nabla h|^2 \tag{13}$$

with $\sigma_1 > 0$. Inserting (13) into (12) and (4), the stability of the general flat solution (2) with respect to a perturbation of the form (6) can be investigated.

The problem is similar to the stability analysis of the growing vicinal surface in Sect. 2.1, but it is more complicated because we are now faced with the interplay between *two* in-plane vectors, the tilt direction $\boldsymbol{m}$ and the electric field direction $\boldsymbol{E}$. This is reflected in the resulting expression for the growth rate of a perturbation with wavevector $\boldsymbol{q}$, which reads [48]

$$\omega(\boldsymbol{q}) = -2\sigma_1 (\boldsymbol{q} \cdot \boldsymbol{E})(\boldsymbol{q} \cdot \boldsymbol{m}). \tag{14}$$

Thus the surface is unstable against perturbations which satisfy the condition $(\boldsymbol{q} \cdot \boldsymbol{E})(\boldsymbol{q} \cdot \boldsymbol{m}) < 0$.

Most experiments on vicinal silicon surfaces have been limited to the situation where the electromigration force points either in the up-step direction ($\boldsymbol{E}$ and $\boldsymbol{m}$ parallel) or in the down-step direction ($\boldsymbol{E}$ and $\boldsymbol{m}$ antiparallel). The expression (14) implies stability for all $\boldsymbol{q}$ in the former case, and instability for all $\boldsymbol{q}$ in the latter. This is in agreement with the classic stability analysis carried out by Stoyanov within a one-dimensional step model based on BCF theory [52], which predicts a step-bunching instability when the force acts in the down-step direction (see [53] for a generalized treatment). For surfaces vicinal to Si(111), four different temperature regimes for electromigration-induced step bunching have been identified [50]. In each regime step bunching occurs either when the electric current is in the up-step or in the down-step direction, but the stable and unstable directions interchange when going from one regime to the next[5]. The step bunching mechanism first predicted by Stoyanov [52], and reproduced by the continuum theory [48], seems to be realized in the lowest temperature regime,

For general relative orientations of $\boldsymbol{E}$ and $\boldsymbol{m}$, the expression (14) implies that the stability depends on the direction of the perturbation wavevector $\boldsymbol{q}$ (Fig. 10). In particular, when $\boldsymbol{E}$ and $\boldsymbol{m}$ are almost, but not quite parallel, the unstable wavevectors accumulate in the direction *perpendicular* to $\boldsymbol{m}$ (Fig. 10a). As was discussed in Sect. 2.1, a perturbation wave vector perpendicular to $\boldsymbol{m}$, i.e., parallel to the steps, is indicative of a step meandering instability. Given that the local miscut typically displays some fluctuations relative to the global miscut $\boldsymbol{m}$, one may expect that step meandering persists also when the global miscut is exactly parallel to the field direction. Such a behavior has indeed been seen in Monte Carlo simulations [48].

The scenario is also qualitatively consistent with experimental observations for Si(111) in the second of the four temperature regimes mentioned above, in which step meandering is found when the electric current is in the down-step direction and step bunching occurs for an up-step current [50]. Experiments carried out with other relative orientations of $\boldsymbol{E}$ and $\boldsymbol{m}$ near the down-step direction display patterns with wavevectors roughly between the vectors perpendicular to $\boldsymbol{E}$ and $\boldsymbol{m}$, as would be expected from (14) [54]. This interpretation of the experiments would imply, however, that the electromigration force is directed against the current in this temperature regime (i.e. that the effective valence Z^* in (12) is negative), which is known not to be the case from other measurements [55]. The actual mechanism acting in this temperature regime, as well as the origin of the additional transitions occurring at higher temperature, remains largely unexplained at present.

3 Wavelength Selection in the Linear Regime

When observing a regular array of ripples or mounds on a surface, the first, most obvious question to address concerns the origin of the characteristic length scale of the pattern. In the initial stages of the morphological instability, when the amplitude of modulation is small, this question can be answered on the basis of the linearized version of the evolution equation (1). The linear evolution equations derived in Sect. 2 do not select a characteristic wavelength, because they favor the growth of perturbations of arbitrarily small length scale: The growth rate $\omega(\boldsymbol{q})$ of the perturbation (6) is a quadratic function of $\boldsymbol{q}$, which grows without bounds as $|\boldsymbol{q}|$ increases.

---

[5] The dependence of the stability of the surface on the current direction is a clear indication that a directional effect like electromigration must be involved, rather than just thermal effects related to Ohmic heating.

Wavelength selection therefore requires that the instability mechanisms described in Sect. 2 are counteracted by a smoothening mechanism on short length scales, which reduces the growth rate $\omega(\boldsymbol{q})$ for large $|\boldsymbol{q}|$. Smoothening is generally driven by the increase of the surface free energy caused by the modulation. According to the Gibbs-Thomson relation, the chemical potential near a curved surface is enhanced by an amount $\Delta\mu = \gamma\kappa$, where γ is the surface free energy and κ is the curvature. The relaxation towards thermodynamic equilibrium thus drives a mass transfer from the maxima ($\Delta\mu > 0$) to the minima ($\Delta\mu < 0$) of the topography. The smoothening mechanism depends on the available kinetic pathways for mass transport, as was first analyzed by Mullins in a continuum setting [56]. For the conditions of interest here, the dominant smoothening mechanism is by surface diffusion. The surface diffusion current is proportional to the gradient of the chemical potential, and can thus be written in the form

$$\boldsymbol{j}_{\mathrm{smooth}} = -\sigma\nabla(\gamma\kappa) \approx K\nabla(\nabla^2)h, \tag{15}$$

where the small slope approximation $\kappa \approx -\nabla^2 h$ has been used, σ denotes the adatom mobility [compare to (12)] and $K = \sigma\gamma$.

We illustrate the wavelength selection mechanism using the Villain instability as an example. Adding the smoothening current (15) in the divergence on the right hand side of (4) and linearizing around the horizontal, flat surface, we now obtain the equation

$$\frac{\partial\varepsilon}{\partial t} = -f(0)\nabla^2\varepsilon - K(\nabla^2)^2\varepsilon\ , \tag{16}$$

which leads to the wavenumber-dependent growth rate

$$\omega(\boldsymbol{q}) = f(0)|\boldsymbol{q}|^2 - K|\boldsymbol{q}|^4\ . \tag{17}$$

This shows that the surface is stable ($\omega < 0$) against perturbations of wavenumbers $|\boldsymbol{q}| > q_c = \sqrt{f(0)/K}$, while the amplification is maximal for perturbations of wavenumber $q^* = q_c/\sqrt{2}$. The corresponding wavelength

$$\lambda^* = 2\pi/q^* = 2\pi\sqrt{2K/f(0)} \tag{18}$$

will therefore dominate the morphology in the early stages of growth. Analogous considerations apply to the other instabilities discussed in Sect. 2.

In order for (18) to attain predictive power, it is necessary to express the coefficients $f(0)$ and K in terms of microscopic quantities. This problem is discussed in detail elsewhere [4, 6, 57, 58]; here we only summarize the main results. When the step edge barrier is large, the initial length scale in mound growth is essentially determined by the spacing l_D of the submonolayer islands which form during the growth of the first layer, and which act as templates for the further evolution. This is the case in the growth experiment on Pt(111) shown in Fig. 3, and it is the typical behavior observed in metal homoepitaxy [4]. For weak barriers the continuum theory predicts that $\lambda^* \gg l_D$, and that mound formation is delayed such that the pattern appears only after the deposition of a large number of layers. This situation appears to be realized in homoepitaxial growth on Ge(100) [59, 60]. In this case the experimental measurement of the initial mound spacing can be used to infer quantitative information about the magnitude of the step edge barrier; at long times the mounds coarsen and steepen (see Sect. 4.1), which eventually leads to epitaxial breakdown (the formation of an amorphous film).

For the growth-induced step meandering instability on vicinal Cu(100) surfaces described in Sect. 2.1 (see Fig. 5), the quantitative analysis of the meander wavelength has been used to distinguish between two possible instability mechanisms, the Bales-Zangwill instability and the one-dimensional (1D) Villain instability [17, 20, 21]. As will be explained below in Sect. 4.1, in this case the meander wavelength does not change during growth, and therefore the experimentally observed wavelength of the fully developed pattern can be determined by linear stability analysis. For the Bales-Zangwill instability the meander wavelength $\lambda_\perp$ is predicted to scale with the deposition flux as $\lambda_\perp \sim F^{-1/2}$, while for the 1D Villain instability with strong kink rounding barriers the initial wavelength is set by the spacing between one-dimensional nuclei along the step, which scales as $F^{-1/4}$. The latter is consistent with the experimentally observed flux dependence, while the former is not [17]. Also the effective activation energy governing the temperature dependence of $\lambda_\perp$ is markedly different for the two mechanisms, because in the case of the Bales-Zangwill instability it involves energy barriers related to terrace diffusion and step crossing, whereas the 1D Villain instability is governed by step edge diffusion and kink crossing barriers. The assumption of the 1D Villain mechanism leads to a consistent description[6] of the experimental data, and allows (along with additional experimental information) to estimate the additional energy barrier for kink rounding [21, 22].

4 Scenarios of Nonlinear Evolution

Within the linearized theory, unstable modes grow exponentially according to (6). Thus after a time of order $\omega(q^*)^{-1}$ the modulation amplitude becomes so large that the nonlinear terms in the evolution equation can no longer be neglected. The nature of the subsequent evolution depends on the structure of these nonlinear terms. A priori, several different scenarios are conceivable: (i) The morphology may *coarsen*, in the sense that the characteristic wavelength increases with time. (ii) The topography may evolve at fixed local slope (*slope selection*) or it may *steepen* during evolution; steepening and coarsening may

[6] This statement applies to the case when the average step orientation is along the close-packed direction (left panel of Fig. 5). The origin of the instability at the open steps (right panel of Fig. 5) is presently not well understood [21].

occur together, or steepening proceeds at fixed wavelength. (iii) Finally, the morphology may evolve in an irregular, *chaotic* fashion while locally retaining the characteristic wavelength of the linear regime. All these possibilities can be realized in the context of pattern formation at solid surfaces, and will be described in the following subsections.

4.1 Coarsening

In many cases coarsening involves a power law increase of the characteristic wavelength,

$$\lambda(t) \sim t^{1/z} , \tag{19}$$

which defines the *coarsening exponent* $1/z$. At the same time the modulation amplitude, measured e.g. through the surface width

$$w(t) = \sqrt{\langle (h - \bar{h})^2 \rangle} \tag{20}$$

also increases as

$$w(t) \sim t^{\beta} . \tag{21}$$

In (20), the angular brackets refer to a spatial average and $\bar{h}(t)$ is the mean height at time t. The exponent β in (21) is the *roughening exponent*. The typical slope of the topography is of the order w/λ. Thus coarsening occurs at fixed slope if $\beta = 1/z$, while *steepening* occurs if $\beta > 1/z$. A special case is steepening at fixed lateral length scale, $1/z = 0$ and $\beta > 0$.

Coarsening is a familiar notion in the kinetics of first order phase transitions, where it refers to the evolution of the domain structure of the stable phases during phase separation [61]. Although we are here dealing with structures that do not, properly speaking, approach a state of thermal equilibrium, many of the concepts and methods developed in phase ordering kinetics can be, and have been, applied to kinetic surface instabilities [62, 63, 64, 65, 66, 67, 68]. Some of the pertinent results will be summarized below.

Coarsening of Mounds and Pits

The basic continuum model for the coarsening of mounds generated by the Villain instability can be constructed from the ingredients introduced in Sects. 2.1 and 3. We just need to specify the full slope dependence of the growth-induced current (3) and write it together with the smoothening current (15) on the right hand side of the evolution equation (4). The limiting behaviors of the function $f(m^2)$ for small and large slopes have been determined already: For $m \to 0$ it approaches a constant $f(0) > 0$, and for large m

it vanishes as $1/m^2$ according to (9). A simple extrapolation formula which connects these two limits is [69]

$$f(m^2) = \frac{f(0)}{1 + (m/\hat{m})^2} \; ; \tag{22}$$

for a detailed justification and a discussion of the physical meaning of the parameters $f(0)$ and $\hat{m}$ we refer to [6, 57].

The form (9,22) implies that the uphill surface current remains nonzero for arbitrary values of the slope; as a consequence, the mounds steepen indefinitely. The mound slopes attain a finite limit only if $f(m^2)$ vanishes at some selected slope m^* [70, 71]. A simple functional form describing this situation is

$$f(m^2) = f(0)[1 - (m/m^*)^2]. \tag{23}$$

Once the current function $f(m^2)$ has been fixed, the full nonlinear evolution equation takes the form

$$\frac{\partial h}{\partial t} = -\nabla \cdot f(|\nabla h|^2)\nabla h - K(\nabla^2)^2 h + F \, . \tag{24}$$

To understand mathematically why the morphology described by (24) coarsens, it is useful to introduce the functional

$$\mathcal{L}[h(\boldsymbol{r},t)] = \int d\boldsymbol{r} \left(\frac{1}{2}(\nabla^2 h)^2 + \mathcal{V}(\nabla h) \right) \, , \tag{25}$$

where the *slope potential* $\mathcal{V}(\nabla h)$ is defined by

$$\mathcal{V}(\boldsymbol{m}) = -\frac{1}{2} \int_0^{|\boldsymbol{m}|^2} du \; f(u) \tag{26}$$

for an arbitrary choice of $f(m^2)$. Using the evolution equation (25), it is a simple matter to verify that [64, 65]

$$\frac{\mathrm{d}\mathcal{L}}{\mathrm{d}t} = - \int d\boldsymbol{r} \left(\frac{\partial h}{\partial t} \right)^2 < 0 \, . \tag{27}$$

Thus $\mathcal{L}$ acts as an effective "free energy" functional for this nonequilibrium process, which is minimized during the evolution.

The two terms in the integrand of (25) describe different aspects of this minimization: On the one hand, the value of the slope potential $\mathcal{V}$ should be minimized locally; on the other hand, the square of the surface curvature $\nabla^2 h$ should become small. The minimization of the slope potential drives the process of slope selection. The potential has a minimum only if the function f goes through zero at some nonzero slope; otherwise, $\mathcal{V}$ decreases indefinitely

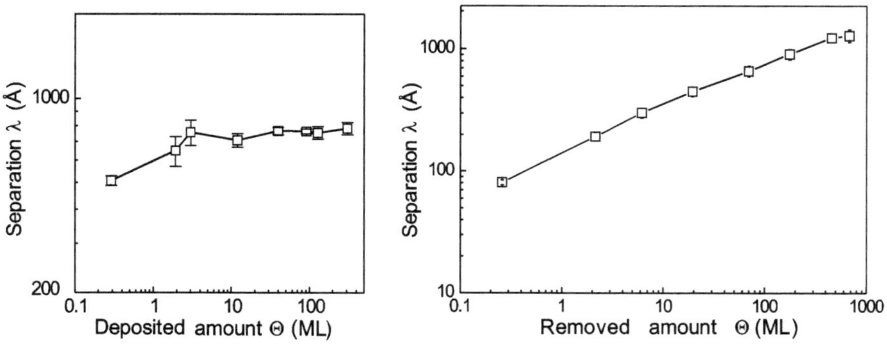

Fig. 11. Left panel: Evolution of the lateral mound separation, determined from the shape of the height-height correlation function, during growth of Pt on Pt(111) at 440 K. Right panel: Evolution of the lateral pit size during ion erosion of Pt(111) at 650 K. Courtesy of Thomas Michely

with increasing $|\boldsymbol{m}|$, and the attempt to minimize it leads to unbounded steepening.

Within the continuum theory, the minimization of the surface curvature term in (25) provides the driving force for coarsening. The coarsening behavior therefore depends on how the curvature is distributed on the surface. In this respect the two models defined by (22) and (23) differ: For the model (22) without slope selection, the curvature is distributed evenly on the scale λ of the mounds, while for the model (23) with slope selection it is concentrated in the edges separating adjacent facets at which the selected slope $|\boldsymbol{m}| = m^*$ is attained. The width of these edge regions introduces an additional length into the problem, which invalidates naive dimensional analysis [61]. Based on these observations, scaling argument can be developed [4, 5] which show that the scaling exponents defined in (19,21) take the values $1/z = 1/4$, $\beta = 1/2$ for model (22) [62, 63] and $1/z = \beta = 1/3$ for model (23) [65]. For model (23) a weak version of the inequality $1/z \leq 1/3$ has been rigorously established [72].

Additional features of the model which may affect the coarsening exponents include in-plane anisotropy of the current function [which can then no longer be written in the form (3)] [19, 64, 65, 71, 73], fluctuations due to the shot noise in the deposition beam [74] and nonlinear terms which break the $h \to -h$ symmetry of the evolution equation (24). In the latter case one must distinguish between terms that also change the conserved nature of the evolution equation [75] and those that do not [67, 76]. Terms of the first kind lead to a significant speedup of coarsening, and will be discussed further in Sect. 4.3.

Experimentally, coarsening mound morphologies have been observed for a large number of homoepitaxial growth systems [4, 6]. The coarsening exponents typically fall into the range $1/6 \leq 1/z \leq 1/3$. The growth of Pt on

Pt(111) is an exception: As is seen in Fig. 3 and demonstrated quantitatively in Fig. 11, for this system the mound morphology evolves at constant lateral length scale. At the same time the surface roughness increases with the square root of the deposit thickness, leading to the scaling exponents $1/z = 0$ and $\beta = 1/2$ [11]. This behavior can be well described by a simple model for the growth of a single mound which does not exchange mass with its neighbors [4, 5, 77, 78]. The model provides a detailed prediction for the shape of the individual mounds, which can be used to infer the value of the step edge barrier [79].

In contrast to the growth at 440 K, the pit morphology formed during ion erosion of Pt(111) at 650 K clearly coarsens, with a coarsening exponent of $1/z = 0.30 \pm 0.02$ [34, 35] (see Figs. 7 and 11). This difference in behavior shows that the microscopic mass transport process needed for coarsening, the generation of step adatoms at kinks, becomes thermally activated between 440 and 650 K [4, 80]. Coarsening of surface structures formed by ion erosion has also been reported for pits on Au(111) [36] and ripples on Cu(110) [38].

Coarsening in One Dimension

A detailed mathematical analysis of coarsening is possible for one-dimensional evolution equations. In one dimension, the growth equation (24) with the current function (23) becomes equivalent to the Cahn-Hilliard equation of phase ordering kinetics [61]. For this system it was shown long ago by Langer that the domain size grows logarithmically, $\lambda \sim \ln t$ [81]. Langers derivation is based on the analysis of the stationary periodic solutions of the equation. These periodic solutions are all linearly unstable, but the characteristic lifetime on which they decay increases exponentially with the wavelength, reflecting the exponential spatial decay of the interactions between the domain walls. During coarsening the system passes through a sequence of almost periodic configurations, and the time it spends in each is given by the lifetime of the corresponding stationary solution. Inverting the relationship between the lifetime and the wavelength thus yields the coarsening law.

Politi and Torcini have applied this approach to the one-dimensional version of the surface evolution equation (24) with a generalization of the current function (22), in which the current decays for large slopes as $j(m) \sim |m|^{-\gamma}$, with $\gamma \geq 1$ [68]. They find that $1/z = 1/4$ for $1 \leq \gamma \leq 3$ and $1/z = (1+\gamma)/(1+5\gamma)$ for $\gamma > 3$. This is in contrast to the scaling arguments employed in the two-dimensional case, which would predict that $1/z = 1/4$ independent of γ. The reason for the discrepancy is that the different, competing contributions which govern the growth of λ and w, and which are assumed in the scaling theory to be of a similar order of magnitude, may precisely cancel, to leading order, in the one-dimensional case [6]. As a consequence, the scaling arguments only give an upper bound on the coarsening exponent.

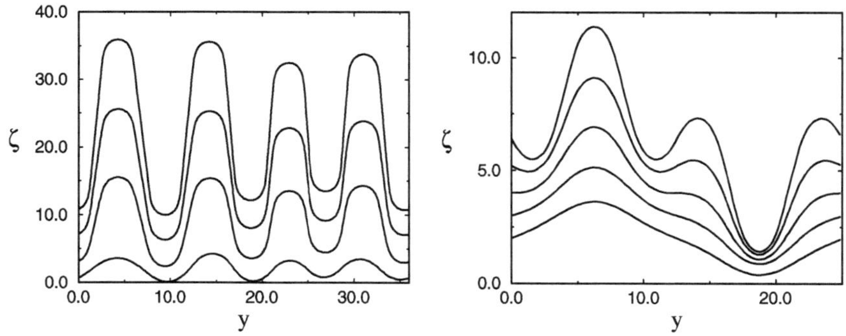

Fig. 12. *Left panel:* Time evolution of the collective step meander according to (28) with $\mu = 1/2$. *Right panel:* When the evolution equation (28) is started with an initial condition of wavelength $\lambda_0 > 2\lambda^*$, an additional meander appears; here $\lambda_0 = 2.8\lambda^*$

One-dimensional growth equations appear naturally in the description of the nonlinear evolution of vicinal surfaces undergoing the Bales-Zangwill instability [82, 83, 84]. The linear stability analysis shows that the different steps meander in phase in the early time regime [85]; within the continuum theory of Sect. 2.1, this is reflected in the fact that, because $\nu_\parallel > 0$ and $\nu_\perp < 0$, the growth rate $\omega(\boldsymbol{q})$ is maximal for perturbations with $q_x = 0$. Therefore a single, one-dimensional function $x = \zeta(y,t)$ is sufficient to describe the displacement of the common step profile from the flat straight reference configuration $\zeta = 0$. A solvability condition arising from a multiscale expansion of the BCF equations for the problem yields an evolution equation of the form [84]

$$\zeta_t = -\left\{ \frac{A\,\zeta_y}{1+\zeta_y^2} + \frac{B}{(1+\zeta_y^2)^\mu} \left[\frac{\zeta_{yy}}{(1+\zeta_y^2)^{3/2}} \right]_y \right\}_y , \tag{28}$$

where subscripts denote partial derivatives and A and B are positive constants. The first term inside the curly brackets is destabilising and proportional to the deposition flux, while the second term is a one-dimensional version of the stabilizing smoothening current introduced in (15). The expression inside the square brackets is the curvature of the step, and the term multiplying the square brackets is a kind of mobility, which depends on the dominant channel for mass transport; the exponent $\mu = 1$ when mass transport is over the terraces (the coefficient B is then proportional to the adatom diffusion coefficient on the terrace), and $\mu = 1/2$ when mass transport is along the step (with B proportional to the step edge diffusion coefficient) [83]. The geometric nonlinearities in (28) arise because, in contrast to (15), here the shape gradient ζ_y is not assumed to be small.

For the physically relevant values of μ (more precisely, for $\mu > -1/2$) the step patterns generated by (28) show an unbounded growth of the meander amplitude as $\sqrt{t}$, but no lateral coarsening (Fig. 12). This finding is consistent with the experimental observation that the wavelength of the step meander does not change during growth [17]. Asymptotically for large t, the equation admits separable solutions of the form $\zeta(y,t) = \sqrt{t}\, g(y)$, where the function $g(y)$ can be constructed to have *any* wavelength larger than a minimal value λ_{min}. Starting from an initial condition with small random fluctuations, the wavelength of the pattern is determined by the maximum of the linear growth rate, as explained in Sect. 3. In the notation of (28), the maximally unstable wavelength is $\lambda^* = 2\pi\sqrt{2B/A}$, and $\lambda^* > \lambda_{\text{min}}$ for $\mu > 0.2283$. Patterns with wavelengths larger than λ^* can be created by starting with an appropriate initial condition; however the initially chosen wavelength λ_0 persists only up to $\lambda_0 \approx 2\lambda^*$, as for larger values an additional meander can grow between the maxima and minima of the initial profile (Fig. 12).

In general, the presence or absence of coarsening for an evolution equation like (28) depends on the relationship between the wavelength λ and the amplitude $\mathcal{A}$ of its periodic stationary solutions: Coarsening requires that $\mathrm{d}\lambda/\mathrm{d}\mathcal{A} > 0$ [86]. This condition is satisfied in (28) for arbitrarily large λ when $\mu < -1/2$ [83]. A physical mechanism which changes the form of the nonlinearities in (28) such as to induce coarsening is the elastic interaction between steps [87]. Somewhat surprisingly, it is not obvious whether a one-dimensional surface evolution equation with stationary, fixed wavelength solutions which neither coarsen nor steepen can be written down [86]; this question is of interest in relation to sand ripples formed under an oscillatory flow of water [88].

Coarsening and Scaling of Step Bunches

A clear experimental example for power law coarsening is provided by the electromigration-induced step bunching of Si(111), where the distance L between step bunches is found to increase with time as $L \sim t^{1/2}$ in two of the temperature regimes mentioned in Sect. 2.4 [89]. This behavior has been reproduced in Monte Carlo simulations [48] as well as in a step-dynamical model based on BCF theory [90, 91]. In addition, the scaling relations

$$N \sim W^{\alpha}, \quad l_{\text{min}} \sim N^{-\gamma} \tag{29}$$

connecting the number of steps in a bunch, N, the minimal distance between steps in the bunch, l_{min}, and the lateral width W of the bunch have been proposed [90, 91, 92] and experimentally verified [93]. The minimal step distance decreases with increasing N, because in a large bunch the steps are pushed more strongly together against the repulsive step-step interaction. Depending on the temperature regime, the values $\gamma = 0.60 \pm 0.04$ and $\gamma = 0.68 \pm 0.03$ were obtained experimentally for Si(111) [93].

In order of magnitude, $l_{\text{min}} \sim W/N$, which yields the exponent identity $\gamma = 1 - 1/\alpha$. The steepening of the bunch with increasing size implies $\gamma > 0$

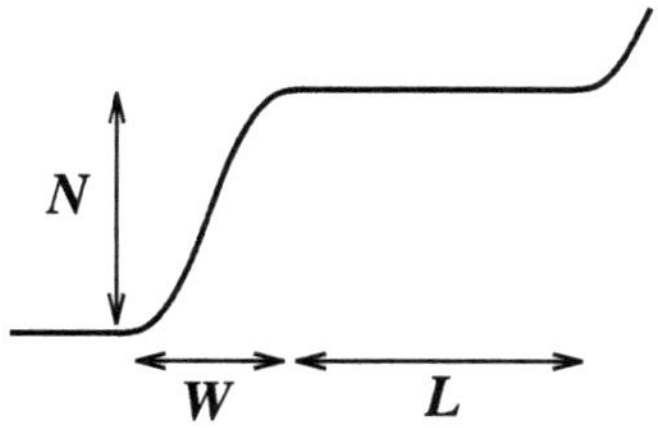

Fig. 13. Characteristic length scales of a step bunch

and hence $\alpha > 1$. Since the overall surface slope is fixed globally, we also have $L \sim N$. It is important to note that W and L define two distinct lateral length scales characterizing the bunch morphology (see Fig. 13). Introducing a coarsening exponent $1/z$ through $W \sim t^{1/z}$, it follows that $L \sim t^{\alpha/z}$, which grows faster than W when $\alpha > 1$.

A unified framework for describing the scaling properties of step bunch morphologies has recently been proposed by Pimpinelli and coworkers [94]. It is based on the continuum equation

$$\frac{\partial h}{\partial t} = -\frac{\partial}{\partial x} j(m) - K \frac{\partial}{\partial x} |m|^{-k} \frac{\partial^2}{\partial x^2} |m|^n \tag{30}$$

for the one-dimensional surface profile $h(x,t)$ perpendicular to the steps (in the x-direction, see Fig. 4). Here $m = \partial h / \partial x$ is the slope in the x-direction[7], $j(m)$ is the destabilising surface current, and the second term on the right hand side describes the smoothening of the vicinal surface by surface diffusion, with $K > 0$ being proportional to the surface diffusion coefficient. For a vicinal surface, the driving force for smoothening is the repulsive interaction between the surface steps. The step-step repulsion is encoded in the exponent n, which describes the decay of the interaction potential with step distance l as l^{-n}; for the most common type of entropic and elastic step-step interactions, $n = 2$ [7]. The exponent k in (30) is related to the diffusion kinetics on the vicinal surface[8]. If diffusion on the terraces is fast compared to the attachment-detachment processes at the steps we have $k = 1$, while in the opposite limit $k = 0$ [95, 96].

In addition, it is assumed in [94] that the destabilising current $j(m)$ can be represented by its leading order term in a gradient expansion, $j(m) \approx Bm^\rho$, where $B\rho > 0$ to satisfy the condition $\mathrm{d}j/\mathrm{d}m > 0$ required for a linear instability (compare to Sect. 2.1). Then the scaling exponents α and z can be determined by requiring that (30) should be invariant under the scale transformation

$$h(x,t) \Rightarrow b^{-\alpha} h(bx, b^z t) \tag{31}$$

[7] We take $m > 0$ without loss of generality.

[8] In [94], only the case $k = 0$ was considered.

for an arbitrary scale factor b. Comparing the three terms on the right hand side of (30) yields two conditions for the unknown exponents α and z, which are solved by the expressions

$$\alpha = 1 + \frac{2}{n-k-\rho}, \quad \gamma = \frac{2}{2+n-k-\rho}, \quad z = \frac{2(1+n-k-2\rho)}{n-k-\rho}. \tag{32}$$

The authors of [94] argue that all known examples of step bunching are covered by three *universality classes* corresponding to $\rho = 1$, $\rho = -1$ and $\rho = -2$.

Let us see how this framework applies to the problem of electromigration-induced step bunching. For detachment-limited dynamics, a continuum equation can be derived which has the form (30) with $\rho = -1$ and $k = 1$ [90]. According to (32), this implies $\gamma = 2/(2+n)$, in particular $\gamma = 1/2$ for the canonical case $n = 2$. This result is at variance with numerical simulations of step dynamical models [91, 92], which yield $\gamma = 2/(1+n)$, as well as with the experimental results quoted above [93], which are consistent with $\gamma = 2/3$.

The reason for the discrepancy is revealed by a careful analysis of the stationary solutions of (30) [97]. The shape of a stationary bunch is obtained by setting the total current on the right hand side of (30) equal to a constant j_0. As explained in [90], this current must be determined by the microscopic boundary conditions at the step, and it turns out to be *independent* of the bunch size. As the bunches steepen with increasing size, this implies that the destabilising current $j(m) \sim 1/m$ becomes *irrelevant* relative to the mean current j_0 for large bunches. Their shape is instead determined by the balance between j_0 and the stabilizing term. In this sense the correct value of ρ to chose for the scaling of the stationary bunch shape is $\rho = 0$. Together with $k = 1$ this gives $\gamma = 2/(1+n)$, in agreement with simulations [91, 92] and experiments [93].

Turning next to the dynamic scaling properties, we have already noted that experiments on Si(111) yield $\alpha/z = 1/2$ [89]. Step dynamical simulations show that this result is actually independent[9] of the step interaction exponent n [91]. Inspection of the expressions (32) reveals that $\alpha/z = 1/2$ independent of n only if $\rho = -1$. Thus, as far as the dynamical properties are concerned, our initial, naive choice $\rho = -1$ is appropriate. This probably reflects the fact that the coarsening dynamics is governed by current differences, which are not affected by the mean current j_0.

While somewhat preliminary and incomplete, the considerations in this subsection show that the step bunch equation (30) must be interpreted with caution. It must be kept in mind that the derivation of a continuum equation from discrete step dynamics can be rigorously justified [98] only when the terrace size is a slowly varying function of the terrace index. This condition

[9] An appealing explanation for the robustness of the value 1/2 for the coarsening exponent is given in [90]; the argument assumes, however, that the bunch spacing L is the only lateral length scale in the problem, which is not strictly true.

clearly breaks down at the boundaries of the step bunch, where the steeply sloped part joins the large terraces that separate it from neighboring bunches.

4.2 Chaotic Bubbling

Coarsening is the typical long-time behavior for surface evolution equations that are of conservation type, in the sense that the terms on the right hand side can be written (apart from an additive constant) as the divergence of a mass current. The Villain instability in epitaxial growth and the instabilities related to surface electromigration (in the absence of desorption) fall into this class of *conserved* dynamics, but the steering instability and the Bradley-Harper instability in ion erosion of amorphous surfaces do not. As a representative example of the class of *non-conserved* surface dynamics, we discuss here the nonlinear evolution of the Bradley-Harper instability.

Building on the work of Bradley and Harper [41] and of Sigmund [42], Cuerno and Barabási derived the leading order nonlinear contributions in the expansion of the sputtering yield Y in terms of the gradients of the topography [33, 99]. The resulting evolution equation is of the form

$$\frac{\partial h}{\partial t} = \nu_\| \frac{\partial^2 h}{\partial x^2} + \nu_\perp \frac{\partial^2 h}{\partial y^2} + \frac{\lambda_\|}{2} \left(\frac{\partial h}{\partial x} \right)^2 + \frac{\lambda_\perp}{2} \left(\frac{\partial h}{\partial y} \right)^2 - K(\nabla^2)^2 h \,, \tag{33}$$

where the first two terms on the right hand side of (11) have been eliminated by going to a comoving frame, and the smoothening current (15) has also been added. Clearly the nonlinear terms $\sim (\partial h/\partial x)^2$, $(\partial h/\partial y)^2$ cannot be written as the divergence of a conserved current.

Equation (33) is an anisotropic, two-dimensional generalization [100] of the *Kuramoto-Sivashinsky* (KS) equation, which was originally derived in the context of flame front propagation [101] and phase dynamics in spatially extended oscillatory systems [102]. The KS equation has become a paradigm of *spatio-temporal chaos* [103, 104]. It is the prime example of an interface evolution equation where the nonlinear terms stabilize the linear instability not by forcing the system into a near-periodic morphology which evolves more and more slowly through coarsening, but instead by establishing a dynamic steady state of irregular but bounded fluctuations. Two snapshots from the evolution of the isotropic, two-dimensional KS equation are shown in Fig. 14. The surface consists of convex cells of a characteristic size given by the most unstable wavelength λ^* of the linear instability (see Sect. 3). These cells split and merge in a random fashion, leading to an appearance of chaotic bubbling[10], similar, perhaps, to the surface of a boiling mudpool. On large length scales, the chaotic fluctuations can be described by an effective stochastic interface equation [105, 106, 107].

The introduction of anisotropy [100] opens up the possibility of qualitatively different modes of evolution, apart from the isotropic chaotic state

[10] This term was suggested to me by Per Jögi.

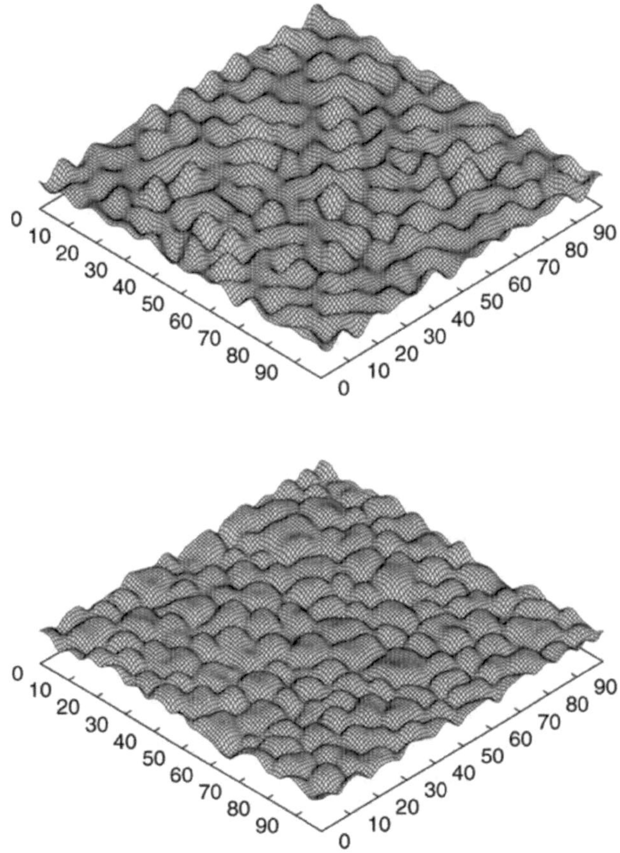

Fig. 14. Snapshots from the evolution of the isotropic, two-dimensional Kuramoto-Sivashinsky equation. The upper panel shows the early time regime governed by the linear instability, the lower panel shows the fully developed chaotic state. Note that the lateral length scale is the same in both images, but in the chaotic regime a cellular structure with clear up-down asymmetry has developed. Courtesy of Martin Rost

shown in Fig. 14. For example, when $\nu_\parallel > 0$ and $\nu_\perp < 0$, as in the Bradley-Harper instability at near grazing ion beam incidence, and $\lambda_\parallel$ and $\lambda_\perp$ have the same sign, then the initial pattern of ripples parallel to the x-axis undergoes a secondary pinching instability, and the surface develops an anisotropic chaotic pattern with a finite coherence length along the ripples. The most dramatic anisotropy effect occurs when $\lambda_\parallel/\lambda_\perp < 0$ and $\nu_\parallel/\nu_\perp > \lambda_\parallel/\lambda_\perp$ (with at least one of the two linear coefficients $\nu_{\parallel,\perp}$ negative). Then (33) can be shown to possess one-dimensionally modulated solutions for which the nonlinear terms precisely cancel, and which therefore show unbounded exponential growth. These *cancellation modes* correspond to tilted ripple patterns in which the ripples form an angle $\pm \arctan \sqrt{-\lambda_\parallel/\lambda_\perp}$ with respect to the y-axis. Because the angle can take two different values, there are actually two symmetry-related ripple patterns which compete for domination of the surface. As a consequence, a kind of coarsening scenario is observed which seems to prevent the exponential blowup of the modulation amplitude. This remarkable prediction of the continuum theory has so far not been verified experimentally, but it has been shown numerically that the phenomenon persists in the presence of external noise [108].

4.3 From Coarsening to Chaos

Having described two qualitatively different scenarios for nonlinear morphological evolution in the preceding two subsections 4.1 and 4.2, we now ask how the transition from one regime to the other occurs when, for example, the height conservation property of the evolution equation is only weakly violated. A physically relevant example is epitaxial growth in the presence of a small amount of desorption. A continuum description of this situation is provided by the evolution equation[11] [75]

$$\frac{\partial h}{\partial t} = -\nabla \cdot [1 - (\nabla h)^2]\nabla h - (\nabla^2)^2 h + F - \frac{\varrho}{1 + (\nabla h)^2}\,, \tag{34}$$

which combines the (suitably rescaled) mound evolution equation (24) [with slope selection, i.e. $f(m)$ given by (23)] with an additional nonlinear term modeling the desorption flux. The form of this term can be motivated from BCF theory. Qualitatively, it expresses the fact that the desorption flux decreases with increasing tilt (increasing step density), because the atoms are captured by steps before they can desorb; the coefficient ϱ in (34) is proportional to the desorption rate on a flat terrace.

The desorption term breaks the up-down symmetry of the evolution equation (24) and leads to a morphology of conical mounds (Fig. 15). More importantly, desorption is found to significantly speed up the coarsening behavior, with the coarsening exponent increasing from $1/z = 1/3$ for $\varrho = 0$ (see Sect. 4.1) to $1/z = 1/2$ for $\varrho \neq 0$ and long times [75]. This result can be understood as follows. Desorption occurs mostly from flat surface regions, i.e. from hilltops and valleys. Because of the up-down asymmetry, hilltops are pointlike whereas valleys form a network of lines. The fraction of the surface occupied by valleys is of order $1/\lambda$, much larger than the fraction $\sim 1/\lambda^2$ occupied by hilltops. As more material desorbs from the valleys than from the hilltops, the peak-to-valley height difference grows at rate ϱ/λ. Due to the slope selection property of the evolution equation, the lateral mound size has to increase at the same rate. Thus we have

[11] Similar equations have been used to describe the faceting of thermodynamically unstable surfaces in the presence of a growth flux [109, 110].

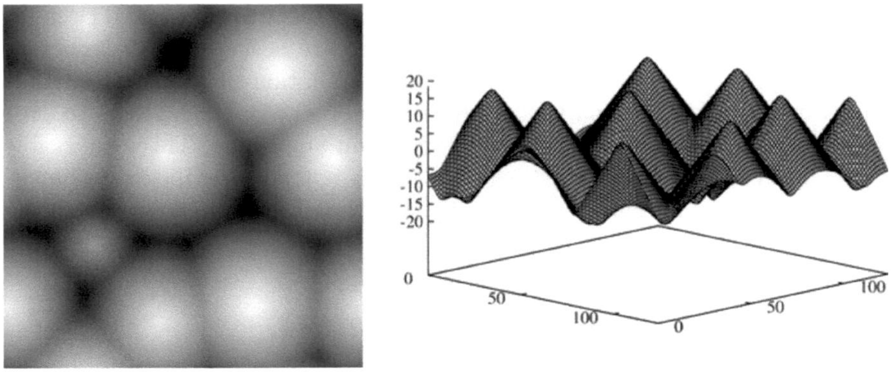

Fig. 15. Mounded surface generated by the evolution equation (34) with $\varrho \approx 0.32$. Conical mounds (*right*) form a cellular structure shown in gray-scale representation on the *left*

$$\mathrm{d}\lambda/\mathrm{d}t \approx \varrho/\lambda \quad \Rightarrow \quad \lambda \approx \sqrt{\varrho t}\ . \tag{35}$$

A model system for which the speedup of coarsening and the transition from coarsening to chaotic dynamics has been studied in detail is the one-dimensional *convective Cahn-Hilliard equation* [110, 111]

$$\frac{\partial m}{\partial t} + \frac{\partial^2}{\partial x^2}(m - m^3) + \frac{\partial^4 m}{\partial x^4} + 2\varrho m \frac{\partial m}{\partial x} = 0\ . \tag{36}$$

This is just the one-dimensional version of (34), written in terms of the slope $m = \partial h/\partial x$, in which the desorption term has been replaced by its leading gradient expansion. For $\varrho = 0$ (36) becomes the one-dimensional Cahn-Hilliard equation [61], which shows logarithmic coarsening, as was mentioned above in Sect. 4.1. The term proportional to ϱ has a formal similarity with the nonlinear convection term in the equations of hydrodynamics; it implies that fluctuations in m are convected at a speed proportional to m. Equation (36) has been introduced as a description of phase ordering kinetics in the presence of an external field [112, 113] and in the context of step growth with crystal anisotropy [114].

Rewriting (36) in terms of the rescaled variable $\tilde{m} = \varrho m$ and taking $\varrho \to \infty$, one finds that (36) reduces to the one-dimensional KS equation, which displays spatio-temporal chaos (see Sect. 4.2). Thus it is clear that (36) must show a transition from coarsening to chaos as ϱ is increased. The analysis of the stationary ($\partial m/\partial t = 0$) solutions of (36) indeed shows that weakly unstable periodic patterns, through which the coarsening process passes, exist only for $2\varrho < \sqrt{2}/3$. For larger values of ϱ stationary and traveling wave solutions with a fixed periodicity and amplitude are found, while around $2\varrho \approx 7$ spatio-temporal chaos sets in [110].

The coarsening behavior of (36) for small but nonzero ϱ has been analyzed in [111]. In this work the phenomenologically derived [113] scaling law

(35) has been firmly established for times $1/\varrho \ll t \ll 1/\varrho^3$. For even longer times a crossover to logarithmic coarsening, as in the standard Cahn-Hilliard equation, is found.

5 Conclusion

An important lesson from the examples described in this article is that similar patterns can be formed through widely different mechanisms. For solid surfaces driven out of equilibrium, patterns of mounds or pits typically appear if no in-plane direction is distinguished, while in the presence of a direction singled out by, e.g., the surface miscut, an electric current or an erosion beam, ripples perpendicular or parallel to this direction can be expected. The existence of a small number of archetypical patterns implies that the appearance of a particular pattern on a surface does not tell us much about the processes that are involved in its formation.

We should therefore be cautious in postulating analogies between widely different systems just on the basis of qualitative similarities in the observed patterns; this applies even more when patterns on the nanometer scale are compared to macroscopic patterns like sand ripples and dunes (although in some cases such an analogy can be well-founded, see [48]). It is only through a detailed quantitative analysis of the microscopic physics, in close cooperation between theory and experiment, that an understanding of the mechanisms responsible for pattern formation, and thus, the ability to manipulate the features of the patterns can be achieved. In this article I have described some of the recent progress in this direction; but clearly much work remains to be done.

Acknowledgements

The work on continuum equations began in collaboration with Michael Plischke and Martin Siegert, and continued with Claudio Castellano, Harvey Dobbs, Jouni Kallunki, Miroslav Kotrla, Paolo Politi, Martin Rost and Pavel Šmilauer. I am most grateful to Thomas Michely for many enlightening discussions and a critical reading of the manuscript, and I thank him as well as Hans-Joachim Ernst and Martin Rost for providing figures. Stoyan Stoyanov has promoted my understanding of electromigration-induced step bunching. Financial support has been provided by DFG within SFB 237 *Unordnung und grosse Fluktuationen* and SFB 616 *Energiedissipation an Oberflächen*, and by Volkswagenstiftung.

References

1. M.C. Cross, P.C. Hohenberg: Rev. Mod. Phys. **65**, 851 (1993)
2. J.L. Hansen, M. van Hecke, C. Ellegaard, K.H. Andersen, T. Bohr, A. Haaning, T. Sams: Phys. Rev. Lett. **87**, 204301 (2001)

3. K.H. Andersen, M. Abel, J. Krug, C. Ellegaard, L.R. Søndergaard, J. Udesen: Phys. Rev. Lett. **88**, 234302 (2002)
4. T. Michely, J. Krug: *Islands, Mounds and Atoms. Patterns and Processes in Crystal Growth Far From Equilibrium* (Springer, Berlin 2003)
5. J. Krug: Physica A **313**, 47 (2002)
6. P. Politi, G. Grenet, A. Marty, A. Ponchet, J. Villain: Phys. Rep. **324**, 271 (2000)
7. A. Pimpinelli, J. Villain: *Physics of Crystal Growth* (Cambridge University Press, Cambridge 1998)
8. C. Teichert: Phys. Rep. **365**, 335 (2002)
9. G. Ehrlich, F. Hudda: J. Chem. Phys. **44**, 1039 (1966)
10. J. Villain: J. Phys. I France **1**, 19 (1991)
11. M. Kalff, P. Šmilauer, G. Comsa, T. Michely: Surf. Sci. **426**, L447 (1999)
12. J. Krug, M. Schimschak: J. Phys. I **5**, 1065 (1995)
13. M. Rost, P. Šmilauer, J. Krug: Surf. Sci. **369**, 393 (1996)
14. R.L. Schwoebel, E.J. Shipsey: J. Appl. Phys. **37**, 3682 (1966)
15. G.S. Bales, A. Zangwill: Phys. Rev. B **41**, 5500 (1990)
16. W. Burton, N. Carbrera, F. C. Frank: Phil. Trans. Roy. Soc. A **243**, 299 (1951)
17. T. Maroutian, L. Douillard, H.-J. Ernst: Phys. Rev. B **64**, 165401 (2001)
18. O. Pierre-Louis, M.R. D'Orsogna, T. Einstein: Phys. Rev. Lett. **82**, 3661 (1999)
19. P. Politi, J. Krug: Surf. Sci. **446**, 89 (2000)
20. M. Rusanen, I.T. Koponen, J. Heinonen, T. Ala-Nissila: Phys. Rev. Lett. **86**, 5317 (2001)
21. J. Kallunki, J. Krug, M. Kotrla: Phys. Rev. B **65**, 205411 (2002)
22. J. Kallunki, J. Krug: Surf. Sci. Lett. **523**, L53 (2003)
23. D.E. Sanders, D.M. Halstead, A.E. DePristo: J. Vac. Sci. Technol. A **10**, 1986 (1992)
24. F. Montalenti, A.F. Voter: Phys. Rev. B **64**, 081401 (2001)
25. J. Yu, J.G. Amar: Phys. Rev. Lett. **89**, 286103 (2002)
26. S. van Dijken, L.C. Jorritsma, B. Poelsema: Phys. Rev. B **61**, 14047 (2000)
27. J. Krug (unpublished)
28. H. Park, A. Provata, S. Redner: J. Phys. A **24**, L1391 (1991)
29. A. Mazor, D.J. Srolovitz, P.S. Hagan, B.G. Bukiet: Phys. Rev. Lett. **60**, 424 (1988)
30. D.J. Srolovitz, A. Mazor, B.G. Bukiet: J. Vac. Sci. Technol. A **6**, 2371 (1988)
31. M. Raible, S.G. Mayr, S.J. Linz, M. Moske, P. Hänggi, K. Samwer: Europhys. Lett. **50**, 61 (2000)
32. G. Carter, B. Navinšek, J.L. Whitton: 'Heavy Ion Sputtering Induced Surface Topography Development' In: *Sputtering by Particle Bombardment I*, Topics in Applied Physics Vol. 47, ed. by R. Behrisch (Springer, Berlin 1981) pp. 231–269
33. M.A. Makeev, R. Cuerno, A.-L. Barabási: Nucl. Instr. and Meth. B **197**, 185 (2002)
34. T. Michely, M. Kalff, G. Comsa, M. Strobel, K.-H. Heinig: Phys. Rev. Lett. **86**, 2589 (2001)
35. M. Kalff, G. Comsa, T. Michely: Surf. Sci. **486**, 103 (2001)
36. M.V.R. Murty, T. Curcic, A. Judy, B.H. Cooper, A.R. Woll, J.D. Brock, S. Kycia, R.L. Headrick: Phys. Rev. Lett. **80**, 4713 (1998)

37. S. Rusponi, G. Costantini, C. Boragno, U. Valbusa: Phys. Rev. Lett. **81**, 2735 (1998)
38. S. Rusponi, G. Costantini, C. Boragno, U. Valbusa: Phys. Rev. Lett. **81**, 4184 (1998)
39. G. Costantini, S. Rusponi, R. Gianotti, C. Boragno, U. Valbusa: Surf. Sci. **416**, 245 (1998)
40. C. Busse, H. Hansen, U. Linke, T. Michely: Phys. Rev. Lett. **85**, 326 (2000)
41. R.M. Bradley, J.M.E. Harper: J. Vac. Sci. Technol. A **6**, 2390 (1988)
42. P. Sigmund, J. Mater. Sci. **8**, 1545 (1973)
43. R. Friedrich, G. Radons, T. Ditzinger, A. Henning: Phys. Rev. Lett. **85**, 4884 (2002)
44. S. Habenicht, W. Bolse, K.P. Lieb, K. Reimann, U. Geyer: Phys. Rev. B **60**, R2200 (1999)
45. I. Koponen, M. Hautala, O.P. Sievänen, Phys. Rev. Lett. **78**, 2612 (1997)
46. D.N. Bly, P.J. Rous: Phys. Rev. B **53**, 13909 (1996)
47. R.S. Sorbello: Solid State Phys. **51**, 159 (1998)
48. H. Dobbs, J. Krug: J. Phys. I France **6**, 413 (1996)
49. R.P. Johnson: Phys. Rev. A **54**, 459 (1938)
50. K. Yagi, H. Minoda, M. Degawa: Surf. Sci. Rep. **43**, 45 (2001)
51. G. Frohberg, P. Adam: Thin Solid Films **25**, 525 (1975)
52. S. Stoyanov: Jpn. J. Appl. Phys. **30**, 1 (1991)
53. O. Pierre-Louis: Surf. Sci. **529**, 114 (2003)
54. M. Degawa, H. Minoda, Y. Tanishiro, K. Yagi: Phys. Rev. B **63**, 045309 (2001)
55. M. Degawa, H. Minoda, Y. Tanishiro, K. Yagi: Surf. Sci. Lett. **461**, L528 (2000)
56. W.W. Mullins: J. Appl. Phys. **30**, 77 (1959)
57. P. Politi, J. Villain: Phys. Rev. B **54**, 5114 (1996)
58. J. Krug: Physica A **263**, 170 (1999)
59. J.E. Van Nostrand, S.J. Chey, D.G. Cahill: Phys. Rev. B **57**, 12536 (1998)
60. K.A. Bratland, Y.L. Foo, J.A.N.T. Soares, T. Spila, P. Desjardins, J.E. Greene: Phys. Rev. B **67**, 125322 (2003)
61. A.J. Bray: Adv. Phys. **43**, 357 (1994)
62. L. Golubović: Phys. Rev. Lett. **78**, 90 (1997)
63. M. Rost, J. Krug: Phys. Rev. E **55**, 3952 (1997)
64. M. Siegert: Physica A **239**, 420 (1997)
65. D. Moldovan, L. Golubovic: Phys. Rev. E **61**, 6190 (2000)
66. C. Castellano, J. Krug: Phys. Rev. B **62**, 2879 (2000)
67. P. Politi: Phys. Rev. E **58**, 281 (1998)
68. P. Politi, A. Torcini: J. Phys. A **33**, L77 (2000)
69. M.D. Johnson, C. Orme, A.W. Hunt, D. Graff, J. Sudijono, L.M. Sander, B.G. Orr: Phys. Rev. Lett. **72**, 116 (1994)
70. J. Krug, M. Plischke, M. Siegert: Phys. Rev. Lett. **70**, 3271 (1993)
71. M. Siegert, M. Plischke: Phys. Rev. Lett. **73**, 1517 (1994)
72. R.V. Kohn, X. Yan: Comm. Pure Appl. Math. **56** 1549 (2003)
73. M. Siegert: Phys. Rev. Lett. **81**, 5481 (1998)
74. L.-H. Tang, P. Šmilauer, D.D. Vvedensky: Eur. Phys. J. B **2**, 409 (1998)
75. P. Šmilauer, M. Rost, J. Krug: Phys. Rev. E **59**, R6263 (1999)
76. J.A. Stroscio, D.T. Pierce, M.D. Stiles, A. Zangwill, L.M. Sander: Phys. Rev. Lett. **75**, 4246 (1995)
77. J. Krug: J. Stat. Phys. **87**, 505 (1997)

78. J. Krug, P. Kuhn: 'Second Layer Nucleation and the Shape of Wedding Cakes' In: *Atomistic Aspects of Epitaxial Growth*, ed. by M. Kotrla, N.I. Papanicolaou, D.D. Vvedensky, L.T. Wille (Kluwer, Dordrecht 2002) pp. 145–163
79. J. Krug, P. Politi, T. Michely: Phys. Rev. B **61**, 14037 (2000)
80. T. Michely, M. Kalff, G. Comsa, M. Strobel, K.-H. Heinig: 'Coarsening Mechanisms in Surface Morphological Evolution'. In: *Atomistic Aspects of Epitaxial Growth*, ed. by M. Kotrla, N.I. Papanicolaou, D.D. Vvedensky, L.T. Wille (Kluwer, Dordrecht 2002) pp. 185–196
81. J.S. Langer, Ann. Phys. **65**, 53 (1971)
82. O. Pierre-Louis, C. Misbah, Y. Satio, J. Krug, P. Politi: Phys. Rev. Lett. **80**, 4221 (1998)
83. J. Kallunki, J. Krug: Phys. Rev. E **62**, 6229 (2000)
84. F. Gillet, O. Pierre-Louis, C. Misbah: Eur. Phys. J. B **18**, 519 (2000)
85. A. Pimpinelli, I. Elkinani, A. Karma, C. Misbah, J. Villain: J. Phys. Condens. Matter **6**, 2661 (1994)
86. P. Politi, C. Misbah: Phys. Rev. Lett. **92**, 090601 (2004)
87. S. Paulin, F. Gillet, O. Pierre-Louis, C. Misbah: Phys. Rev. Lett. **86**, 5538 (2001)
88. J. Krug: Adv. Compl. Sys. **4**, 353 (2001)
89. Y.-N. Yang, E.S. Fu, E.D. Williams: Surf. Sci. **356**, 101 (1996)
90. D.-J. Liu, J.D. Weeks: Phys. Rev. B **57**, 14891 (1998)
91. M. Sato, M. Uwaha: Surf. Sci. **442**, 318 (1999)
92. S. Stoyanov, V. Tonchev: Phys. Rev. B **58**, 1590 (1998)
93. K. Fujita, M. Ichikawa, S.S. Stoyanov: Phys. Rev. B **60**, 16006 (1999)
94. A. Pimpinelli, V. Tonchev, A. Videcoq, M. Vladimirova: Phys. Rev. Lett. **88**, 206103 (2002)
95. P. Nozières: J. Physique **48**, 1605 (1987)
96. D.-J. Liu, E.S. Fu, M.D. Johnson, J.D. Weeks, E.D. Williams: J. Vac. Sci. Technol. B **14**, 2799 (1996)
97. J. Krug, V. Tonchev, S. Stoyanov, A. Pimpinelli (submitted to Phys. Rev. B)
98. J. Krug: 'Continuum Equations for Step Flow Growth'. In: *Dynamics of Fluctuating Interfaces and Related Phenomena*, ed. by D. Kim, H. Park, B. Kahng (World Scientific, Singapore 1997) pp. 95–113
99. R. Cuerno, A.-L. Barabási: Phys. Rev. Lett. **74**, 4746 (1995)
100. M. Rost, J. Krug: Phys. Rev. Lett. **75**, 3894 (1995)
101. G.I. Sivashinsky: Ann. Rev. Fluid Mech. **15**, 179 (1983)
102. Y. Kuramoto: *Chemical Oscillations, Waves and Turbulence* (Springer, Berlin 1984)
103. P. Manneville: *Dissipative Structures and Weak Turbulence* (Academic Press, San Diego 1990)
104. T. Bohr, M.H. Jensen, G. Paladin, A. Vulpiani: *Dynamical Systems Approach to Turbulence* (Cambridge University Press, Cambridge, UK, 1998)
105. V. Yakhot: Phys. Rev. A **24**, 642 (1981)
106. K. Sneppen, J. Krug, M.H. Jensen, C. Jayaprakash, T. Bohr: Phys. Rev. A **46**, R7351 (1992)
107. B.M. Boghosian, C.C. Chow, T. Hwa: Phys. Rev. Lett. **83**, 5262 (1999)
108. S. Park, B. Kahng, H. Jeong, A.-L. Barabási: Phys. Rev. Lett. **83**, 3486 (1999)
109. A.A. Golovin, S.H. Davis, A.A. Nepomnyashchy: Phys. Rev. E **59**, 803 (1999)
110. A.A. Golovin, A.A. Nepomnyashchy, S.H. Davis, M.A. Zaks: Phys. Rev. Lett. **86**, 1550 (2001)

111. S.J. Watson, F. Otto, B.Y. Rubinstein, S.H. Davis: Physica D **178**, 127 (2003)
112. K. Leung: J. Stat. Phys. **61**, 345 (1990)
113. C.L. Emmott, A.J. Bray: Phys. Rev. E **54**, 4568 (1996)
114. Y. Saito, M. Uwaha: J. Phys. Soc. Jpn. **65**, 3576 (1996)

Nonlinear Dynamics and Pattern Formation in Semiconductor Systems

Eckehard Schöll

Institut für Theoretische Physik, Technische Universität Berlin,
10623 Berlin, Germany
schoell@physik.tu-berlin.de

In the first part of this review article the basic concepts of self-organized pattern formation and spatio-temporal dynamics of the charge carriers in semiconductors are introduced. In the second part these concepts are applied to two model systems of semiconductor nanostructures which are of particular current interest: double-barrier resonant-tunnelling diodes and superlattices. We study time delayed feedback control (time delay autosynchronization) of chaotic spatio-temporal patterns in these semiconductor models. Different control schemes, e.g., a diagonal control matrix, or global control, or combinations of both, are introduced and compared.

1 Complex Nonlinear Spatio-temporal Dynamics in Semiconductors

1.1 Semiconductor Nanostructures as a Nonlinear Dynamic System

This review deals with complex nonlinear spatio-temporal dynamics, pattern formation, and chaotic behavior in semiconductors [1]. Its aim is to build a bridge between two well-established fields: the theory of dynamic systems, and nonlinear charge transport in semiconductors.

Semiconductors are complex many-body systems whose physical, e.g., electrical or optical, properties are governed by a variety of nonlinear dynamic processes. In particular, modern semiconductor structures whose structural and electronic properties vary on a nanometer scale provide an abundance of examples for nonlinear transport processes. In these structures nonlinear transport mechanism are given, for instance, by quantum mechanical tunnelling through potential barriers, or by thermionic emission of hot electrons which have enough kinetic energy to overcome the barrier. A further important feature connected with potential barriers and quantum wells in such semiconductor structures is the ubiquitous presence of space charges. This, according to Poisson's equation, induces a further feedback between the charge carrier distribution and the electric potential distribution governing the transport. This mutual nonlinear interdependence is particularly pronounced in the cases of semiconductor heterostructures (consisting of layers of different materials) and low-dimensional structures where abrupt junctions between different materials on an atomic length scale cause conduction band discontinuities resulting in potential barriers and wells. The local charge accumulation in these potential wells, together with nonlinear transport processes across the barriers have been found to provide a number of nonlinearities.

The view of a semiconductor as a nonlinear dynamic system is a fairly recent development. Such nonlinear dynamic systems can exhibit a variety of complex behaviour such as bifurcations, phase transitions, spatio-temporal pattern formation, self-sustained oscillations, and deterministic chaos. Semiconductors are *dissipative* dynamic systems, i.e., a steady state can only be maintained by a continuous flux of energy, and possibly matter, through them. Mathematically, this is described by the feature that volume elements in a suitable space of dynamic variables - the *phase space* - shrink with increasing time. In the language of thermodynamics, this represents an *open* system which is driven by external fluxes and forces so far from thermodynamic equilibrium that linear dynamic laws no longer hold. Due to the driving forces and the inherent nonlinearities of these systems, they may spontaneously evolve into a state of highly ordered spatial or temporal structures, so-called *dissipative structures*. Unlike an isolated, closed system, which after a perturbation always returns to a thermodynamic equilibrium state characterized by maximum entropy, an open dissipative nonlinear system may exhibit a process of *self-organization*, in which the entropy is locally decreased. Such processes usually involve qualitative changes in the state of the system, similar to phase transitions. Nonequilibrium phase transitions and dissipative structures have been noted in a great number of very different dissipative systems occurring in physics, chemistry, biology, ecology [2, 3, 4], and even economics and social sciences, but the observed phenomena are similar.

In the field of semiconductor physics, these matters have become an active field of research only within the last decade, although there exists some singular early work, for example on the bifurcation of current filaments in connection with dielectric breakdown [5], or phase portrait analysis of field domains in CdS crystals [6]. The analogy of an overheating instability of the electron gas with an equilibrium phase transition was pointed out by Volkov and Kogan [7], and Pytte and Thomas [8] drew this analogy in the case of the Gunn instability of the electron drift velocity at about the same time. The early theory of domain instabilities in semiconductors was reviewed by Bonch-Bruevich [9]. Generation-recombination induced phase transitions in semiconductors were first noted by Landsberg and Pimpale [10]. Impact ionization of electrons or holes by hot carriers across the bandgap or from localized levels was recognized as the main autocatalytic process which is necessary for phase transitions [11, 12], and current filamentation was treated as a process of self-organization in a system far from equilibrium [13]. The interest in this field was greatly enhanced by the experimental discovery of deterministic chaos in semiconductors [14, 15]. The introduction of concepts and methods from

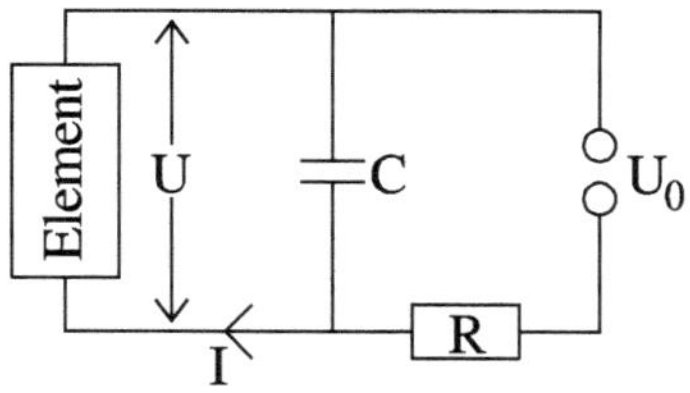

Fig. 1. A semiconductor element operated in a circuit with load resistor R and capacitor C, and applied bias voltage U_0

nonlinear dynamics subsequently stimulated a large amount of experimental and theoretical work on chaos and spatio-temporal self-organized pattern formation in a variety of semiconducting materials [16, 17, 18, 19, 20, 21, 1].

The electric transport properties of a semiconductor show up most directly in its current-voltage characteristic. It is determined in a complex way by the microscopic properties of the material, which specify the current density j as a function of the local electric field $\mathcal{E}$, and by the contacts. A local, static, scalar $j(\mathcal{E})$ relation need not always exist, but in fact does in many cases. Close to thermodynamic equilibrium, i.e., at sufficiently low bias voltage, the $j(\mathcal{E})$ relation is linear (Ohm's Law), but under practical operating conditions it will generally become nonlinear and may even display a regime of *negative differential conductivity* $\sigma_{diff} = dj/d\mathcal{E} < 0$. The *global* current-voltage characteristic $I(U)$ of a semiconductor can in principle be calculated from the *local* $j(\mathcal{E})$ relation by integrating the current density j over the cross section A of the current flow ($I = \int_A j df$) and the electric field $\mathcal{E}$ over the length L of the sample ($U = \int_0^L \mathcal{E} dz$). Unlike the $j(\mathcal{E})$ relation, the $I(U)$ characteristic is not only a property of the semiconductor material, but depends also on the geometry, the boundary conditions, and the contacts of the sample. Only for the idealized case of spatially uniform states, the $j(\mathcal{E})$ and the $I(U)$ characteristics are identical, up to re-scaling. The $I(U)$ relation is said to display *negative differential conductance* if $dI/dU < 0$.

In the case of negative differential conductance, the current decreases with increasing voltage, and vice versa, which normally corresponds to an unstable situation. The actual electric response depends upon the attached circuit which in general contains – even in the absence of external load resistors – unavoidable resistive and reactive components like lead resistances, lead inductances, package inductances, and package capacitances. These reactive components give rise to additional degrees of freedom which are described by Kirchhoff's equations of the circuit. If, for instance, the circuit is as shown in Fig. 1, where the capacitance C parallel to the device is given by the sum of the device capacitance, external and parasitic wire capacitances, R is a load resistance, and U_0 is the bias voltage, then Kirchhoff's laws lead to

$$U_0 = RI_0(t) + U(t) \tag{1}$$

$$I_0(t) = I(t) + C\frac{dU}{dt}. \tag{2}$$

Hence the temporal behavior of the voltage is determined by the circuit equation

$$\frac{dU(t)}{dt} = \frac{1}{C}\left(\frac{U_0 - U}{R} - I\right). \tag{3}$$

If a semiconductor element with negative differential conductance is operated in a reactive circuit, oscillatory instabilities may be induced by these reactive components, even if the relaxation time of the semiconductor is much smaller than that of the external circuit so that the semiconductor can be described by its stationary $I(U)$ characteristic and simply acts as a nonlinear resistor. Self-sustained semiconductor oscillations, where the semiconductor itself introduces an internal unstable temporal degree of freedom, must be distinguished from those circuit-induced oscillations. The self-sustained oscillations under time-independent external bias will be discussed in the following. Examples for internal degrees of freedom are the charge carrier density, or the electron temperature, or a junction capacitance within the device. Equation (3) is then supplemented by a dynamic equation for this internal variable.

1.2 Current Filaments, Field Domains, Fronts

Two important cases of negative differential conductivity (NDC) are described by an *N*-shaped or an *S*-shaped $j(\mathcal{E})$ characteristic, and denoted by *NNDC* and *SNDC*, respectively (Fig. 2). However, more complicated forms like Z-shaped, loop-shaped, or disconnected characteristics are also possible. NNDC and SNDC are associated with voltage- or current-controlled instabilities, respectively. In the NNDC case the current density is a single-valued function of the field, but the field is multivalued: the $\mathcal{E}(j)$ relation has three branches in a certain range of j. The SNDC case is complementary in the sense that $\mathcal{E}$ and j are interchanged. In case of *NNDC*, the NDC branch is often but not always - depending upon external circuit and boundary conditions - unstable against the formation of nonuniform field profiles along the charge transport direction (*electric field domains*), while in the *SNDC* case *current filamentation* generally occurs, i.e. the current density becomes nonuniform over the cross-section of the current flow and forms a conducting channel (Fig. 3). The elementary structures which make up these self-organized patterns are stationary or moving *fronts* representing the boundaries of the high-field domain or high-current filament. These primary self-organized spatial patterns may themselves become unstable in secondary bifurcation leading to periodically or chaotically breathing, rocking, moving, or spiking filaments or domains, or even solid-state turbulence and spatio-temporal chaos. Alternatively, the spatially uniform steady state may already become unstable with respect to uniform oscillations in a Hopf bifurcation.

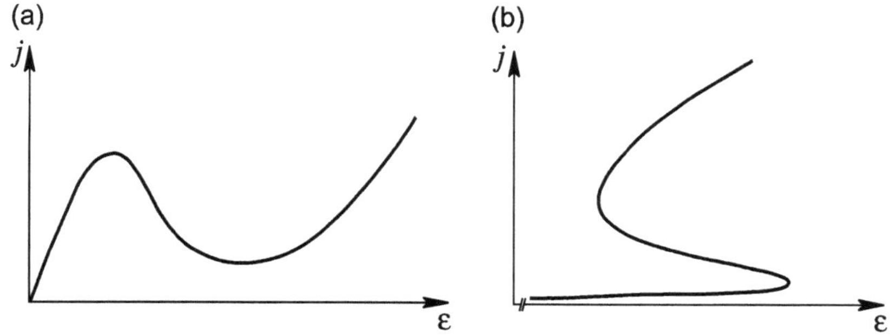

Fig. 2. Current density j versus electric field $\mathcal{E}$ for two types of negative differential conductivity (NDC): (**a**) NNDC; (**b**): SNDC (schematic)

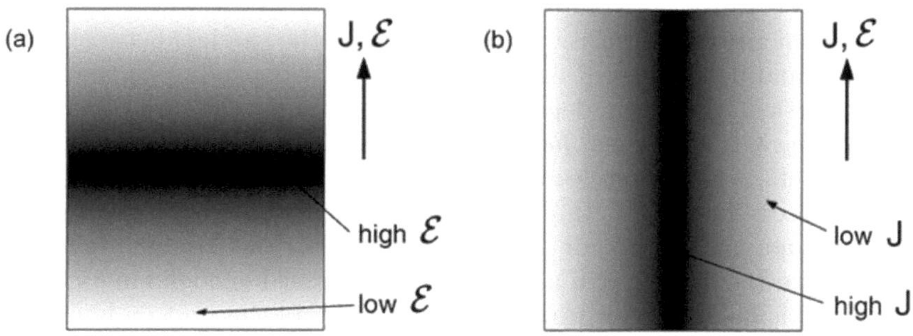

Fig. 3. Sketch of (**a**) a high-field domain, (**b**) a high-current filament, where j is the current density and $\mathcal{E}$ is the electric field

1.3 Mechanisms for Instabilities and Pattern Formation

Semiconductor instabilities may be classified as dominated by a bulk mechanism (e.g. drift instability; generation–recombination instability), or by heterojunctions and potential barriers and wells (e.g. resonant tunnelling across, or thermionic emission over, barriers in nanostructures) [17]. The first class of semiconductor oscillators includes the case of a *drift instability*. In the simplest extension of the Drude model, the current density is given by $j(\mathcal{E}) = -env(\mathcal{E})$ where $e > 0$ is the electron charge, n is the electron density, and $v(\mathcal{E})$ is the field dependent drift velocity, which may give rise to negative differential conductivity if $d|v|/d|\mathcal{E}| < 0$. The best known example is the *Gunn effect* [22]. The mechanism is based upon intervalley transfer of electrons in k-space from a state of high mobility to a state of low mobility under a strong electric field in direct semiconductors like GaAs or other III-V compounds. At low electric fields the electrons are essentially in the minimum of the central conduction band valley, which has a low effective mass m^* and hence a high mobility. As the field is increased, the electrons are heated up, and gain enough energy to be transferred to the satellite valley with a higher minimum energy, but larger effective mass, and hence lower mobility. As more and more electrons are transferred, the averaged mobility μ decreases strongly so that the current density $j = en\mu(\mathcal{E})\mathcal{E}$ decreases with increasing field, as a result of negative differential mobility $d(\mu\mathcal{E})/d\mathcal{E} < 0$. When most electrons are in the upper valley, j increases again. Thus an NNDC $j(\mathcal{E})$ characteristic is produced. Depending upon the cathode boundary condition, this may lead to the formation of travelling high field domains which show up as transit time oscillations of the current. They are used in real devices (Gunn diodes) to generate and amplify microwaves at frequencies typically beyond 1 GHz.

The large class of generation–recombination (GR) instabilities is distinguished by a nonlinear dependence of the steady-state carrier concentration n upon the field $\mathcal{E}$ that yields a non-monotonic relation $j = en(\mathcal{E})\mu\mathcal{E}$ of either NNDC or SNDC type [16]. This dependency is due to a redistribution of electrons between the conduction band and bound states with increasing field. The microscopic transition probabilities of the carriers between different states, and hence the GR coefficients, generally depend upon the electric field. A particularly strong dependence is expected for the rate constants of the following GR processes: *field-enhanced trapping* and emission from impurity centers, which often leads to NNDC, and *impact ionization*, which is the key process for SNDC. Impact ionization of carriers from impurity levels (shallow donors, acceptors, or deep traps) or across the bandgap (avalanching) is due to the following. If a free carrier has gained enough kinetic energy in the electric field, it may transfer this energy in a collision to a bound carrier which is then released to the conduction band; thereby an additional free carrier is generated which may, in turn, impact-ionise other carriers. Such a positive feedback (*autocatalysis*) leads to a rapid increase of the free-carrier density. Models of this type are relevant for a variety of materials and in various temperature ranges, and can explain SNDC and current filamentation in the regime of low-temperature impurity breakdown, and self-generated current oscillations including chaotic behavior, as observed experimentally.

A variety of instabilities can arise due to the specific transport properties of semiconductor heterostructures. One mechanism for NNDC, which is the real space analog of the k-space intervalley transfer in the Gunn effect, uses electron transfer between a high-mobility layer and a low-mobility layer in a modulation doped semiconductor heterostructure under a time-independent bias applied parallel to the layers [23]. In the NNDC regime current oscillations of $2 - 200$ MHz have been experimentally observed and theoretically explained. Another class of oscillatory instabilities occurs under *vertical* electrical transport in layered semiconductor structures, e.g. in the heterostructure hot-electron diode (HHED), or the double-barrier resonant tunnelling diode (DBRT), which are associated with S-shaped and Z-shaped current–voltage characteristics, respectively. A *resonant tunnelling structure* [24] is composed of alternating layers of two different semiconductor materials with different bandgaps (Fig. 4). The energy diagram shows a modulation of the conduction band edge on a nanometer scale, forming potential barriers and

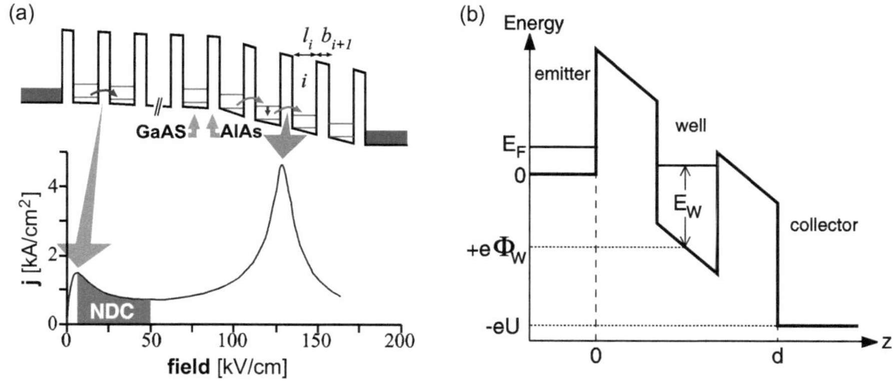

Fig. 4. (**a**) Superlattice exhibiting domain formation. The associated current density (j) versus field ($\mathcal{E}$) characteristic shows negative differential conductivity (NDC). The low-field domain corresponds to sequential tunnelling between equivalent levels of adjacent quantum wells (low-field peak of the $j(\mathcal{E})$ characteristic), while the high-field domain corresponds to resonant tunnelling between different levels of adjacent wells (high-field peak). (**b**) Schematic potential profile of the double barrier resonant tunnelling structure (DBRT). E_F and E_w denote the Fermi level in the emitter, and the energy level in the quantum well, respectively. U is the voltage applied across the structure

quantum wells. The current density across the barrier between two wells is due to quantum mechanical tunnelling and exhibits a strongly nonlinear dependency upon the electric field. It is maximum if there is maximum overlap between the occupied states in one well and the available unoccupied states in the other, i.e., if the energies are in resonance. For low fields equivalent levels in adjacent wells are approximately in resonance. With increasing field the energies of the two wells are shifted with respect to each other, and the available states in the collecting well are lowered with respect to the emitting well, and hence the current density drops as the overlap between the energy levels decreases, thereby displaying NDC. Upon further increase of the field the current density rises again up to a sharp resonance peak when the ground energy level in one quantum well is aligned with the second level in the neighbouring well. Thus resonant tunnelling produces NNDC.

The simplest system of this type consists of a double-barrier structure with one embedded quantum well in between, sandwiched between a highly doped emitter and collector region (DBRT), see Fig. 4b. However, the situation becomes more complicated if the nonlinear feedback between space charges and transport processes is taken into account. The built-up charge in the well leads to an electrostatic feedback mechanism which increases the energy of the well state supporting resonant tunnelling conditions for higher applied voltages. This may result in bistability and hysteresis where a high current and a low current state coexist for the same applied voltage and the current–voltage characteristic becomes Z-shaped. Switching between the two stable states as well as self-sustained current oscillations may occur under appropriate external circuit conditions. The bistability also provides a basis for lateral pattern formation (current filamentation) and spatio-temporal bifurcation scenarios including chaotic breathing and spiking.

Sequential resonant tunnelling in a periodic structure of multiple quantum wells, a *semiconductor superlattice*, likewise displays NNDC (Fig. 4a). Now, along the growth direction, the uniform field distribution may break up into a low-field domain where the field is near the first peak of the $j(\mathcal{E})$ characteristic, and a high-field domain where the field is close to the second, resonant-tunnelling peak. Depending upon the doping, the applied voltage, the structural parameters and the emitter contact conductivity, stationary or travelling domains occur, the latter leading to self-sustained current oscillations up to 150 GHz.

2 Chaos Control of Spatio-temporal Patterns

2.1 Time–Delay Autosynchronziation

Control of complex chaotic dynamics has evolved as one of the central issues in applied nonlinear science over the last decade [25]. One intriguing aspect is the possibility of noninvasive control. This refers to the stabilisation of a target state which is not changed by the control term, and the control force vanishes once the target state has been reached. A natural choice for the target state are unstable periodic orbits (UPOs), since they are dense in the chaotic attractor of the uncontrolled system. Chaos control makes use of noninvasive control methods applied to UPOs [26]. A particularly simple and efficient scheme uses time–delayed signals $s(t) - s(t-\tau)$ to generate control forces for stabilizing τ–periodic states [27] (time–delay autosynchronization or "Pyragas method"). It is simple to implement, quite robust, and has been applied successfully in real experiments. An extension to multiple time-delays (extended time–delay autosynchronization) has been proposed by Socolar et al. [28]:

$$\begin{aligned} F(t) &= (s(t) - s(t-\tau)) + RF(t-\tau) \\ &= \sum_{\nu=0}^{\infty} R^{\nu} \left(s(t-\nu\tau) - s(t-(\nu+1)\tau) \right) \end{aligned} \tag{4}$$

where F is the control force, and R is a memory parameter. Analytical insight into those schemes has been gained only recently [29, 30, 31, 32, 33, 34].

Various forms of the control force of time–delay autosynchronization have been applied to several models of semiconductor oscillators, viz impact ionization induced breakdown [35], real-space transfer [36], heterostructure hot-electron diodes [37, 38, 39], resonant tunnelling diodes [40], and superlattices [41]. Such self–stabilizing control schemes only require a time-delayed feedback, and should be straight-forward to implement in practical semiconductor devices, although this has not been realized so far. In the following we will demonstrate that it is possible to stabilize spatio-temporal patterns in semiconductor heterostructures by time–delay autosynchronization, using two models of semiconductor nanostructures which are of great current interest [1]:

(i) Charge accumulation in the quantum-well of a double-barrier resonant-tunnelling diode (DBRT) may result in lateral spatio-temporal patterns of the current density. Various oscillatory instabilities in form of periodic or chaotic breathing and spiking modes may occur [42]. We demonstrate that unstable current density patterns, e.g., periodic breathing oscillations, can be stabilized in a wide parameter range by means of a simple delayed feedback loop [40]. We compare the control domains of different control schemes, relating them to their Floquet spectrum.

(ii) Electron transport in semiconductor superlattices shows strongly nonlinear spatio-temporal dynamics. Complex scenarios including chaotic motion of multiple fronts have been found under time-independent bias conditions [43]. Unstable periodic orbits corresponding to travelling field domain modes can be stabilized by time delayed feedback control. We apply a novel control scheme using a low-pass filtered global feedback [41].

2.2 Spatio-temporal Oscillations in Resonant Tunnelling Diodes

As a first example we consider a double–barrier resonant tunnelling diode (DBRT) as sketched in Fig. 4b, which exhibits a Z-shaped (bistable) current–voltage characteristic [1]. We include the lateral re–distribution of electrons in the quantum well plane (x coordinate) giving rise to filamentary current flow [44, 45]. Complex chaotic scenarios including spatio-temporal breathing and spiking oscillations have been found in a simple reaction–diffusion model of activator–inhibitor type [42]. We extend this model to include control terms, and obtain the following equations, where we use dimensionless variables throughout:

$$\frac{\partial a}{\partial t} = \frac{\partial}{\partial x}\left(D(a)\frac{\partial a}{\partial x}\right) + f(a,u) - KF_a(x,t) \tag{5}$$

$$\frac{du}{dt} = \frac{1}{\varepsilon}\left(U_0 - u - r\langle j\rangle\right) - KF_u(t) \tag{6}$$

Here $u(t)$ is the inhibitor and $a(x,t)$ is the activator variable. In the semiconductor context $u(t)$ denotes the (normalized) voltage drop across the device and $a(x,t)$ is the (normalized) electron density in the quantum well. Equation (5) is the continuity equation for the charge in the well. The nonlinear, nonmonotonic function $f(a,u)$ describes the balance of the incoming and outgoing current densities of the quantum well, and $D(a)$ is an effective, electron

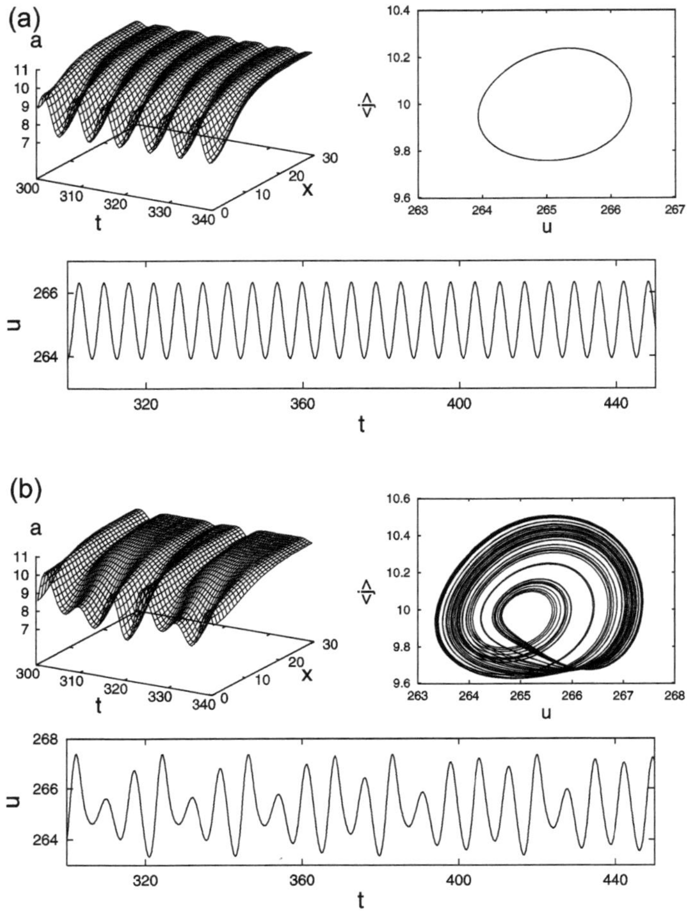

Fig. 5. Spatio-temporal breathing patterns in the DBRT: electron density evolution, phase portrait, and voltage time series for (**a**)$\varepsilon = 7.0$: periodic breathing, (**b**) $\varepsilon = 9.1$: chaotic breathing ($r = -35$,$U_0 = 84.2895$,$L = 30$)

density dependent transverse diffusion coefficient. The local current density in the device is $j(a, u) = \frac{1}{2}(f(a, u) + 2a)$, and $\langle j \rangle \equiv \frac{1}{L}\int_0^L j dx$ is associated with the global current. Equation (6) represents Kirchhoff's law of the circuit (3) in which the device is operated. The external bias voltage U_0, the dimensionless load resistance r, and the time-scale ratio $\varepsilon = rC/\tau_a$ (where C is the capacitance of the circuit and τ_a is the tunnelling time) act as control parameters. By choosing $r < 0$, which can be realized by an active circuit, i.e. applying an additional control voltage proportional to the device current $\langle j \rangle$ in series with the bias U_0 [46], it is in principle possible to stabilize the middle branch of the stationary $j(u)$ characteristic and access it experimentally. The one–dimensional spatial coordinate x corresponds to the direction transverse to the current flow. We consider a system of width L with Neumann boundary conditions $\partial_x a = 0$ at $x = 0, L$, corresponding to no charge transfer through the lateral boundaries.

Equations (5),(6) contain control forces F_a and F_u for stabilizing time periodic spatial patterns. The strength of the control terms is proportional to the control amplitude K, which gives one important parameter of each control scheme. In the semiconductor context these forces can be implemented by appropriate electronic feedback circuits [37].

The dynamics of the free system, i.e. $K = 0$, develops temporally chaotic and spatially nonuniform states (spatio–temporal breathing or spiking) in appropriate parameter regimes [42], as shown in Fig. 5. In the semiconductor context the time and length scales of our dimensionless variables are typically given by picoseconds and 100 nanometers, respectively. A typical bifurcation diagram exhibiting a period-doubling route to chaos is shown in Fig. 6.

We are concerned with controlling unstable time periodic patterns $u_p(t) = u_p(t + \tau)$, $a_p(x,t) = a_p(x, t + \tau)$ which are embedded in a chaotic attractor. For that purpose we apply control forces F_a and F_u which are derived from time–delayed differences of the voltage and the charge density. For example we may choose $F_u = F_{\mathrm{vf}}$ with the voltage feedback force

$$F_{\mathrm{vf}}(t) = u(t) - u(t - \tau) + RF_{\mathrm{vf}}(t - \tau) \tag{7}$$

(extended time–delay autosynchronization). The control forces are designed such that they vanish exactly on a target orbit of period τ.

Here we concentrate on the question how the coupling of the control forces to the internal degrees of freedom influences the performance of the control. For our model we consider two different choices for the control force F_a. On the one hand we use a force which is based on the local charge density according to

$$F_{\mathrm{loc}}(x,t) = a(x,t) - a(x, t - \tau) + RF_{\mathrm{loc}}(x, t - \tau) \quad . \tag{8}$$

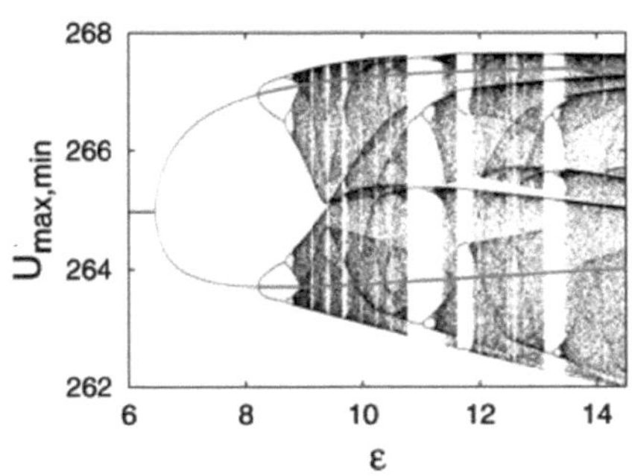

Fig. 6. Chaotic bifurcation diagram of the resonant tunnelling diode. The maxima and minima of the voltage oscillations are plotted versus the time-scale parameter ε

whereas on the other hand we propose a construction which is only based on its spatial average

$$F_{\mathrm{glo}}(t) = \langle a\rangle(t) - \langle a\rangle(t-\tau) + RF_{\mathrm{glo}}(t-\tau) \quad . \tag{9}$$

We call the choice $F_a = F_{\mathrm{loc}}$ a *local* control scheme in contrast to the *global* control scheme $F_a = F_{\mathrm{glo}}$ which requires only the global average and does not depend explicitly on the spatial variable. The second option has considerable experimental advantages since the spatial average is related to the total charge in the quantum well and does not require a spatially resolved measurement.

In general the analysis of the control performance of time–delayed feedback methods results in differential–difference equations which are hard to tackle. Stability of the orbit is governed by eigenmodes and the corresponding complex valued growth rates Λ (Floquet exponents). There exists a simple case (which we call *diagonal* control) where analytical results are available [30, 47], namely for $F_a = F_{\mathrm{loc}}$ and $F_u = F_{\mathrm{vf}}$. It is a straightforward extension to a spatially extended system of an identity matrix for the control of discrete systems of ordinary differential equations (cf. [29]). Fig. 7a-c shows successful control of a chaotic breathing oscillation after the control force is switched on.

In Fig. 7d, e the regime of successful control in the K–R parameter plane and the real part of the Floquet spectrum $\Lambda(K)$ for $R = 0$ is depicted. The control domain has a typical triangular shape bounded by a flip instability (period doubling bifurcation: Re $\Lambda = 0$, Im $\Lambda = \pi/\tau$) to its left and by a Hopf (torus) bifurcation to its right. Inclusion of the memory parameter R increases the range of K over which control is achieved. We observe that the numerical result fits very well with the analytical prediction.

To confirm the bifurcations at the boundaries we consider the real part of the Floquet spectrum of the orbit subjected to control. Complex conjugate Floquet exponents show up as doubly degenerate pairs. The largest nontrivial exponent decreases with increasing K and collides at negative values with a branch coming from negative infinity. As a result a complex conjugate pair develops and real parts increase again. The real part of the exponent finally crosses the zero axis giving rise to a Hopf bifurcation. Our numerical simulations are in agreement with the analytical result.

Let us now replace the local control force $F_a = F_{\mathrm{loc}}$ by the global control $F_a = F_{\mathrm{glo}}$. Figure 8a, b shows the corresponding control regime and Floquet spectrum. The control domain looks similar in shape as for diagonal control, although the domain for the global scheme is drastically reduced. The shift in the control boundaries is due to different branches of the Floquet spectrum crossing the (Re $\Lambda = 0$)-axis.

It is now interesting to note that if we keep $F_a = F_{\mathrm{glo}}$ as before but remove the voltage feedback completely, the control domain is shifted to higher K values and at the same time is dramatically increased (Fig. 8c).

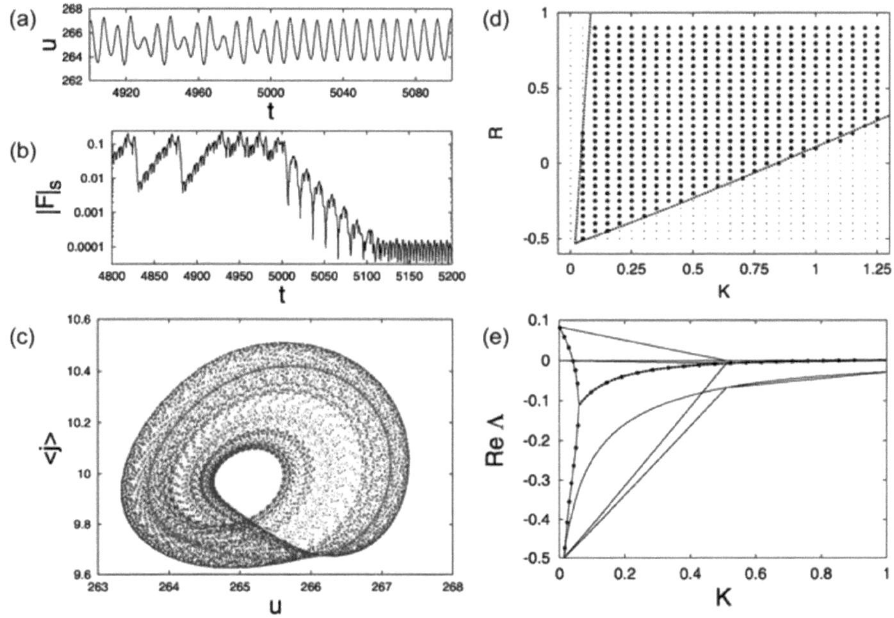

Fig. 7. Diagonal control in the DBRT, where the control force is switched on at $t = 5000$. (**a**) Voltage u vs. time, (**b**) Supremum of the control force vs. time, (**c**) Phase portrait (global current vs. voltage) showing the chaotic breathing attractor and the embedded stabilized periodic orbit. Parameters: $r = -35$, $\varepsilon = 9.1$, $\tau = 7.389$, $K = 0.137$, $R = 0$. (**d**) Control domain in the K–R parameter plane for diagonal control of the unstable periodic orbit with period $\tau = 7.389$. $*$ successful control in the numerical simulation, $\cdot$ no control, lines: analytical result according to [47]. (**e**) Leading real parts Λ of the Floquet spectrum for diagonal control in dependence on K ($R = 0$)

From the Floquet spectrum we see that after the flip bifurcation the largest Floquet exponent does not immediately hybridize into a complex conjugate pair, but the Hopf bifurcation is caused by another complex conjugate pair which is not connected to the largest Floquet exponent. Thereby the Hopf bifurcation is suppressed and the control regime is increased. This behavior is very similar to the one observed in a different reaction–diffusion model (modeling a heterostructure hot electron diode, HHED) [38], where it was found that additional control of the global variable u may gradually reduce the control regime to zero.

Finally, we note that the period-one orbit can be stabilized by our control scheme throughout the whole bifurcation diagram including chaotic bands and windows of higher periodicity, as marked by two solid lines in Fig. 6 for diagonal control. Thus our method represents a way of obtaining stable self-sustained voltage oscillations independently of parameter fluctuations.

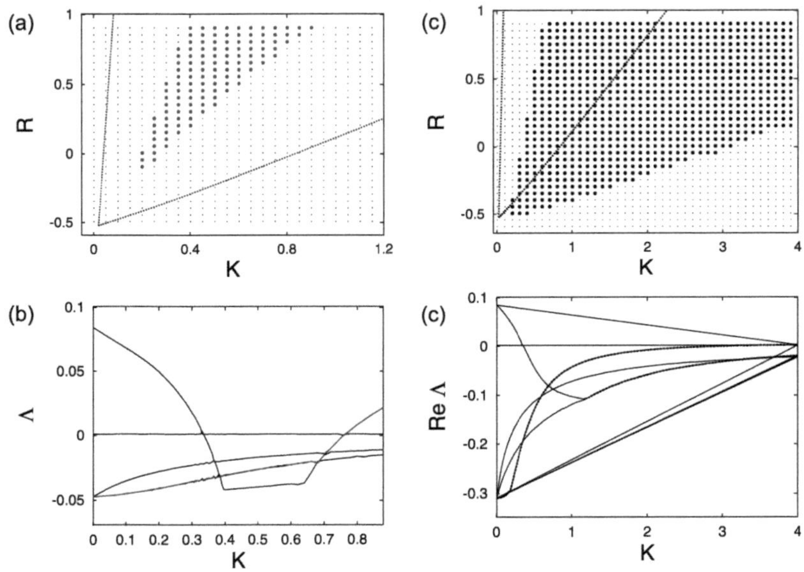

Fig. 8. Same as Fig. 7d-e in (**a**) and (**b**) for global control and in (**c**) and (**d**) for global control without voltage feedback ((**b**) $R = 0.7$, (**d**) $R = 0.1$)

2.3 Domains and Fronts in Superlattices

Semiconductor superlattices [48] have been demonstrated to give rise to self-sustained current oscillations ranging from several hundred MHz [49, 50, 51] to 150 GHz at room temperature [52]. Chaotic scenarios have been found experimentally [53, 54, 55] and described theoretically in periodically driven [56] as well as in undriven systems [43]. For a reliable operation of a superlattice as an ultra-high frequency oscillator such unpredictable and irregular conditions should be avoided, which might not be easy in practice.

Here we focus on simulations of dynamic scenarios for superlattices under fixed time-independent external voltage in the regime where self-sustained dipole waves are spontaneously generated at the emitter. The dipole waves are associated with traveling field domains, and consist of electron accumulation and depletion fronts which in general travel at different velocities and may merge and annihilate. Depending on the applied voltage and the contact conductivity, this gives rise to various oscillation modes as well as different routes to chaotic behavior [43].

Our model of a superlattice is based on sequential tunnelling of electrons, see Fig. 4a. In the framework of this model the quantum wells are assumed to be only weakly coupled, and electrons are localized at these wells. The tunnelling rate to the next well is lower than the typical relaxation rate between the different energy levels within one well. The electrons within one well are then in quasi–equilibrium and transport through the barrier is incoherent.

The resulting tunnelling current density $J_{m\to m+1}(F_m, n_m, n_{m+1})$ from well m to well $m+1$ depends only on the electric field F_m between both wells and the electron densities n_m and n_{m+1} in the respective wells (in units of cm^{-2}). A detailed microscopic derivation of the model has been given elsewhere [57].

The rate of variation of electron density in well m is governed by the continuity equation

$$e\frac{dn_m}{dt} = J_{m-1\to m} - J_{m\to m+1} \quad \text{for } m = 1, \dots N \tag{10}$$

and the discrete version of Gauss's law

$$\epsilon_r \epsilon_0 (F_m - F_{m-1}) = -e(n_m - N_D) \quad \text{for } m = 1, \dots N, \tag{11}$$

where N is the number of wells in the superlattice, ϵ_r and ϵ_0 are the relative and absolute permittivities, $e > 0$ is the electron charge, N_D is the donor density, and F_0 and F_N are the fields at the emitter and collector barrier, respectively. The total applied voltage U between emitter and collector imposes a global constraint

$$U = -\sum_{m=0}^{N} F_m d, \tag{12}$$

where d is the superlattice period. This, together with (11), allows us to eliminate the field variables $F_m(n_1, ..., n_N, U)$ from the dynamic equations.

At the contacts Ohmic boundary conditions $J_{0\to 1} = \sigma F_0$, $J_{N\to N+1} = \sigma F_N n_N / N_D$ are chosen, where σ is the Ohmic contact conductivity, and the factor n_N/N_D is introduced in order to avoid negative electron densities at the collector. The value of σ essentially determines the oscillation mode [43].

If the contact conductivity σ is chosen appropriately, electron accumulation and depletion fronts are generated at the emitter. Those fronts form a traveling high field domain, with leading electron depletion front and trailing accumulation front. This leads to self-generated current oscillations. A fixed voltage U imposes a constraint on the lengths of the high-field domains and thus on the front velocities. If N_a accumulations fronts and N_d depletion fronts are present, the respective front velocities v_a and v_d must obey $v_d/v_a = N_a/N_d$. If the accumulation and depletion fronts have different velocities, they may collide in pairs and annihilate. At certain combinations of contact conductivity σ and voltage U, chaotic motion arises, when the annihilation of fronts of opposite polarity occurs at irregular positions within the superlattice [43], leading to complex bifurcation diagrams. In Fig. 9a the density plot of the positions (well numbers) at which two fronts annihilate is shown as a function of the voltage. We see that for low voltage the annihilation takes place at a definite position in the superlattice with a variation of only a few wells. This distribution broadens for increasing voltage in characteristic bifurcation scenarios reminiscent of period doubling, leading to chaotic

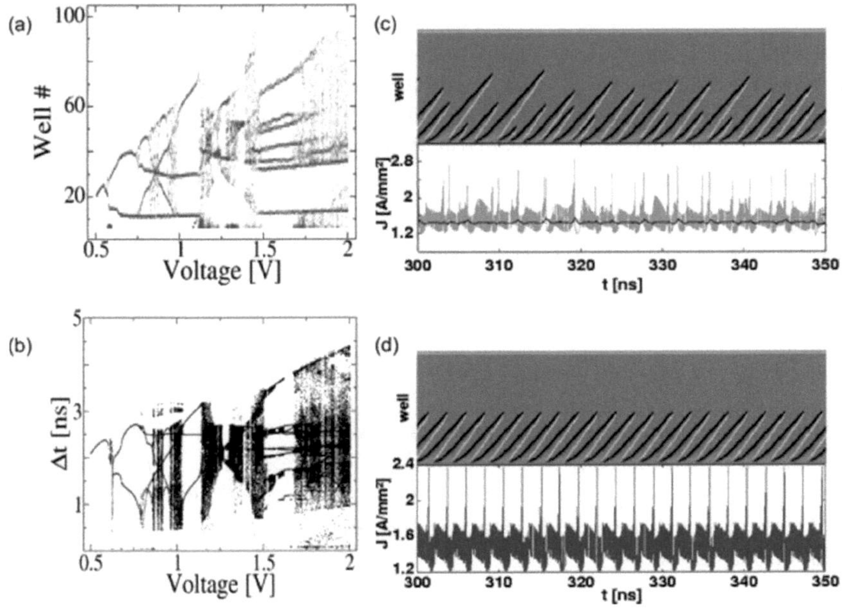

Fig. 9. (**a**), (**b**) One-parameter bifurcation diagram: (a) Positions where accumulation and depletion fronts annihilate vs voltage at $\sigma = 0.5 \quad \Omega^{-1}\mathrm{m}^{-1}$. The grayscale indicates high (black) and low (white) numbers of annihilations at a given well. (**b**) Time differences between consecutive maxima of the electron density in well no. 20 vs voltage. Time series of length 600 ns have been used for each value of the voltage. Simulation of an $N = 100$ superlattice with $\mathrm{Al_{0.3}Ga_{0.7}As}$ barriers of width $b = 5$nm and GaAs quantum wells of width $w = 8$nm, doping density $N_D = 1.0 \times 10^{11}\mathrm{cm}^{-2}$ at $T = 20$K. (**c**) Space-time plot of the uncontrolled charge density, and current density J vs. time. (**d**) Same for extended time-delay autosynchronization with global voltage control with exponentially weighted current density (denoted by the black curve). $U = 1.15$ V, $\tau = 2.29$ ns, $K = 3 \times 10^{-6} Vmm^2/A$, $R = 0.2$, $\alpha = 10^9 s^{-1}$. Light and dark regions denote electron accumulation and depletion fronts in the space-time plots of the charge densities, respectively

regimes. In Fig. 9b the time differences Δt between two consecutive maxima of the electron density in a specified well are plotted versus U. Chaotic bands and periodic windows can be clearly seen.

The transition from periodic to chaotic oscillations is enlightened by considering the space-time plot for the evolution of the electron densities (Fig. 9c). At $U = 1.15$V chaotic front patterns with irregular sequences of annihilation of front pairs at varying positions within the superlattice occur. The largest Lyapunov exponent is calculated as $1.1 \times 10^9 s^{-1}$ which is a clear indication of chaos.

We shall now introduce a time delayed feedback loop to control the chaotic front motion and stabilize a periodic oscillation mode which is inherent in

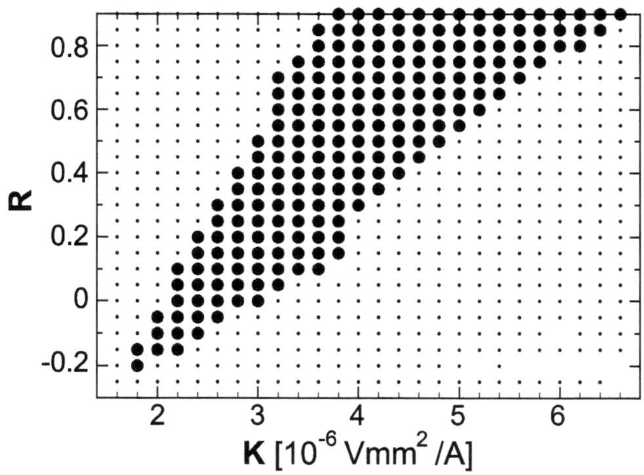

Fig. 10. Control domain for global voltage control with exponentially weighted current density. Full circles denote successful control, small dots denote no control. Parameters as in Fig. 9

the chaotic attractor. As we have seen in the previous section, local coupling schemes usually lead to efficient control in a large control domain, but they are not easily implemented in real systems since local, spatially resolved measurements are necessary. Therefore, here we use a much simpler global scheme. In our problem, as a global output signal that is coupled back in the feedback loop, it is natural to use the total current density J defined as follows: $J = \frac{1}{N+1}\sum_{m=0}^{N} J_{m\to m+1}$ [57]. For the uncontrolled chaotic oscillations, J is given in Fig. 9c by grey, showing irregular spikes at those times when two fronts annihilate. Note that the grey current time trace is modulated by fast small-amplitude oscillations (due to well-to-well hopping of depletion and accumulation fronts in our discrete model) which are not resolved in the plot. However, as the variable J is fed back to the system for the purposes of control, these high–frequency oscillations render the control loop unstable. They need to be filtered out by using, e.g., the following low–pass filter:

$$\overline{J}(t) = \alpha \int_0^t J(t')e^{-\alpha(t-t')}\mathrm{d}t', \tag{13}$$

with a cut-off frequency α.

The multiple time delays of the resulting signal $\overline{J}$ (Fig. 9c, black curve) are then used to modulate the voltage U across the superlattice:

$$\begin{aligned} U &= U_0 + U_c(t) \\ U_c(t) &= -K\left(\overline{J}(t) - \overline{J}(t-\tau)\right) + RU_c(t-\tau) \end{aligned} \tag{14}$$

where U_0 is a time–independent external bias, and U_c is the control voltage. Such a global control scheme is easy to implement experimentally. It is non-invasive in the sense that the control force vanishes when the target state of period τ has been reached. This target state is an unstable periodic orbit of the uncontrolled system. The period τ can be determined by observing the resonance-like behavior of the mean control force versus τ. The result of the control is shown in Fig. 9d. The front dynamics exhibits annihilation of front pairs at fixed positions within the superlattice, and stable periodic oscillations of the current are obtained. In Fig. 10 the control domain is depicted in the parameter plane of the amplitude K of the control force and and the memory parameter R. A typical horn-like control domain similar to the ones known from other coupling schemes (see Sect. 2.2) is found.

In conclusion, we have reviewed self-organized spatio-temporal pattern formation in semiconductor systems. It has been demonstrated that time-delay autosynchronization represents a convenient and simple scheme for the self-stabilisation of high-frequency current oscillations associated with space–time patterns in semiconductor nanostructures. This approach lacks the drawback of synchronization by external forcing, since it requires nothing but delaying of the global electrical system output by the specified time lag. The proposed low-pass filtering of the output signal presents a solution of the problem one necessarily encounters when trying to control a nano-system with a crucially discrete quantum structure leading to superimposed fast well-to-well hopping oscillations in case of the superlattice.

This work was supported by DFG in the framework of Sfb 555. I am indebted to stimulating collaboration and discussion with A. Amann, N. Baba, O. Beck, N. Janson, W. Just, S. Popovich, P. Rodin, J. Schlesner, J. E. S. Socolar, G. Stegemann, J. Unkelbach, A. Wacker.

References

1. E. Schöll: *Nonlinear spatio-temporal dynamics and chaos in semiconductors* (Cambridge University Press, Cambridge, 2001).
2. H. Haken: *Synergetics, Introduction and Advanced Topics* (Springer, Berlin, Heidelberg 2004)
3. A.S. Mikhailov: *Foundations of Synergetics Vol. I* (Springer, Berlin, 1994), 2nd ed.
4. F.H. Busse and S.C. Müller: *Evolution of spontaneous structures in dissipative continuous systems* (Springer, Berlin, 1998).
5. H. Lueder, W. Schottky, and E. Spenke: *Zur technischen Beherrschung des Wärmedurchschlags*, Naturwiss. **24**, 61 (1936).
6. K.W. Böer and P.L. Quinn: *Inhomogeneous field distribution in homogeneous semiconductor s having an N-shaped negative differential conductivity*, phys. stat. sol. **17**, 307 (1966).
7. A.F. Volkov and S.M. Kogan: *Physical phenomena in semiconductors with negative differential conductivity*, Sov. Phys. Usp. **11**, 881 (1969), [Usp. Phys. Nauk **96**, 633 (1968)].
8. E. Pytte and H. Thomas: *Soft modes,critical fluctuations,and optical properties for a two-valley model of Gunn- instability semiconductors*, Phys. Rev. **179**, 431 (1969).

9. V.L. Bonch-Bruevich, I.P. Zvyagin, and A.G. Mironov: *Domain electrical instabilities in semiconductors* (Consultant Bureau, New York, 1975).
10. P.T. Landsberg and A. Pimpale: *Recombination-induced nonequilibrium phase transitions in semiconductors*, J. Phys. C **9**, 1243 (1976).
11. P.T. Landsberg, D.J. Robbins, and E. Schöll: *Threshold switching as a generation-recombination induced non-equilibrium phase transition*, phys. status solidi (a) **50**, 423 (1978).
12. E. Schöll and P.T. Landsberg: Proc. Roy. Soc. A **365**, 495 (1979).
13. E. Schöll: *Current layers and filaments in a semiconductor model with an impact ionization induced instability*, Z. Phys. B **48**, 153 (1982).
14. K. Aoki, T. Kobayashi, and K. Yamamoto: J. Phys. Soc. Jpn. B **51**, 2372 (1982).
15. S.W. Teitsworth, R.M. Westervelt, and E.E. Haller: *Nonlinear oscillations and chaos in electrical breakdown in Ge*, Phys. Rev. Lett. **51**, 825 (1983).
16. E. Schöll: *Nonequilibrium Phase Transitions in Semiconductors* (Springer, Berlin, 1987).
17. M.P. Shaw, V.V. Mitin, E. Schöll, and H.L. Grubin: *The Physics of Instabilities in Solid State Electron Devices* (Plenum Press, New York, 1992).
18. J. Peinke, J. Parisi, O. Rössler, and R. Stoop: *Encounter with Chaos* (Springer, Berlin, Heidelberg, 1992).
19. B.S. Kerner and V.V. Osipov: *Autosolitons* (Kluwer Academic Publishers, Dordrecht, 1994).
20. F.-J. Niedernostheide (Editor): *Nonlinear Dynamics and Pattern Formation in Semiconductors and Devices* (Springer, Berlin, 1995).
21. K. Aoki: *Nonlinear dynamics and chaos in semiconductors* (Institute of Physics Publishing, Bristol, 2000).
22. J.B. Gunn: *Instabilities of current in III-V semiconductors*, IBM J. Res. Develop. **8**, 141 (1964).
23. Z.S. Gribnikov, K. Hess, and G. Kosinovsky: *Nonlocal and nonlinear transport in semiconductors: Real-space transfer effects*, J. Appl. Phys. **77**, 1337 (1995).
24. L. Esaki and L.L. Chang: *New transport phenomenon in a semiconductor superlattice*, Phys. Rev. Lett. **33**, 495 (1974).
25. H.G. Schuster: *Handbook of chaos control* (Wiley-VCH, Weinheim, 1999).
26. E. Ott, C. Grebogi, and J.A. Yorke: *Controlling chaos*, Phys. Rev. Lett. **64**, 1196 (1990).
27. K. Pyragas: *Continuous control of chaos by self-controlling feedback*, Phys. Lett. A **170**, 421 (1992).
28. J.E.S. Socolar, D.W. Sukow, and D.J. Gauthier: *Stabilizing unstable periodic orbits in fast dynamical systems*, Phys. Rev. E **50**, 3245 (1994).
29. M.E. Bleich and J.E.S. Socolar: *Stability of periodic orbits controlled by time-delay feedback*, Phys. Lett. A **210**, 87 (1996).
30. W. Just, T. Bernard, M. Ostheimer, E. Reibold, and H. Benner: *Mechanism of time-delayed feedback control*, Phys. Rev. Lett. **78**, 203 (1997).
31. H. Nakajima: *On analytical properties of delayed feedback control of chaos*, Phys. Lett. A **232**, 207 (1997).
32. K. Pyragas: *Analytical properties and optimization of time-delayed feedback control*, Phys. Rev. E **66**, 26207 (2002).
33. W. Just, S. Popovich, A. Amann, N. Baba, and E. Schöll: *Improvement of time–delayed feedback control by periodic modulation: analytical theory of Floquet mode control scheme*, Phys. Rev. E **67**, 026222 (2003).

34. W. Just, H. Benner, and E. Schöll: *Control of chaos by time–delayed feedback: a survey of theoretical and experimental aspects*, in *Advances in Solid State Phyics*, edited by B. Kramer (Springer, Berlin, 2003).
35. E. Schöll and K. Pyragas: *Tunable semiconductor oscillator based on self-control of chaos in the dynamic hall effect*, Europhys. Lett. **24**, 159 (1993).
36. D.P. Cooper and E. Schöll: *Tunable real space transfer oscillator by delayed feedback control of chaos*, Z. f. Naturforsch. **50a**, 117 (1995).
37. G. Franceschini, S. Bose, and E. Schöll: *Control of chaotic spatiotemporal spiking by time-delay autosynchronisation*, Phys. Rev. E **60**, 5426 (1999).
38. O. Beck, A. Amann, E. Schöll, J.E.S. Socolar, and W. Just: *Comparison of time-delayed feedback schemes for spatio-temporal control of chaos in a reaction-diffusion system with global coupling*, Phys. Rev. E **66**, 016213 (2002).
39. N. Baba, A. Amann, E. Schöll, and W. Just: *Giant improvement of time-delayed feedback control by spatio-temporal filtering*, Phys. Rev. Lett. **89**, 074101 (2002).
40. J. Unkelbach, A. Amann, W. Just, and E. Schöll: *Time–delay autosynchronization of the spatio-temporal dynamics in resonant tunneling diodes*, Phys. Rev. E **68**, 026204 (2003).
41. J. Schlesner, A. Amann, N.B. Janson, W. Just, and E. Schöll: *Self-stabilization of high–frequency oscillations in semiconductor superlattices by time–delay autosynchronization*, Phys. Rev. E **68**, 066208 (2003).
42. E. Schöll, A. Amann, M. Rudolf, and J. Unkelbach: *Transverse spatio-temporal instabilities in the double barrier resonant tunneling diode*, Physica B **314**, 113 (2002).
43. A. Amann, J. Schlesner, A. Wacker, and E. Schöll: *Chaotic front dynamics in semiconductor superlattices*, Phys. Rev. B **65**, 193313 (2002).
44. A. Wacker and E. Schöll: *Criteria for stability in bistable electrical devices with S- or Z-shaped current voltage characteristic*, J. Appl. Phys. **78**, 7352 (1995).
45. V. Cheianov, P. Rodin, and E. Schöll: *Transverse coupling in bistable resonant-tunneling structures*, Phys. Rev. B **62**, 9966 (2000).
46. A.D. Martin, M.L.F. Lerch, P.E. Simmonds, and L. Eaves: *Observation of intrinsic tristability in a resonant tunneling structure*, Appl. Phys. Lett. **64**, 1248 (1994).
47. W. Just, E. Reibold, H. Benner, K. Kacperski, P. Fronczak, and J. Holyst: *Limits of time-delayed feedback control*, Phys. Lett. A **254**, 158 (1999).
48. L. Esaki and R. Tsu: *Superlattice and negative differential conductivity in semiconductors*, IBM J. Res. Develop. **14**, 61 (1970).
49. Y. Kawamura, K. Wakita, H. Asahi, and K. Kurumada: *Observation of room temperature current oscillation in InGaAs/InAlAs MQW pin diodes*, Jpn. J. Appl. Phys. **25**, L928 (1986).
50. J. Kastrup, R. Klann, H.T. Grahn, K. Ploog, L.L. Bonilla, J. Galán, M. Kindelan, M. Moscoso, and R. Merlin: *Self-oscillations of domains in doped GaAs-AlAs superlattices*, Phys. Rev. B **52**, 13761 (1995).
51. K. Hofbeck, J. Grenzer, E. Schomburg, A.A. Ignatov, K.F. Renk, D.G. Pavel'ev, Y. Koschurinov, B. Melzer, S. Ivanov, S. Schaposchnikov, and P.S. Kop'ev: *High-frequency self-sustained current oscillation in an Esaki-Tsu superlattice monitored via microwave emission*, Phys. Lett. A **218**, 349 (1996).
52. E. Schomburg, R. Scheuerer, S. Brandl, K.F. Renk, D.G. Pavel'ev, Y. Koschurinov, V. Ustinov, A. Zhukov, A. Kovsh, and P.S. Kop'ev: *InGaAs/InAlAs superlattice oscillator at 147 GHz*, Electronics Letters **35**, 1491 (1999).

53. Y. Zhang, J. Kastrup, R. Klann, K.H. Ploog, and H.T. Grahn: *Synchronization and chaos induced by resonant tunneling in GaAs/AlAs superlattices*, Phys. Rev. Lett. **77**, 3001 (1996).
54. K.J. Luo, H.T. Grahn, K.H. Ploog, and L.L. Bonilla: *Explosive bifurcation to chaos in weakly coupled semiconductor superlattices*, Phys. Rev. Lett. **81**, 1290 (1998).
55. O.M. Bulashenko, K.J. Luo, H.T. Grahn, K.H. Ploog, and L.L. Bonilla: *Multifractal dimension of chaotic attractors in a driven semiconductor superlattice*, Phys. Rev. B **60**, 5694 (1999).
56. O.M. Bulashenko and L.L. Bonilla: *Chaos in resonant-tunneling superlattices*, Phys. Rev. B **52**, 7849 (1995).
57. A. Wacker: *Semiconductor superlattices: A model system for nonlinear transport*, Phys. Rep. **357**, 1 (2002).

Group Theoretic Methods in the Theory of Pattern Formation

Rudolf Friedrich

Institute of Theoretical Physics, Westfälische Wilhelms-Universität Münster, Wilhelm-Klemm-Str. 9, 48149 Münster, Germany
fiddir@uni-muenster.de

1 Introduction

Nonequilibrium systems like lasers, fluid motions, and chemical reactions exhibit a rich variety of instabilities. A common feature of these transitions is symmetry breaking [1], [2], [3], [4]. In contrast to equilibrium systems, whose final states are determined by the minimum of a thermodynamic potential, nonequilibrium systems can allow for the emergence of temporal, spatial and spatio-temporal structures whose properties have to be described in terms of dynamical systems. Close to transitions between different states, these equations are evolution equations for a set of *order parameters*, which explicitly determine the temporal behaviour [1], [2]. In pattern forming systems, the order parameters usually are amplitudes of spatial patterns. These, in turn, can be characterized by their symmetries. Therefore, group theoretic methods are extremely helpful in the characterization of nonequilibrium phase transitions. The present article gives an overview on the application of group theoretic methods to pattern formation. It can not cover all aspects and we refer the reader to the excellent review articles and monographs [6], [7], [8] [9], [10], [11]. We shall put our attention onto slightly different topics: After a brief summary of the mathematical treatment of the theory of pattern formation we introduce necessary group theoretic notations. We discuss the structure of order parameter equations and several useful group theoretic notions for their investigation. Then we turn to natural patterns in large aspect ratio systems, i.e. patterns which include defects, grain boundaries and domain walls. A subsequent section is devoted to higher instabilities in complex systems and indicate why Goldstone modes play a crucial role. Furthermore, a discussion of the impact of fluctuations on the behaviour of dissipative solitons close to drift bifurcations is included.

2 Pattern Formation in Nonequilibrium Systems

The theory of pattern formation in hydrodynamical, optical, chemical, and biological systems is based on continuum theoretical field equations for a state vector $\mathbf{q}(\mathbf{r},t) = \{q_\alpha(\mathbf{r},t)\}$, which consist of velocity fields, light fields, chemical concentrations etc.. It evolves according to an evolution equation of the form

$$\frac{\partial}{\partial t}\mathbf{q}(\mathbf{r},t) = \tilde{N}[\mathbf{q}(\mathbf{r},t),\nabla,\boldsymbol{\sigma}] + \mathbf{F}(\mathbf{q},t)\,. \tag{1}$$

Fluctuations, which are a signature of the microscopic constituents of the systems under considerations, are denoted by $\mathbf{F}(\mathbf{q},t)$. $\mathbf{N}$ is a nonlinear functional of $\mathbf{q}$ depending on control parameters $\boldsymbol{\sigma}$, which account for a variation of experimental conditions etc..

In the following we consider the deterministic case $\mathbf{F}=0$ assuming that the evolution equation allows for a time-independent basic state $\mathbf{q}^0(\mathbf{r})$ This state should exist for a compact region in the space of control parameters $\boldsymbol{\sigma}$.

It is convenient to take the basic state as a reference state and introduce deviations $\mathbf{w}(\mathbf{r},t)$:

$$\mathbf{q}(\mathbf{r},t) = \mathbf{q}^0(\mathbf{r}) + \mathbf{w}(\mathbf{r},t)\ . \tag{2}$$

An evolution equation for the deviation $\mathbf{w}(\mathbf{r},t)$ is obtained by inserting this ansatz into the evolution equation (1):

$$\dot{\mathbf{w}}(\mathbf{r},t) = N[\mathbf{q}^0(\mathbf{r}) + \mathbf{w}(\mathbf{r},t),\nabla,\boldsymbol{\sigma}]\ . \tag{3}$$

Since $N[\mathbf{q}^0(\mathbf{r})\nabla,\boldsymbol{\sigma}]=0$ an expansion of the nonlinearity yields:

$$\begin{aligned} N[\mathbf{q}^0(\mathbf{r}) + \mathbf{w}(\mathbf{r},t),\nabla,\boldsymbol{\sigma}] &= L[\nabla,\boldsymbol{\sigma}]\mathbf{w}(\mathbf{r},t) \\ &\quad + \Gamma[\nabla,\boldsymbol{\sigma}] : \mathbf{w}(\mathbf{r},t) : \mathbf{w}(\mathbf{r},t) + \ldots\ . \end{aligned} \tag{4}$$

In the following we consider the evolution equation

$$\frac{\partial}{\partial t}\mathbf{w}(\mathbf{r},t) = L[\nabla,\boldsymbol{\sigma}]\mathbf{w}(\mathbf{r},t) + \Gamma[\nabla,\boldsymbol{\sigma}] : \mathbf{w}(\mathbf{r},t) : \mathbf{w}(\mathbf{r},t) \qquad , \tag{5}$$

where only nonlinearities up to quadratic order in the vector $\mathbf{w}(\mathbf{r},t)$ are made explicit. $L[\nabla,\boldsymbol{\sigma}]$ denotes a linear operator. The notation for the quadratic terms is a shorthand notation. In general, it is a functional quadratic in $\mathbf{w}(\mathbf{r},t)$. The inclusion of higher order terms is evident.

2.1 Linear Stability Analysis of the Basic State

An important question related with the basic state concerns its stability. Stability with respect to infinitesimal disturbances is investigated on the basis of the linearized evolution equation (5):

$$\frac{\partial}{\partial t}\mathbf{w}(\mathbf{r},t) = L[\nabla,\boldsymbol{\sigma}]\mathbf{w}(\mathbf{r},t)\ . \tag{6}$$

This linear equation is solved by a normal mode ansatz:

$$\mathbf{q}(\mathbf{r},t) = \sum_j c_j \Phi_j(\mathbf{r}) exp[\lambda_j t] \ . \tag{7}$$

The normal modes $\Phi_j(\mathbf{r})$ and its corresponding complex growth rates are determined by the linear eigenvalue problem:

$$\lambda_j(\boldsymbol{\sigma})\Phi_j(\mathbf{r}) = L[\nabla, \boldsymbol{\sigma}]\Phi_j(\mathbf{r}) \ . \tag{8}$$

Since the linear operator contains spatial derivatives of the state vector $\mathbf{w}(\mathbf{r}, t)$ one actually has to solve a linear boundary value problem.

The consideration of this problem leads us to discuss some mathematical notions. We first define a scalar product (the asterisk denotes complex conjugation)

$$\langle \mathbf{u}(\mathbf{r},t)|\mathbf{w}(\mathbf{r},t)\rangle = \int_V d^3\mathbf{r} \sum_\alpha u_\alpha^*(\mathbf{r},t) w_\alpha(\mathbf{r},t) \ . \tag{9}$$

Furthermore, we assume that each state vector can be represented by a linear superposition of a countable infinity of basis vectors

$$\phi_j(\mathbf{r}), \quad j = 1, 2, ..., \infty \ . \tag{10}$$

In other words, the state vectors $\mathbf{w}(\mathbf{r}, t)$ are elements of a separable Hilbert space.

Using the scalar product (9) one can define a linear operator which is called the adjoint operator $L^\dagger[\nabla, \boldsymbol{\sigma}]$:

$$\langle \mathbf{u}(\mathbf{r},t)|L[\nabla,\boldsymbol{\sigma}]\mathbf{w}(\mathbf{r},t)\rangle = \left\langle L^\dagger[\nabla,\boldsymbol{\sigma}]\mathbf{u}(\mathbf{r},t)|\mathbf{w}(\mathbf{r},t)\right\rangle \ . \tag{11}$$

The adjoint linear eigenvalue problem defines the adjoint normal modes according to

$$\lambda_j^{\dagger *}(\boldsymbol{\sigma})\Phi_j^\dagger(\mathbf{r}) = L^\dagger[\nabla, \boldsymbol{\sigma}]\Phi_j^\dagger(\mathbf{r}) \ . \tag{12}$$

Since quite often the linear operator turns out to be non-selfadjoint the linear eigenvalue problem may have generalized eigenvectors. That means that for some eigenvalues generalized eigenvectors $\Phi_{j,\alpha}(\mathbf{r})$, which fulfill the relations

$$L[\nabla, \boldsymbol{\sigma}]\Phi_{j,\alpha} = \lambda_j \Phi_{j,\alpha} + \Phi_{j,\alpha-1} \tag{13}$$

have to be included. In a finite dimensional space L can only transformed to a Jordan normal form.

The (generalized) eigenvectors fulfill the following orthogonality and completeness relations:

$$\left\langle \Phi_{i\alpha}^\dagger(\mathbf{r})|\Phi_{j\beta}(\mathbf{r})\right\rangle = \delta_{\alpha,\beta}\delta_{ij}, \qquad \sum_{j,\alpha} |\Phi_{j\alpha}(\mathbf{r})\rangle \left\langle \Phi_{j\alpha}^\dagger(\mathbf{r}')\right| = I\delta(\mathbf{r}-\mathbf{r}') \ . \tag{14}$$

2.2 Mode Equations

If one or several eigenvalues of the normal modes get positive real parts disturbances (including contributions of the corresponding normal modes) start to grow exponentially. In order to deal with the newly evolving state it is necessary to take into account the nonlinear terms.

If one expects the newly evolving states to be close to the basic state it is convenient to perform a normal mode ansatz:

$$\mathbf{w}(\mathbf{r},t) = \sum_j \xi_j(t)\Phi_j(\mathbf{r}) \; . \tag{15}$$

Inserting this ansatz into the basic equation (5) we obtain

$$\sum_j [\frac{d}{dt}\xi_j(t) - \lambda_j \xi_i(t)]\Phi_j(\mathbf{r}) \\ = \sum_{j,k} \xi_j(t)\xi_k(t)\Gamma[\nabla,\sigma] : \Phi_j(\mathbf{r}) : \Phi_k(\mathbf{r}) \; . \tag{16}$$

Performing the scalar products with the adjoint modes $\Phi_i^\dagger(\mathbf{r})$ we are led to the mode equations for the amplitudes $\xi_i(t)$

$$\dot{\xi}_i(t) = \lambda_i \xi_i(t) + \sum_{j,k} \Gamma_{i;j,k}\xi_j(t)\xi_k(t) \; , \tag{17}$$

where we have made use of the biorthogonality of the normal modes. Furthermore, we have defined the scalar products

$$\Gamma_{i;j,k} = \left\langle \Phi_i^\dagger(\mathbf{r}) | \Gamma[\nabla,\sigma] : \Phi_j(\mathbf{r}) : \Phi_k(\mathbf{r}) \right\rangle \; . \tag{18}$$

These coefficients specify the nonlinear interaction between the different normal modes. If the linear eigenvalue problem possesses generalized eigenvectors λ_i is a matrix in Jordan normal form.

Since the set of normal modes $\Phi_j(\mathbf{r})$ is assumed to be complete the infinite dimensional system of ordinary differential equations for the mode amplitudes $\xi_j(t)$ are completely equivalent to the evolution equation (5).

2.3 Order Parameter Equations

An instability of the basic state is indicated by the eigenvalues of the linear eigenvalue problem. Close to instability there are modes with vanishing real part of their eigenvalues. These modes are denoted as critical or unstable modes. Furthermore, there are modes whose corresponding eigenvalues have negative real parts. In the linear case, these modes would simply decay and are denoted as stable modes. Accordingly, we decompose the amplitudes ξ of these modes into stable and unstable amplitudes ξ_s, ξ_u, respectively.

$$\mathrm{Re}\lambda_j(\sigma) \approx 0 \to \xi_u(t) , \tag{19}$$
$$\mathrm{Re}\lambda_j(\sigma) < 0 \to \xi_s(t) . \tag{20}$$

Since the amplitudes of the stable modes are linearly fast relaxing variables, their long time behaviour is entirely determined by the behaviour of the amplitudes of the stable modes:

$$\xi_s(t) = \xi_s[\xi_u(t)] . \tag{21}$$

As a result, one obtains a closed set of ordinary differential equations for the order parameters $\xi_u(t)$:

$$\dot{\xi}_u(t) = \lambda_u[\sigma]\xi_u(t) + H_u[\xi_u(t), \sigma] . \tag{22}$$

The state vector is then constructed according to

$$\mathbf{q}(\mathbf{r}, t) = \mathbf{q}^0(\mathbf{r}) + \sum_u \xi_u(t)\Phi_u(\mathbf{r}) + \sum_s \xi_s[\xi_u(t)]\Phi_s(\mathbf{r}) , \tag{23}$$

which explicitly demonstrates that the dynamical behaviour is entirely determined (after a transient period) exclusively by the order parameters $\boldsymbol{\xi}_u(t)$.

The relationship $\xi_s = \xi_s[\xi_u]$ can be evaluated approximately as a power series in $\xi_u(t')$. Let us briefly indicate the necessary steps. The evolution equation for the stable modes can be solved:

$$\xi_s(t) = \int_{-\infty}^{t} e^{\lambda_s(t-t')} \sum_{u,u'} \Gamma_{s;uu'}\xi_u(t')\xi_{u'}(t') . \tag{24}$$

Only quadratic terms in the amplitudes of the critical modes are retained since the amplitudes of the stable modes are expected to be of the order of $\xi_u\xi_{u'}$. Since the exponential function is a fast decaying function in $t - t'$, we can perform the following approximation

$$\begin{aligned} \xi_u(t') &= e^{\mathrm{Im}(\lambda_u)t'}\tilde{\xi}_u(t') \\ &\approx e^{\mathrm{Im}(\lambda_u)t'}\tilde{\xi}_u(t) = e^{\mathrm{Im}(\lambda_u)(t'-t)}\xi_u(t) . \end{aligned} \tag{25}$$

Here, we have taken into account that the order parameter $\xi_u(t)$ may be fastly oscillating with a frequency given by the imaginary part of the eigenvalue λ_u. However, $\tilde{\xi}_u(t')$ can be expected to be a slowly varying function and, in turn, can be approximated by $\tilde{\xi}_u(t)$. As a result we obtain (to quadratic order in ξ_u):

$$\xi_s(t) = \sum_{uu'} \frac{\Gamma_{s;u,u'}}{[\mathrm{Im}(\lambda_u + \lambda_{u'}) - \lambda_s]}\xi_u(t)\xi_{u'}(t) . \tag{26}$$

This result yields the order parameter equations (up to cubic nonlinearities):

$$\dot{\xi}_u(t) = \lambda_u[\sigma]\xi_u(t) + \sum_{u,u'} \Gamma_{u;u',u''}\xi_{u'}(t)\xi_{u''}(t)$$
$$+ \sum_{u',u'',u'''} \Gamma^3_{u;u',u'',u'''}\xi_{u'}(t)\xi_{u''}(t)\xi_{u'''}(t) \ . \tag{27}$$

The nonlinear coupling coefficient $\Gamma^3_{u;u',u'',u'''}$ is explicitly determined according to

$$\Gamma^3_{u;u',u'',u'''} = \sum_s \frac{[\Gamma_{u;s,u'} + \Gamma_{u;u',s}]}{[Im(\lambda_u + \lambda_{u'}) - \lambda_s]} \Gamma_{s;u',u''} \ . \tag{28}$$

The calculation of higher order terms is straightforward.

Instabilities can be classified by the spectrum of the linear eigenvalue problem (6). Typical instabilities are steady state bifurcations and Hopf bifurcations.

3 Group Theoretic Methods

The traditional field of application of group theoretic methods is atomic and molecular physics. Here, group theory is used to investigate atomic spectra by a classification of the eigenstates of the Hamiltonian based on irreducible representations of underlying symmetry groups. However, group theory is not able to provide quantitative results, i.e. it does not help in calculating numbers of the energy levels.

The situation is quite similar in the study of instabilities in nonequilibrium systems. The main points can be outlined by the following simple example (c.f. Fig.1). Consider a system like a Rayleigh-Bénard experiment: A fluid in a cell is heated from below and cooled from above. For small temperature gradients a purely heat conductive state exists, which has reflectional symmetry $x \to -x$. For higher temperature gradients convective motions set in. These motions consist of one or several convection cells.

We characterize the system by its temperature field $T(x, y, z, t)$. There is a group of symmetry transformations, which consist of the parity transformation P and the identity $E = P^2$:

$$PT(x, y, z, t) = T(-x, y, z, t) \ . \tag{29}$$

The basic (purely heat conductive) state $T_o(x, y, z) = PT_0(x, y, z)$ is invariant with respect to this parity transformation.

We now assume the existence of a steady state bifurcation, i.e. an instability where a single real eigenvalue becomes positive at the instability point. The corresponding critical mode can then be classified by its symmetry properties: It can either be a mode with positive $PT_c(x, y, z) = T_c(x, y, z)$ or negative parity $PT_c(x, y, z) = -T_c(x, y, z)$, respectively. Close to instability the newly evolving state has the form

$$T(x,y,z,t) = T_0(x,y,z) + \xi(t)T_c(x,y,z) + h.o.$$
$$\dot{\xi}(t) = \lambda\xi(t) + h(\xi(t)) \,, \tag{30}$$

where the amplitude $\xi(t)$ is determined by an order parameter equation.

A parity breaking transition means $PT(x,y,z,t) \neq T(x,y,z,t)$. However, if $(T(x,y,z,t)$ is a possible state, due to symmetry, also $PT(x,y,z)$ has to be realizable. As a result the order parameter equation has to allow for solutions $-\xi(t)$. This immediately leads to

$$h(\xi) = -h(-\xi) = \lambda\xi - a\xi^3 - b\xi^5 \ldots \,. \tag{31}$$

In turn, this implies the existence of a supercritical transition, or, in physical terms, a nonequilibrium phase transition of second order.

The second case is a parity conserving transition. Here, one has to expect a subcritical bifurcation or a nonequilibrium phase transition of first order, since the nonlinearity of the order parameter equation $h(\xi)$ has the form

$$h(\xi) = -h(-\xi) = \lambda\xi + \alpha\xi^2 - a\xi^3 + \ldots \,. \tag{32}$$

The main signature of such a transition is the existence of hysteresis effects.

The physical reason for the differences of the both instabilities is evident (c.f. fig. (1): In case of a parity breaking transition the states

$$T(x,y,z) = T_0(x,y,z) + \xi T_c(x,y,z)$$
$$T(x,y,z) = T_0(x,y,z) - \xi T_c(x,y,z) \tag{33}$$

are physically equivalent, whereas in case of a parity conserving transition they are physically inequivalent (they differ e.g. by the fact that in the first case fluid particles close to both lateral boundaries flow in upwards direction, whereas it flows in downwards direction for the second case).

This simple example already demonstrates the main points of a group theoretic treatment of nonequilibrium phase transitions: It can be used in a qualitative manner for

- classifying spatial modes according to representations of underlying symmetry groups
- fixing the number of unstable modes (order parameters) and characterizing their symmetry properties
- determining the structure of nonlinear couplings among modes and, in turn, the structure of the order parameter equations
- classifying the newly evolving states by their symmetries, i.e. classifying possible symmetry breakings.

However, in order to obtain quantitative results on the behaviour of the order parameter equations, the coefficients Γ arising in the order parameter equations have to be calculated explicitly. Thereby, not all coefficients have to be evaluated. Symmetry implies certain relationships among the nonlinear coupling coefficient (28), which is quite important for practical calculations.

In the following we shall make these points more explicit.

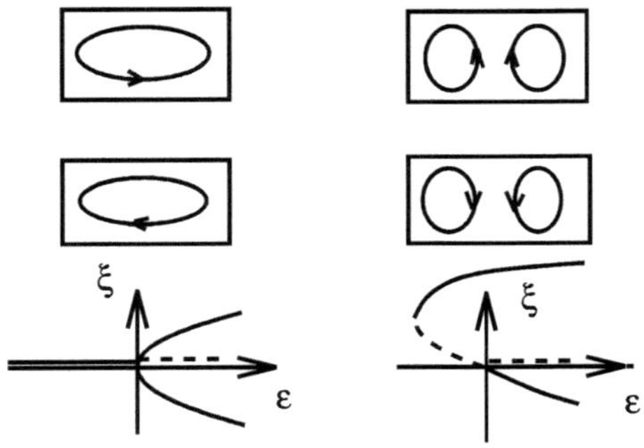

Fig. 1. Steady state bifurcations in a small Rayleigh-Bénard cell: Super- and subcritical instabilities and its relation to reflectional symmetry (exemplified by instabilities in a small aspect ration Rayleigh-Bénard cell). *Left hand side:* Parity breaking transition leads to two physically equivalent states through a supercritical bifurcation. *Right hand side:* Parity conserving transition results in inequivalent states through a subcritical bifurcation

3.1 Symmetry Groups and Equivariant Evolution Equations

We consider systems which allow for a group G of symmetry transformations (e.g. reflections, translations, rotations etc.). The symmetry operation g transforms the state vector $\mathbf{q}(\mathbf{r},t)$ into a new vector $g \circ \mathbf{q}(\mathbf{r},t)$ and, according to the theory of group representations, induces the existence of a linear operator $D(g)$:

$$g \circ \mathbf{q}(\mathbf{r},t) = D(g)\mathbf{q}(\mathbf{r},t) \ . \tag{34}$$

The system possesses the symmetry of the symmetry group G if, for a state $\mathbf{q}(\mathbf{r},t)$ of the system, the state vector $D(g)\mathbf{q}(\mathbf{r},t)$ is also an admissible state for all times t and all symmetry operations g of the symmetry group G. In that case,

$$\frac{\partial}{\partial t}\mathbf{q}(\mathbf{r},t) = \tilde{N}[\mathbf{q}(\mathbf{r},t), \nabla, \sigma] \tag{35}$$

implies that the transformed state vector $D(g)\mathbf{q}(\mathbf{r},t)$ is a solution of the evolution equation, too:

$$\frac{\partial}{\partial t}D(g)\mathbf{q}(\mathbf{r},t) = D(g)N[\mathbf{q}(\mathbf{r},t), \nabla, \sigma] = N[D(g)\mathbf{q}(\mathbf{r},t), \nabla, \sigma] \ . \tag{36}$$

The evolution equation is called to be equivariant with respect to symmetry operations of the group G.

3.2 Instability of a Totally Symmetric Stationary State

In analogy to the previous section we assume the existence of a time independent solution $\mathbf{q}_0(\mathbf{r})$. This basic state is supposed to be invariant with respect to all symmetry operations g:

$$D(g)\mathbf{q}_0(\mathbf{r}) = \mathbf{q}_0(\mathbf{r}) \quad \forall g \in G\,. \tag{37}$$

This condition will be released below, where higher instabilities are treated.

In order to examine pattern formation close to an instability of the basic state we introduce again deviations $\mathbf{w}(\mathbf{r}, t)$ from the basic state obeying the evolution equation (5).

Equivariance of the evolution equation for the deviation $\mathbf{w}(\mathbf{r}, t)$ has the following two consequences.

I. All operators $D(g)$ commute with the linear operator $L[\nabla, \boldsymbol{\sigma}]$:

$$D(g)L[\nabla, \boldsymbol{\sigma}] = L[\nabla, \boldsymbol{\sigma}]D(g)\,. \tag{38}$$

In fact, the modes $\Phi_j(\mathbf{r})$ of the linear operator $L[\nabla, \boldsymbol{\sigma}]$ can be used to define a representation $D_{ij}(g)$ of the group element g:

$$D_{ij}(g) = \left\langle \Phi_i^\dagger(\mathbf{r})|D(g)|\Phi_j(\mathbf{r})\right\rangle = \left\langle \Phi_i^\dagger(\mathbf{r})|g \circ \Phi_j(\mathbf{r})\right\rangle\,. \tag{39}$$

Since we have decomposed the modes into stable and unstable ones, the representation D(g) possesses the following decomposition into matrices operating in the space of the unstable and stable modes:

$$\begin{aligned} D(g) &= D^u(g) + D^s(g)\,, \\ D(g)\Phi_u(\mathbf{r}) &= \sum_{u'} D^u_{uu'}(g)\Phi_{u'}(\mathbf{r})\,, \\ D(g)\Phi_s(\mathbf{r}) &= \sum_{s'} D^s_{ss'}(g)\Phi_{s'}(\mathbf{r})\,. \end{aligned} \tag{40}$$

The physical reason for the decomposition is evident: A symmetry transformation may only transform unstable (stable) modes into a linear combination of unstable (stable) modes.

II. The nonlinear terms have to fulfill

$$\Gamma[\nabla, \boldsymbol{\sigma}] : \mathbf{w}(\mathbf{r}, t) : \mathbf{w}(\mathbf{r}, t) = D(g^{-1})\Gamma[\nabla, \boldsymbol{\sigma}] : [D(g)\mathbf{w}(\mathbf{r}, t)] : [D(g)\mathbf{w}(\mathbf{r}, t)] \tag{41}$$

This condition immediately takes over to the invariance of the matrix elements $\Gamma_{i;jk}$:

$$\Gamma_{i;jk} = D(g^{-1})_{ii'}\Gamma_{i';j'k'}D(g)_{j'j}D(g)_{k'k}\,. \tag{42}$$

3.3 Structure of the Order Parameter Equations for Systems with Symmetry

The state vector close to instability has the following form:

$$\mathbf{q}(\mathbf{r}, t) = \mathbf{q}_0(\mathbf{r}) + \sum_u \xi_u(t)\Phi_u(\mathbf{r}) + \sum_s \xi_s[\xi_u(t)]\Phi_s(\mathbf{r})\,. \tag{43}$$

A transformed vector is then given by

$$D(g)\mathbf{q}(\mathbf{r},t) = \mathbf{q}_0(\mathbf{r}) + \sum_u \xi_u(t) D(g)\Phi_u(\mathbf{r}) + \sum_s \xi_s[\xi_u(t)] D(g)\Phi_s(\mathbf{r}) \,. \quad (44)$$

This shows that the transformed amplitudes ξ'_u

$$\xi'_u = \sum_{u'} D_{u'u}(g)\xi_{u'} = \sum_{u'} D_{u,u'}(g^{-1})\xi_{u'} \quad (45)$$

have to be a solution of the order parameter equations, too. As a consequence, the order parameter equations have to be equivariant with respect to the transformations $D_{u',u}(g)$. Let us examine this in more detail.

We combine the order parameter amplitudes ξ_u into the vector $\boldsymbol{\xi} = (\xi_1, .., \xi_n)$ and use the following notation for the order parameter equation:

$$\frac{d}{dt}\boldsymbol{\xi} = \Lambda\boldsymbol{\xi} + \mathbf{H}[\boldsymbol{\xi},\boldsymbol{\sigma}] \,. \quad (46)$$

These equations are equivariant with respect to the symmetry transformations $D^u(g) = D_{uu'}(g)$ of the symmetry group G:

$$\frac{d}{dt}D^u(g)\boldsymbol{\xi} = D^u(g)\Lambda\boldsymbol{\xi} + D^u(g)\mathbf{H}[\boldsymbol{\xi},\boldsymbol{\sigma}]$$
$$= \Lambda D^u(g)\boldsymbol{\xi} + \mathbf{H}[D^u(g)\boldsymbol{\xi},\boldsymbol{\sigma}] \,. \quad (47)$$

The nonlinearity $H[\boldsymbol{\xi},\boldsymbol{\sigma}]$ is approximated by a Taylor series where, due to equivariance, the coefficients $\Gamma^n_{u;u_1,..,u_n}$ of the n-th order polynomials have to fulfill the following relations:

$$\Gamma^n_{u;u_1,..,u_n} = \sum_{v,v_1,..,v_n} D^u_{u,v}(g^{-1})\Gamma^n_{v;v_1,..,v_n} D^u_{v_1,u_1}(g) D^u_{v_n,u_n}(g) \,. \quad (48)$$

This can be shown by using the explicit expressions for the nonlinear coupling coefficients (28) using the symmetry properties (42).

An immediate consequence of the relations (48) is a restriction of the number of independent coefficients occurring in the polynomials. This can be made explicit by the following two theorems concerning invariant polynomials and equivariant polynomial vector fields [8], [10]:

a) Each invariant polynomial $f[\boldsymbol{\xi}]$,

$$f[\boldsymbol{\xi}] = f[D(g)\boldsymbol{\xi}] \,, \quad (49)$$

can be expressed as a polynomial function of the finite number of generating invariant polynomials,

$$r_i(\boldsymbol{\xi}), \quad i = 1, ..., N_e \,. \quad (50)$$

That means:

$$f[\boldsymbol{\xi}] = F[\{r_i(\boldsymbol{\xi})\}] \,. \quad (51)$$

b) Each equivariant polynomial vector field can be represented by a finite number of generating equivariant polynomial vector fields

$$\mathbf{g}_i(\boldsymbol{\xi}) \quad i = 1, ..., \tilde{N} \geq N_e \tag{52}$$

according to:

$$\mathbf{h}(\boldsymbol{\xi}) = \sum_{i=1}^{\tilde{N}} \mathbf{g}_i(\boldsymbol{\xi}) f_i[r_1(\boldsymbol{\xi}), \ldots, r_{N_e}(\boldsymbol{\xi})] \, . \tag{53}$$

A subset of equivariant polynomial vector fields can be obtained via the gradients of the set of generating invariant polynomials.

Let us now summarize the implications of these facts for the order parameter equations. They take the following form

$$\frac{d}{dt}\boldsymbol{\xi} = \Lambda\boldsymbol{\xi} + \sum_{i=1}^{\tilde{N}} \mathbf{g}_i(\boldsymbol{\xi}) f_i[r_1(\boldsymbol{\xi}), \ldots, r_{N_e}(\boldsymbol{\xi}] \, . \tag{54}$$

From these equations, one can derive a closed set of evolution equations for the generating polynomials $r_i(\boldsymbol{\xi})$:

$$\dot{r}_i = R_i[r_1(\boldsymbol{\xi}), \ldots, r_{N_e}(\boldsymbol{\xi})] \, . \tag{55}$$

Normal Form Transformations

Normal form transformations are nonlinear transformations among the order parameters $\boldsymbol{\xi}$:

$$\boldsymbol{\xi}' = \boldsymbol{\xi} + \mathbf{h}(\boldsymbol{\xi}) \, . \tag{56}$$

They are chosen such that the nonlinear terms achieve a simple form. We do not want to go into the details and refer the reader to [13]. However, it is evident that the normal form transformations preserve symmetries. As a consequence, one may start from order parameter equations which are already in normal form.

3.4 Continuous Symmetry Groups

In the case of equivariance with respect to a continuous symmetry group, for instance the symmetry group SO(3), the representations $D(\boldsymbol{\alpha})$ can be expressed by their generators J_i [6]:

$$D(\boldsymbol{\alpha}) = e^{i\boldsymbol{\alpha}\cdot\mathbf{J}} \, , [J_i, J_j] = i\epsilon_{ijk} J_k \, . \tag{57}$$

(Here, we have considered a Lie group with structure constants ϵ_{ijk}.) The advantage using the infinitesimal generators is based on a formulation of the equivariance condition in differential form:

$$J_i\mathbf{H}[\boldsymbol{\xi},\boldsymbol{\sigma}] = (J_i\boldsymbol{\xi}) \cdot \frac{\partial}{\partial \boldsymbol{\xi}}\mathbf{H}[\boldsymbol{\xi},\boldsymbol{\sigma}] \ . \tag{58}$$

This differential formulation determines a simple algorithm for the evaluation of invariant polynomials and equivariant polynomial vector fields [6], [15].

3.5 Fixed Point Subspaces and Isotropy Subgroups

Due to symmetry, the dimensionality of the order parameter equations may be quite large and an analysis of these equations may be rather difficult. Therefore, one looks for invariant subspaces of these equations. It is evident, that these invariant subspaces are related to subgroups G_i of the *large* group G. A fixed point subspace of a subgroup G_i is defined as follows:

$$Fix(G_i) = \{\boldsymbol{\xi} : D(g)\boldsymbol{\xi} = \boldsymbol{\xi}, \forall g \in G_i\} \ . \tag{59}$$

All symmetry transformations g of the subgroup G_i leave the state $\boldsymbol{\xi}$ invariant. Usually, one says *the state* $\boldsymbol{\xi}$ *possess the symmetries of the subgroup* G_i.

The isotropy subgroup $\Gamma_{\boldsymbol{\xi}}$ of a state $\boldsymbol{\xi}$ are all transformations g of the group G, which leave the state $\boldsymbol{\xi}$ invariant (i.e. it is the largest subgroup leaving the state invariant):

$$\Gamma_{\boldsymbol{\xi}} = \{g \in G, D(g)\boldsymbol{\xi} = \boldsymbol{\xi}\} \ . \tag{60}$$

The isotropy subgroup, therefore, characterizes the symmetry of the state $\boldsymbol{\xi}$. A knowledge of all isotropy subgroups enables one to characterize the symmetries of the newly evolving states. Simultaneously, the isotropy subgroups define invariant subspaces in the set of order parameter equations. Therefore, one may determine solutions in these lower dimensional invariant subspaces.

3.6 Symmetry Classification of Instabilities

From the above considerations it becomes evident that one can proceed to a classification of instabilities in systems with symmetries. This classification is based on combining a class of instabilities (steady state bifurcation, Hopf bifurcation, mode interactions etc..) with a certain symmetry group:

$$typ \ \ of \ \ instability \times symmetry \ \ group \ . \tag{61}$$

Great efforts have been put into this task and a relatively complete picture of the important cases has emerged. There have been investigations for the two cases of codimension I bifurcations (steady state, Hopf bifurcation in systems with O(2), O(3) symmetries), codimension II bifurcations in systems with O(2), O(3) symmetries and mode interactions in systems with O(2) and O(3) symmetries. For a summary we refer the reader to [14]. In general the investigation of these problems lead to interesting spatio-temporal behaviour. Quite often, homoclinic and heteroclinic chaos occurs. A reason for that is the fact that equivariant dynamical systems are not generic and therefore allow for saddle-saddle connections around which chaotic behaviour is organized. One of the first examples is $l-(l+1)$ mode interactions in systems with O(3)-symmetry [15].

Planform Selection

Experiments like Rayleigh-Bénard convection in large aspect ratio systems lead to patterns, which achieve a two dimensional spatial periodicity far from the boundaries. The question arises what kinds of patterns are selected, e.g. whether roll, quadratic or hexagonal patterns are realized. From a mathematical point of view this problem can be investigated by prescribing a certain type of lattice, e.g. a square lattice or a hexagonal lattice by imposing periodic boundary conditions with respect to the lattice. This restricts the (noncompact) symmetry group E of the infinite plane to subgroups with quadratic or hexagonal symmetries. As a consequence one can perform a classification of the problem of planform selection according to the combination of the following criteria:

$$type \quad of \quad instability \times lattice \,. \tag{62}$$

A list of possible bifurcation scenarios with detailed references are given in [14].

This approach allows one to describe perfect, i.e. strictly periodic patterns. Natural patterns, however, are characterized by the occurrence of defects, grain boundaries and domain walls. This will be the topic of the next section.

4 Order Parameter Equations for Natural Patterns in Large Aspect Ratio Systems

Experiments on Rayleigh-Bénard convection or experiments in nonlinear optics have been performed in so-called large aspect ratio systems. For Rayleigh-Bénard convection this means that the horizontal dimensions are large compared to the vertical height. In such systems a large number of critical modes exist close to onset. In fact, infinitely extended systems with uniform basic state possess translational symmetry and the modes take the form

$$\Phi_{j,\mathbf{k}}(\mathbf{r}) = \phi_j(\mathbf{k}, z)e^{i\mathbf{k}\cdot\mathbf{x}} \,, \tag{63}$$

where $\mathbf{x} = (x, y)$ denotes the horizontal directions. The modes are plane waves with wave vector $\mathbf{k}$ and the corresponding eigenvalues $\Lambda_j(\mathbf{k})$ form bands of modes (c.f. fig. (2)). If the bands of modes are widely separated the modes belonging to the stable bands can be eliminated [1], [2]. The order parameter $\xi(\mathbf{k},t)$ depend on the continuous wave vector $\mathbf{k}$ and its evolution equation takes the form

$$\dot{\xi}(\mathbf{k},t) = \lambda_u(\mathbf{k})\xi(\mathbf{k},t) + \int d\mathbf{k}' \int d\mathbf{k}'' \Gamma(\mathbf{k};\mathbf{k}',\mathbf{k}'')\xi(\mathbf{k}',t)\xi(\mathbf{k}'',t)$$
$$+ \int d\mathbf{k}' \int d\mathbf{k}'' \int d\mathbf{k}''' \Gamma(\mathbf{k};\mathbf{k}',\mathbf{k}'',\mathbf{k}''')\xi(\mathbf{k}',t)\xi(\mathbf{k}'',t)\xi(\mathbf{k}''',t) + \ldots . \tag{64}$$

It is evident, that symmetry arguments constrain the nonlinear coupling terms. In fact, one may formally extend the conditions, eq. (48), to the case of continuously many modes characterized by the wave vector $\mathbf{k}$.

Let us make this transparent for the quadratic terms (Higher order terms are treated similarly).

Translational invariance implies equivariance with respect to the transformation $\xi(\mathbf{k},t) \to \xi(\mathbf{k},t)e^{i\mathbf{k}\cdot\alpha}$,

$$\Gamma(\mathbf{k};\mathbf{k}',\mathbf{k}'') = \delta(\mathbf{k}-\mathbf{k}'-\mathbf{k}'')\gamma(\mathbf{k};\mathbf{k}',\mathbf{k}'') \quad . \tag{65}$$

Rotational symmetry implies invariance with respect to the transformation $\xi(\mathbf{k},t) \to \xi(U\mathbf{k},t)$ (U is a unitary matrix),

$$\gamma(\mathbf{k}_1;\mathbf{k}_2,\mathbf{k}_3) = \gamma(\mathbf{k}_i \cdot \mathbf{k}_j, \mathbf{e}_z \cdot [\mathbf{k}_i \times \mathbf{k}_j) . \tag{66}$$

The nonlinear coupling terms can only depend on the invariants $\mathbf{k}_i \cdot \mathbf{k}_j$, $\mathbf{e}_z \cdot [\mathbf{k}_i \times \mathbf{k}_j]$ ($\mathbf{e}_z$ denotes the unit vector perpendicular to the layer). If the system additionally has reflectional symmetry, the dependency on the cross products vanish.

Furthermore, close to instability one can assume that only modes with critical wave vector close to a critical one (k_c) can be excited (c.f. Fig. (2)). This allows one to expand the coupling coefficients with respect to $\mathbf{k}_i^2 - k_c^2$ which leads to a kind of normal form [16]. It is convenient to formulate this equation in real space by introducing the Fourier transform

$$\psi(\mathbf{x},t) = \int d^2\mathbf{k} e^{i\mathbf{k}\cdot\mathbf{x}}\xi(\mathbf{k},t) . \tag{67}$$

This defines an order parameter field $\psi(\mathbf{x},t)$. The linear term of the order parameter equation simply turns into

$$\lambda_u(-i\nabla)\psi(\mathbf{x},t) . \tag{68}$$

Due to translational invariance the nonlinear terms are convolution integrals and contain nonlocal interactions. In many cases it is possible to expand the convolution integrals to obtain approximate representations involving (low-order) spatial derivatives of the order parameter fields. The procedure consists in an expansion in terms of

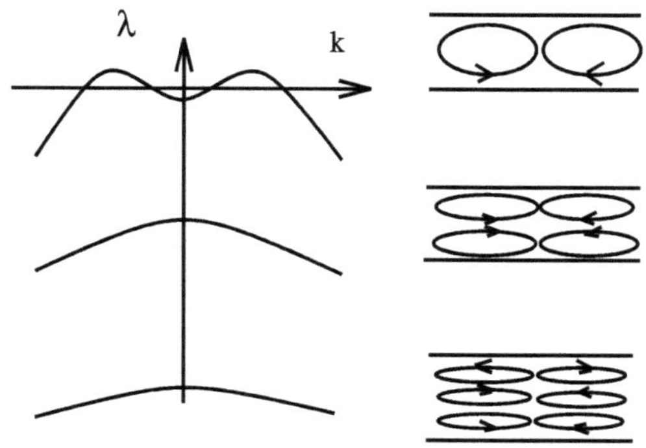

Fig. 2. Eigenvalue bands $\lambda_(k)$ for an instability in an infinitely extended systems. In the case of the Rayleigh-Bénard system the bands belong to modes with increasing number of convection cells in vertical direction. The wave vector **k** determines direction and wavelength of the modes

$$\tilde{\Delta}^k \psi(\mathbf{x},t) = (k_c^2 + \Delta)^k \psi(\mathbf{x},t) \to (k_c^2 - \mathbf{k}^2)^k \xi(\mathbf{k},t) \ . \tag{69}$$

It is possible because $\psi(\mathbf{x},t)$ contains only modes with wave vectors close to k_c. The general procedure is described in [16].

Let us briefly indicate the order parameter equations for some important cases [17].

4.1 Steady State Bifurcation in Reflectionally Invariant Systems

If the unstable band consists of real eigenvalues the order parameter equation takes the form:

$$\begin{aligned}\dot{\psi}(\mathbf{x},t) &= \lambda(-\Delta)\psi(\mathbf{x},t) \\ &+ \sum_{n,ijlm} B_{n;ijlm} \tilde{\Delta}^i [\tilde{\Delta}^j \psi(\mathbf{x},t) \Delta^n [\tilde{\Delta}^l \psi(\mathbf{x},t) \tilde{\Delta}^m \psi(\mathbf{x},t)]] \quad .\end{aligned} \tag{70}$$

The famous Swift-Hohenberg equation is obtained if only the term $B_{0;000}$ is taken into account. The inclusion of further terms allows for a detailed consideration of nonlinear mode interactions [17], [16]. The term with the coefficients $B_{n;000}$, for instance, determine the mode coupling between differently oriented plane waves.

The order parameter equations have been widely used to investigate planform selection in the presence of defects and domain walls between different planforms, i.e. they describe the emergence of natural patterns in large aspect ratio systems close to a steady state bifurcation.

4.2 Steady State Bifurcation in Systems Lacking Reflectional Symmetry

Interesting features of pattern formation occur if reflectional symmetry is broken like in Rayleigh-Bénard systems rotating about a vertical axis. There is an additional control parameter, the Taylor number Ta, which is a dimensionless measure for the rotation rate. At sufficiently high rotation rate the so-called Küpper-Lortz instability sets in [18]. This instability leads to a kind of turbulence already close to onset of convection and can therefore be treated by the order parameter concept.

The class of order parameter equation takes the form [16]

$$\begin{aligned}\dot{\psi}(\mathbf{x},t) &= \lambda(-\Delta)\psi(\mathbf{x},t) \\ &+ \sum_{n,ijlm} B_{n;ijlm}\tilde{\Delta}^i[\tilde{\Delta}^j\psi(\mathbf{x},t)\Delta^n[\tilde{\Delta}^l\psi(\mathbf{x},t)\tilde{\Delta}^m\psi(\mathbf{x},t)]] \\ &+ Ta\sum_{n;ijlm} C_{n;ijlm}\tilde{\Delta}^i[\tilde{\Delta}^j\psi(\mathbf{x},t)\Delta^n\mathbf{e}_z\cdot[\nabla\tilde{\Delta}^l\psi(\mathbf{x},t)\times\nabla\tilde{\Delta}^m\psi(\mathbf{x},t)]] \\ &+ Ta\sum_{n;ijlm} D_{n;ijlm}\tilde{\Delta}^i[\mathbf{e}_z\cdot[\nabla\tilde{\Delta}^j\psi(\mathbf{x},t)\times\nabla\Delta^n[\tilde{\Delta}^l\psi(\mathbf{x},t)\tilde{\Delta}^m\psi(\mathbf{x},t)]]]\end{aligned} \tag{71}$$

Again, using only the first terms of the expansion yields suitable approximations. The corresponding order parameter equations can be used to study especially non-variational aspects in pattern formation [19]-[23]. An example is exhibited in Fig. (3), which shows the spontaneous formation of domain walls and grain boundaries between differently oriented patches of convection rolls as a result of the Küppers-Lortz instability.

4.3 Order Parameters of Pseudoscalar Type

Another interesting pattern forming system is Rayleigh-Bénard convection in fluids with low Prandtl numbers. Here, convection rolls tend to form spiral patterns in a regime denoted as spiral turbulence. As first noted by Siggia and Zippelius [25] there are effects of large scale horizontal drift motions. The large scale vortical motions are only slightly damped and have to be described by a secondary order parameter field. The vortical motions $\mathbf{v}_c$ are expressed by a stream function $\Phi(\mathbf{x},t)$, $\mathbf{v}_c = \nabla\times\mathbf{e}_z\Phi(\mathbf{x},t)$. The field $\Phi(\mathbf{x},t)$ is of pseudoscalar type. The set of order parameter equation takes the general form [16]

$$\begin{aligned}\dot{\psi}(\mathbf{x},t) &= \lambda(-\Delta)\psi(\mathbf{x},t) + M[\psi(\mathbf{x},t)] \\ &\quad + \sum_{n,ijlm} C_{i;jl}\tilde{\Delta}^i\mathbf{e}_z\cdot[\nabla\tilde{\Delta}^j\psi(\mathbf{x},t)\times\nabla\Delta^l\Phi(\mathbf{x},t)] \\ \tau(\Delta)\dot{\Phi}(\mathbf{x},t) &= \gamma(-\Delta)\Phi(\mathbf{x},t) + \sum_{i;jl} D_{i;jl}\Delta^i\mathbf{e}_z\cdot[\nabla\tilde{\Delta}^j\psi(\mathbf{x},t)\times\nabla\tilde{\Delta}^j\psi(\mathbf{x},t)].\end{aligned} \tag{72}$$

If only lowest order terms are retained one obtains the model of Manneville [24]. It has been shown that this model generates spiral type convection patterns [26], [27]. Typical patterns are exhibited in Fig. (4).

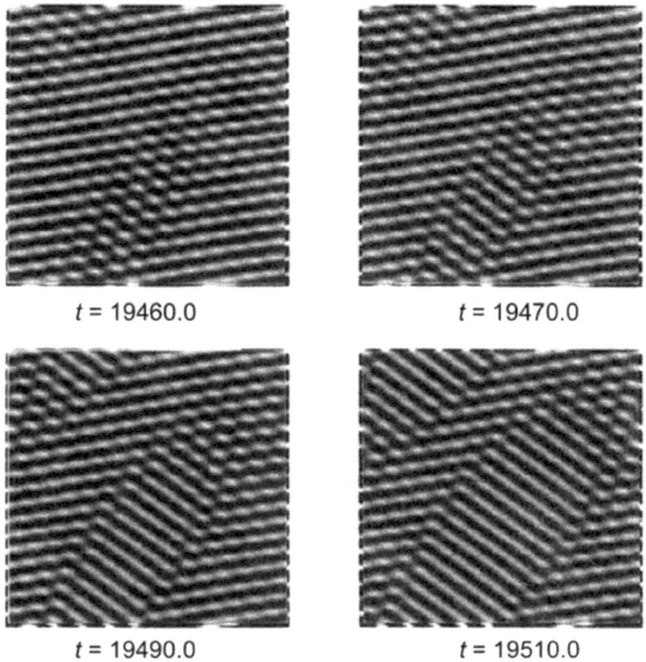

Fig. 3. Küppers-Lortz-instability: Spontaneous formation of domain structures in the order parameter equation (71)

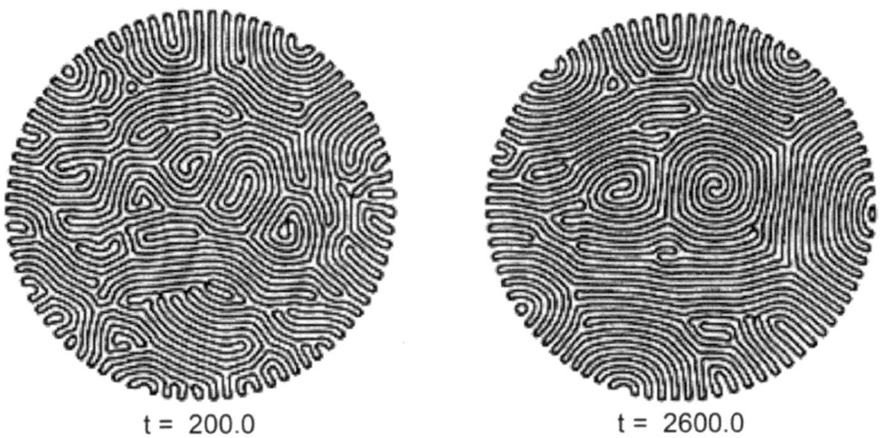

Fig. 4. Spiral turbulence: Numerical solution of an order parameter equations (72) [26]

5 Higher Instabilities in Systems with Continuous Symmetries

Up to now we have considered an instability of a basic state, which is invariant with respect to all symmetry transformations of the system under consideration. In the following we shall relax this assumption. Such a situation arises quite naturally at secondary and higher instabilities [28]. Special features occur for systems which allow for a continuous group of symmetry transformations like rotations or translations.

Continuous symmetries are characterized by their infinitesimal generators $\mathbf{J}$ and are of the following form:

$$D(\boldsymbol{\alpha}) = e^{i\boldsymbol{\alpha}\cdot\mathbf{J}} . \tag{73}$$

The simplest case is a system with SO(2)-symmetry, where one generator has to be considered. Examples for symmetry groups with several generators are systems with spherical symmetry. Here, the generators are the operators of angular momentum

$$\mathbf{J} = \frac{1}{i}\mathbf{r} \times \nabla , \tag{74}$$

where the usual commutator relations $[J_i, J_j] = i\epsilon_{ijk} J_k$ hold.

Additionally, one may also consider infinitely extended systems, which have the symmetries of the Euclidean plane E(2). This symmetry group is important for hydrodynamical large aspect ratio systems and reaction-diffusion equations. Here, the generators of translational symmetries are just

$$\mathbf{J} = \frac{1}{i}\nabla . \tag{75}$$

In mathematical terms, E(2) is a noncompact group.

For instabilities of stationary states $\mathbf{q}_0(\mathbf{r})$ with broken symmetries novel features arise due to the existence of neutral modes, the so called Goldstone modes. In fact, differentiating the equation

$$\mathbf{N}[e^{i\boldsymbol{\alpha}\cdot\mathbf{J}}\mathbf{q}_0(\mathbf{r})] = 0 \tag{76}$$

with respect to α_i (putting $\boldsymbol{\alpha} = 0$) one obtains

$$L[\nabla, \boldsymbol{\sigma}] J_i \mathbf{q}_0(\mathbf{r}) = 0 . \tag{77}$$

This explicitly shows that $\psi_i(\mathbf{r}) = J_i\mathbf{q}_0(\mathbf{r})$ are eigenvectors of the linear stability eigenvalue problem (77) with eigenvalues $\lambda_j = 0$.

In nonequilibrium systems Goldstone modes may play an essential role at higher symmetry breaking instabilities: They may become *active*, which means that the state vector takes the form

$$\mathbf{q}(\mathbf{r},t) = e^{i\boldsymbol{\alpha}(t)\cdot\mathbf{J}}[\mathbf{q}_0(\mathbf{r}) + \mathbf{w}(\mathbf{r},t)] \; , \tag{78}$$

where the operators $e^{i\boldsymbol{\alpha}(t)\cdot\mathbf{J}}$ explicitly become time dependent. The instability induces a motion along the so-called group orbit $e^{i\boldsymbol{\alpha}(t)\cdot\mathbf{J}}\mathbf{q}_0(\mathbf{r})$. If one considers a system with translational invariance, a time dependent parameter $\boldsymbol{\alpha}$ induces a drift of the basic pattern $\mathbf{q}_0(\mathbf{r},t)$.

Let us briefly outline the mathematical treatment of higher instabilities in systems with continuous symmetries. Let us denote the Goldstone modes and their adjoint ones by $\psi_j(\mathbf{r})$, $\psi_j^\dagger(\mathbf{r})$. We define the following projection operators:

$$P_j = |\psi_j(\mathbf{r}\,\rangle\,\langle\,\psi_j^\dagger(\mathbf{r}| \tag{79}$$

and

$$Q = E - \sum_j P_j \; . \tag{80}$$

It is evident from (78) that we can impose the condition

$$P_j \mathbf{w}(\mathbf{r},t) = 0 \; . \tag{81}$$

Using the equivariance condition

$$e^{-i\boldsymbol{\alpha}(t)\cdot\mathbf{J}}\mathbf{N}[e^{i\boldsymbol{\alpha}(t)\cdot\mathbf{J}}\mathbf{q}(\mathbf{r}] = \mathbf{N}[\mathbf{q}(\mathbf{r}] \tag{82}$$

we end up with the following evolution equation:

$$\begin{aligned} e^{-i\boldsymbol{\alpha}(t)\cdot\mathbf{J}} \tfrac{d}{dt} &[e^{i\boldsymbol{\alpha}(t)\cdot\mathbf{J}}][\mathbf{q}_0(\mathbf{r}) + \mathbf{w}(\mathbf{r},t)] \\ &+ \frac{d}{dt}\mathbf{w}(\mathbf{r},t) = L\mathbf{w}(\mathbf{r},t) + \Gamma : \mathbf{w}(\mathbf{r},t) : \mathbf{w}(\mathbf{r},t) \; . \end{aligned} \tag{83}$$

It can be shown that the operator

$$e^{-i\boldsymbol{\alpha}(t)\cdot\mathbf{J}}\frac{d}{dt}e^{i\boldsymbol{\alpha}(t)\cdot\mathbf{J}} = \mathbf{m}(\boldsymbol{\alpha}(t), \frac{d}{dt}\boldsymbol{\alpha}(t)) \cdot \mathbf{J} \tag{84}$$

is a linear combination of the generators $\mathbf{J}$ (This is evident if all operators commute). The resulting equation,

$$\mathbf{m}(\boldsymbol{\alpha}, \frac{d}{dt}\boldsymbol{\alpha})\cdot\mathbf{J}[\mathbf{q}_0(\mathbf{r}) + \mathbf{w}(\mathbf{r},t)] + \frac{d}{dt}\mathbf{w}(\mathbf{r},t) = L\mathbf{w}(\mathbf{r},t) + \Gamma : \mathbf{w}(\mathbf{r},t) : \mathbf{w}(\mathbf{r},t), \tag{85}$$

can be decomposed into two by an application of the projection operators P, Q:

$$m_j = \{1 + P_j J \mathbf{w}\}^{-1}\{P_j L\mathbf{w} + P_j \Gamma : \mathbf{w} : \mathbf{w}\} \tag{86}$$

$$\frac{d}{dt}\mathbf{w} = QL\mathbf{w} + Q\Gamma : \mathbf{w} : \mathbf{w} - Q\mathbf{m}\left(\boldsymbol{\alpha}, \frac{d\boldsymbol{\alpha}}{dt}\right) \cdot \mathbf{J}\mathbf{w} \tag{87}$$

The first equation can be rewritten as an evolution equation for the operator $e^{i\boldsymbol{\alpha}\cdot\mathbf{J}}$.

$$\frac{d}{dt}e^{i\boldsymbol{\alpha}(t)\cdot\mathbf{J}} = e^{i\boldsymbol{\alpha}(t)\cdot\mathbf{J}}\sum_j \{1 + P_j J\mathbf{w}\}^{-1}\{P_j L\mathbf{w} + P_j\Gamma : \mathbf{w} : \mathbf{w}\}J_j \tag{88}$$

The second equation is a closed evolution equation for the deviation $\mathbf{w}(\mathbf{r}, t)$ ($\mathbf{m}$ is given in terms of $\mathbf{w}$). It describes the temporal evolution in a comoving reference frame.

The drift motion $e^{i\boldsymbol{\alpha}(t)\cdot\mathbf{J}}$ depends on the deviations $\mathbf{w}$. If $\mathbf{w}$ relaxes to zero, the basic state $\mathbf{q}_0(\mathbf{r})$ is stable and the drift vanishes. However, a bifurcation leads to a finite value of $\mathbf{w}$ and a drift may be induced.

If $\mathbf{J}$ correspond to a spatial shift ($\mathbf{J} = -i\nabla$) such types of instabilities are called drift bifurcations. They are observed as higher parity breaking transitions of cellular structures and arise under the condition that, at the bifurcation point, the linear operator $L(\nabla, \boldsymbol{\sigma})$ possesses generalized eigenvectors $\Phi_i(\mathbf{r})$ with eigenvalue zero, which are related to the Goldstone modes

$$\begin{aligned} L\psi_i(\mathbf{r}) &= 0 \\ L\Phi_i(\mathbf{r}) &= \psi_i(\mathbf{r}) \ . \end{aligned} \tag{89}$$

For such an instability, the pattern is of the following form

$$\mathbf{q}(\mathbf{r}, t) = \mathbf{q}_0(\mathbf{r} - \alpha(t)) + \sum \xi_j(t)\Phi_j(\mathbf{r} - \alpha(t)) + stable \quad modes \quad . \tag{90}$$

The order parameter equations read:

$$\begin{aligned} \frac{d}{dt}\boldsymbol{\alpha} &= \boldsymbol{\xi} \\ \frac{d}{dt}\boldsymbol{\xi} &= \epsilon\boldsymbol{\xi} + \mathbf{h}(\boldsymbol{\xi}) \ . \end{aligned} \tag{91}$$

Below instability, $\epsilon < 0$, the order parameter ξ relaxes to zero and, in turn, $\boldsymbol{\alpha} \to 0$ indicating a stationary pattern. For a supercritical bifurcation, however, the order parameter $\boldsymbol{\xi}$ gradually increases as a function of ϵ in connection with a slow drift of the pattern. Drift bifurcations have been observed in the Faraday instability [29], Rayleigh-Bénard convection [30], the printer's instability [31], and cellular flame patterns [32].

Up to now we have considered a stationary basic state. However, one also has considered rotating waves as basic states. This is especially important for spiral waves arising in chemical reactions. One may perform a similar bifurcation analysis [33] and show that symmetry properties lead to a meandering motion of the rotating waves in the plane. This approach can explain a lot of the experimental details of meandering spirals in reaction-diffusion systems [34].

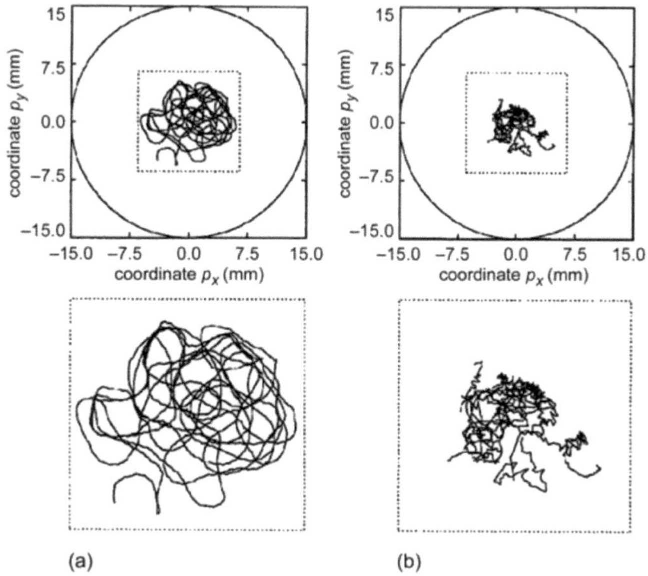

Fig. 5. Trajectories of dissipative solitons in a gas-discharge experiment [41]: (**a**): Regime beyond drift instability. The deterministic drift is covered by fluctuations (**b**): Regime below drift instability. Noise induced motion of dissipative soliton

5.1 Goldstone Modes, Dissipative Solitons, and Fluctuations

Nonequilibium systems like gas-discharge systems or optical systems [35], [36] allow for the existence of *localized states* or *dissipative solitons*. These are solitary objects on an otherwise uniform background. However, in gas-discharge systems, experimentally one never does observe stationary current filaments. They are always drifting and it has been argued that the filaments may have undergone a drift bifurcation [37]. However, the filament drift may also be due to fluctuations. Thus, a bifurcation analysis of the experimental data has to take into account the impact of fluctuations.

The problem has been analysed in the following way [41], [42]. The evolution equation for the localized state position $\boldsymbol{\alpha}$ and the amplitudes $\mathbf{A}$ of the unstable modes are expected to have the following form:

$$\frac{d}{dt}\boldsymbol{\alpha} = \mathbf{A}$$
$$\frac{d}{dt}\mathbf{A} = \mathbf{h}(\mathbf{A}) + \Gamma(t) \,. \tag{92}$$

$\mathbf{A}$ can be interpreted as the velocity of the localized structure. $\Gamma(t)$ denotes fluctuating forces, which stem from the microscopic degrees of freedom of the system. The function $\mathbf{h}$ takes the following form

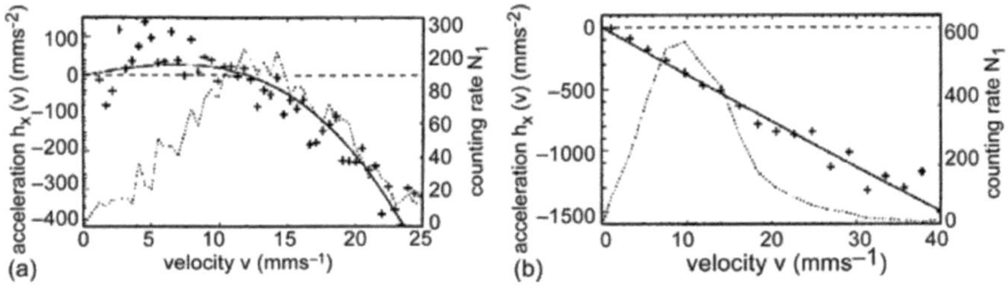

Fig. 6. Estimated deterministic drift of dissipative solitons in a gas-discharge system [41] ($\mathbf{v} = \mathbf{A}$) (**a**): The functional form $h(\mathbf{A}) = \epsilon\mathbf{A} - |\mathbf{A}|^2\mathbf{A}$ indicates a drift bifurcation. (**b**): The functional form $h(\mathbf{A}) = -\gamma\mathbf{A}$ indicates a usual Ornstein-Uhlenbeck process for the velocity of the dissipative soliton.

$$\mathbf{h} = \epsilon\mathbf{A} - a\mathbf{A}^2\mathbf{A} \,. \tag{93}$$

If ϵ is larger than zero, a bifurcation has occurred and the localized structure moves with the velocity $v_0 = \sqrt{\epsilon/a}$. However, it is difficult to observe the drift bifurcation due to the impact of noise. Therefore, one has to use methods from stochastic data analysis in order to disentangle the drift $\mathbf{h}$ from noise.

Such methods have been developed [38], [39], [40] and have successfully been applied to the examination of drift bifurcations in gas-discharge systems [41], [42]. An example is shown in fig. (6), where the velocity v_0, determined by a reconstruction of the nonlinearity $\mathbf{h}$ from velocity data of the localized object is exhibited as a function of the control parameter and indicates the existence of a drift bifurcation. However, if one looks at the realization of paths of the object above and below threshold, one may hardly see any difference. In order to investigate drift bifurcations in real strongly fluctuating systems, one has to apply methods of stochastic data analysis.

References

1. H. Haken, *Synergetics. An Introduction* (Springer-Verlag Berlin, 1983)
2. H. Haken, *Advanced Synergetics* (Springer-Verlag Berlin, 1983)
3. M. Cross, P. Hohenberg, *Pattern formation outside of equilibrium*, Rev. Mod. Phys. **65**, 851 (1993)
4. F.H. Busse, S.C. Müller, *Evolution of spontaneous structures in dissipative continuous systems*, (Springer, Berlin, 1998)
5. P. Manneville, *Dissipative Structures and weak turbulence*, (Academic Press, New York 1990)
6. D.H. Sattinger, P.J. Olver, *Group theoretic methods in bifurcation theory*, (Springer Verlag 1979)
7. M. Field: *Equivariant dynamical systems*, Trans. Am. Math. Soc. **259**(1), 185 (1980)
8. G. Gaeta, *Bifurcation and Symmetry breaking*, Physics Reports **189**, 1 (1990)

9. M. Golubitsky and D.G. Schaeffer, Singularities and Groups in Bifurcation Theory, Vol. I (Springer Verlag 1985)
10. M. Golubitsky, I. Stewart and D.G. Schaeffer, Singularities and Groups in Bifurcation Theory, Vol. II (Springer Verlag 1988)
11. P. Chossat, R. Lauterbach, *Methods in equivariant bifurcations and dynamical systems*, (World Scientific 2000)
12. J.D. Crawford, E. Knobloch, *Symmetry breaking bifurcations in fluid dynamics*, Annu. Rev. Fluid Mech. **23**, 341 (1991)
13. V.I. Arnold, *Geometrical methods in the theory of ordinary differential equations*, (Springer-Verlag Berlin, 1983)
14. G. Dangelmayr and L. Kramer, *Mathematical tools for pattern formation*, in F.H. Busse, S.C. Müller, *Evolution of spontaneous structures in dissipative continuous systems*, (Springer, Berlin, 1998)
15. R. Friedrich and H. Haken: *Stationary, Wavelike, and Chaotic Thermal Convection in Spherical Geometries*, Phys. Rev. A **34**, 2100-2120 (1986)
16. M. Bestehorn, R. Friedrich: *Rotationally invariant order parameter equations for natural patterns in nonequilibrium systems*, Phys. Rev. **E 59**, 2642 (1999)
17. M. Neufeld, R. Friedrich: *Model equations for quasi-twodimensional pattern forming systems*, Ann. Phys. Fr. **19**, 721-733 (1994)
18. G. Küppers, D. Lortz, J. Fluid Mech. **35**, 609 (1969)
19. R. Friedrich, M. Fantz, M. Bestehorn, H. Haken, *Pattern Formation in Rotating Bénard Convection*, Physica D **61**, 147-154 (1992)
20. J.M. Rodriguez, C. Pérez-Garcia, M. Bestehorn, M. Fantz, R. Friedrich, *Pattern formation in convection of rotating fluids with broken vertical symmetry*, Phys. Rev. A **46**, 4729-4735 (1992)
21. M. Neufeld, R. Friedrich, H. Haken, *Order parameter equation and model equation for high Prandtl number convection in rotating systems*, Z. Phys. B **92**, 243-256 (1993)
22. J. Millán-Rodriguez, M. Bestehorn, C. Pérez-Garcia, R. Friedrich, M. Neufeld: *Defect motion in rotating fluids*, Phys. Rev. Lett. **74**, 530-533 (1995)
23. M. Neufeld, R. Friedrich, *Statistical properties of the heat transport in a model of rotating Bénard convection*, Phys. Rev **E 51**, 2033-2045, (1995)
24. P. Manneville, J. Physic (Paris) **44**, 759 (1983)
25. E.D. Siggia, A. Zippelius, Phys. Rev. Lett. **47**, 835 (1981)
26. M. Fantz, M. Bestehorn, R. Friedrich, and H. Haken: *Hexagonal and Spiral Patterns of Thermal Convection*, Physics Letters **A 174**, 48–52 (1993)
27. M. Bestehorn, M. Fantz, R. Friedrich, H. Haken, C. Perez-Garcia:*Spiral Patterns in thermal convection*, Z. Phys. B **88**, 93 (1992)
28. R. Friedrich, *Higher Instabilities in Synergetic Systems with Continuous Symmetries*, Z. Phys. B **90**, 273 (1993)
29. J.P. Gollub, C.W. Meyer, *Symmetry breaking instability on a fluid surface*, Physica **D 6**, 337 (1983)
30. V. Steinberg, G. Ahlers, and D.S. Cannell, *Pattern formation and wavenumber selection by Rayleigh-Bénard convection in a cylindrical container*, Phys. Scr. **32**, 534 (1985)
31. M. Rabaud, Y. Couder, and S. Michalland, *Wavelength selection and transients in the one-dimensional array of cells of the printer's instability*, Eur. J. Mech. B **10**, 253 (1991)
32. G.H. Gunaratne, M. El-Handi, M. Gorman, Asymmetric cells and rotating rings in cellular flames Mod. Phys. Lett. B **10**, 1379 (1996)

33. B. Fiedler, B. Sandstede, A. Scheel, C. Wulf: *Bifurcation from relative equilibria of noncompact group actions: skew products, meanders, and drifts*, Doc. Math. **1**, 479 (1996)
34. S. Müller, *Experiments on excitation waves*, in *Nonlinear Physics of Complex Systems*, eds. J. Parisi, S.C. Müller, W. Zimmermann (Springer-Verlag Berlin, 1996)
35. F.T. Arecchi, S. Boccaletti, P. Ramazza, Phys. Rep. **328**, 1 (1999)
36. T. Ackemann, W. Lange, Appl. Phys. B: Lasers Opt. **72**,21 (2001)
37. M. Bode and H.-G. Purwins, Physica *Pattern formation in reaction-diffusion systems-dissipative solitons in physical systems*, **D 86**, 53 (1995)
38. S. Siegert, R. Friedrich, J. Peinke: *Analysis of data sets of stochastic systems*, Physics Letters **A 234**, 275–280 (1998)
39. R. Friedrich, S. Siegert, J. Peinke, S. Lück, M. Siefert, M. Lindemann, J. Raethjen, G. Deuschl, G. Pfister: *Extracting model equations from experimental data*, Physics Letters **A 271**, 217 (2000)
40. J. Gradisek, S. Siegert, R. Friedrich, I. Grabec: *Analysis of time series from stochastic processes*, Phys. Rev. **E 62**, 3146 (2000)
41. A.W. Liehr, H.U. Bödeker, T.D. Frank, R. Friedrich, H.G. Purwins, *Drift bifurcation detection for dissipative solitons*, New Journ. of Phys. **5**, Art. No. 89 (2003)
42. H.U. Bödeker, M.C. Röttger, A.W. Liehr, T.D. Frank, R. Friedrich, H.G. Purwins, *Noise covered drift bifurcation of dissipative solitons in a planar gas-discharge system*, Phys. Rev. **E 67** (5), 056220 (2003)

On Fundamental Structure-Forming Processes

Consequences for Dynamic Features

Peter Häussler

Technische Universität Chemnitz, Institut für Physik,
09107 Chemnitz, Germany
haeussler@physik.tu-chemnitz.de

1 Introduction

Many of the unsolved severe problems in condensed matter physics today are related to disordered phases such as liquids and glasses[1], including the glass transition [1], energetical low-lying dynamic excitations [2], the physical origin of the so-called Boson-peak in many properties at low temperatures (low-T) [3], as well as the adequate description of their static and dynamic structures themselves [4]. An even more severe problem is that we still do not understand what the fundamental structure-forming processes are when at very early stages structure starts to arise, as in liquid and amorphous systems, from the short-range order only to the long-range one. The understanding of these processes is of particular importance since liquid and amorphous systems are the precursors of any crystalline phase. If one understands how structure is forming in the disordered state, we may get a flavor for how crystals themselves are forming.

A better understanding of the static structure and the fundamental structure-forming processes may contain the key for the above-mentioned unsolved problems. Due to their structural variability, their isotropy, and their broad range of existence, the amorphous (a) and liquid (l) systems are model-like with respect to studies of processes at intermediate steps to the crystal. Structure formation itself is a many-particle effect with strong non-linear aspects.

It is generally believed that local chemical features of the individual atoms play the major role and global as well as collective effects are less important. Accordingly, quantum chemistry is taken as the adequate tool but, on the other hand, is hard to handle whenever the number of particles is large, long-range interactions are acting, or many degrees of freedom of the total system, such as density variations, charge transfer, or hybridization effects, need to be regarded. In large systems the number of parameters is huge and extends the capability of modern computers by far. Only small clusters of atoms can be treated where many atoms are located at their surface. Global and collective effects are hard to include, and they are treated as a perturbation of the local effects or are disregarded entirely.

[1] See contributions in this book.

Bloch's theorem in a crystal itself represents such a global collective effect of many atoms, of those which are located on periodically stacked mirror planes, scattering, e.g. electron waves or dynamic excitations collectively with high efficiency. Its application allows a tremendous reduction of the parameters to a few in a unit cell. But, in disordered systems similar reductions are inapplicable due to the lack of unit cells or translational invariance. Analogous concepts do not exist so far.

Generally, the isotropy of disordered systems gets traced back to the random orientation of pair-interacting nearest neighbours only [4]. Accordingly, global collective effects as in crystals are believed to be of minor influence and are mostly disregarded. Very often, the hard-sphere model serves as an excellent tool and, indeed, is able to explain many features, but, on the other hand, is unable to explain systematic variations with changing electronic conditions [5], and in particular has been shown to fail for especially those elements which are called the glass formers (B, C, Si, Ge, As, P, Sb, S, etc.) [4].

Within the last years there have been increasing indications that collective effects are by no means negligible and may even play the major role [5, 6, 7, 8, 9]. By diffraction experiments, a large scattering contribution was found to be based on the correlated location of nearest neighbours on concentric spherical shells around any adatom. These consecutive nearest-neighbour shells show between them, like mirror planes in the planar crystalline case, a unique spacing. Accordingly, we may talk about *mirror spheres* or *spherical-periodic order* (SPO) [9]. Its collective scattering power gives rise to one of the dominant peaks in the corresponding structure factor, $S(K)$, at $K_{\mathrm{p}e}$, where the index, e, is expressing the electronic influence on their existence. Accordingly, this peak describes the spacing of the shells, analogous to a Bragg peak which describes the spacing between mirror planes. Since the spherical-periodic order exists, in principle, around any adatom, those around nearby adatoms are strongly interpenetrating each other. The first-nearest-neighbour distance itself, which in disordered systems is 25% larger than the spacing between the shells [10], is hardly seen in the diffraction data.

The SPO is triggered by spherical-periodic electron-density oscillations and therefore is a global collective effect [5]; it is the result of a self-organizing resonance between the electrons as one subsystem and the forming static structure as the other one [5, 6, 7, 8, 9]. The SPO is temperature-dependent and has been observed to start within the liquid state well above the melting point, improving with decreasing T [11]. After a rapid quench into the amorphous state it improves further and becomes well pronounced [5].

In crystals, plane electron waves with a wavelength matching the spacing between the mirror planes are scattered with high efficiency and form standing waves between the planes. Quite analogously, in the systems with mirror spheres, standing spherical-periodic electron waves form between them. The better the atoms are organized at the spheres, the better defined are these standing spherical electron waves. As a consequence, in both cases there will be a gap or a pseudogap in the electronic density of states. If they arise at E_F they optimize (maximize) the band-structure energy, U_{bs}, and hence reduce the total energy of the phase under question [12]. Accordingly, the stabilisation effect under consideration is called resonance stabilisation [13].

Due to the pseudogap at E_F, consequences arise for the electronic transport properties [5, 6]. The better the resonance, the deeper the pseudogap and, as a consequence, the less conducting is the system and the larger are deviations to a simple free-electron behaviour. Under certain conditions the resonance and hence the SPO is such that different systems are completely dominated by the resonance and many properties become indistinguishable from one system to the other [5].

While in good crystals the ordered region is huge, containing, e.g. (in a single crystal) all the ions, in the disordered state the SPO lasts only to a few nearest-neighbour shells. Accordingly, the total mass of the atoms involved can be taken as infinite for the crystals, but finite for the SPO, with consequences for the latter in particular on the dispersion of electrons and dynamic excitations. Under these conditions, a new quasiparticle forms, called a *spheron*[2], expressing its existence in spherically periodic systems only [14]. Additional transport anomalies arise at temperatures above a characteristic T_o [15], which will not exist in single crystals. Some of these consequences have been published already elsewhere [5, 15] but are discussed below in Sects. 8/9.

Quite similar to resonances based on momentum, there are indications that resonances can also be based on angular momentum [16, 17]. Whereas the former causes spatial correlations between the atoms (the SPO), the latter may cause angular correlations (AC).

The first of our reviews on this topic reported on glassy metals only [5]. Quite recently, in another contribution [6], we have reviewed new aspects of fundamental structure-forming processes in disordered quasicrystals as well as quasicrystals themselves. Quasicrystals were found to be of additional advantage since they can be prepared by simply heat treating their amorphous precursors, showing improvements of the SPO as well as increasing angular correlations. In some other publications, analogous results on magnetic glassy alloys [18], ionic glasses [7], and glassy semiconductors [16, 19, 20, 21] have been reported. It is the goal of the present contribution to review some aspects of all these systems in one paper only and to discuss the related aspects of dynamic anomalies. We describe two major effects, namely those resonance effects which cause the stability of the disordered state (including different scenarios to optimize the resonance), and those which are responsible for the formation of the spherons and additional transport anomalies.

The paper is organized as follows. In the next section we discuss the resonance stabilisation effect in a simple mechanical analog, as well as, generalized, in disordered condensed matter. The following two sections 3/4, divided

[2] In earlier publications we called them *amorphons*.

into several paragraphs, show some scenarios of how the total system tries to optimize the resonance and hence the stability of the phases. At the end of Sect. 4 and in Sect. 5, experimental indications of angular-momentum-based resonances are shown. In Sect. 6 consequences of resonance effects on the electronic system, such as pseudogaps at the Fermi edge, as well as those far below, are presented and discussed. In Sect. 7 only few remarks on subsequent electronic transport properties are made. More details have already been published elsewhere (references are given below). In Sect. 8 we show that the SPO also influences dynamic excitations. Consequences arise, very special for disordered systems, causing the new type of dynamic excitations, the spherons (Sect. 9), and hence low-T transport anomalies.

2 Resonance Stabilisation in Disordered Matter

As briefly mentioned above, structure formation at an early stage, based on momentum, can be described as affected by a *spherical-periodic resonance* between the self-organizing electrons and static structure (Hume-Rothery- or Peierls-like) [22, 23]. In some cases the static structure is adjusting to the electronic constraints [5, 6, 9], in some others the electronic structure adjusts too by adjusting the electron density via different scenarios [7, 8].

Whenever two systems are coupled, their degenerate states get lifted, and bonding states as well as anti-bonding states exist, with a gap in between [13]. The simplest resonating system, known from physics ground course, is two weakly coupled mathematical pendula. Whenever they get weakly coupled, as a result we get two new states: the *bonding* one, where both are oscillating in phase, and the *anti-bonding* one, where both are oscillating in anti-phase. The former has the lower, the latter the higher energy; the original degenerate state has been split into two with a gap in between. In principle, now the total system can go permanently into the bonding state, if energy loss is allowed.

If the pendula have been resonantly prepared (same mass, same length of the suspension) the splitting between the bonding and the anti-bonding states is maximal (resonance gap) [13]. If both pendula are different (out of resonance) in this mechanical case, there is no way for them to get into resonance, to maximize the splitting and to minimize the energy of the bonding state.

We now transfer this picture into condensed matter physics by taking all the conduction electrons as one *pendulum* and the forming static structure of the ions as the other one, with the electron-ion interaction as the coupling between both. Generally, they would be out of resonance. But, now both subsystems have internal degrees of freedom and may mutually adjust in order to come into resonance, to optimize the splitting between the bonding and the anti-bonding states, and to decrease the total energy as much as possible under the given constraints [5, 7, 8]. There exists a real-space [24] as well as a reciprocal-space [12] interpretation.

2.1 Description of Resonance Stabilisation in r-Space

In real-space (r-space) the ideal, momentum-based resonance in disordered systems is expressed by the equality of two characteristic wavelengths of the interacting subsystems, caused by spherical-periodic electron-density oscillations, and, subsequently, by the so-called Friedel oscillations [25] in the effective pair potential [14, 24, 26]. The Friedel oscillations can be deduced via the screening of a local charge (ion) by the conduction electrons. For systems with a short mean free path of the electrons [10, 27, 28], the effective pair potential deviates from the well known cosine function [25] by a phase shift of $\pi/2$ and therefore is written as

$$\varphi_{\text{eff}}(r) \propto -\sin(2k_{\text{F}}r)/r^3 \quad , \tag{1}$$

with the wavelength $\lambda_{\text{Fr}} = 2\pi/2k_{\text{F}} = 1/2\times\lambda_{\text{F}}$ (λ_{F} the Fermi wavelength) (Fig. 1b). As a first approximation, the Fermi momentum, k_{F}, gets estimated by its free-electron value

$$k_{\text{F}} = \sqrt[3]{3\pi^2\, \bar{n}_{\text{o}} \bar{Z}} \quad , \tag{2}$$

with $\bar{n}_{\text{o}}$ the mean particle density (for alloys) and $\bar{Z}$ the mean valency per atom. Positive ions arrange preferentially in equidistant spherical shells around any adatom at positions where the minima of (1) occur, at approximately

$$r_{\text{n+1}}^{\text{Fr}} = (5/4 + n) \times \lambda_{\text{Fr}}, \qquad n = 0, 1, 2, 3, \cdots \quad . \tag{3}$$

The resonance is fulfilled when the locations of the nearest-neighbour shells are identical with the $r_{\text{n+1}}^{\text{Fr}}$ [10]. Accordingly, the first nearest neighbour should be located at $r_1^{\text{Fr}} = 5/4\times\lambda_{\text{Fr}}$. The further minima cause particle-density oscillations with a wavelength $\lambda_{\text{i}} = 2\pi/K_{\text{pe}}$. K_{pe} represents the location of the resonance-induced structural peak in $S(K)$, and hence describes the spacing between the nearest-neighbour shells. Accordingly, the resonance in the disordered state is based on medium-range order, by the equality of the characteristic wavelengths [26], namely

$$\lambda_{\text{i}} = \frac{2\pi}{K_{\text{pe}}} = \lambda_{\text{Fr}} = \frac{2\pi}{2k_{\text{F}}} \quad . \tag{4}$$

The consecutive nearest-neighbour shells of positive ions are schematically shown in Fig. 1a. Negative ions would preferentially be located in positive regions (between the Friedel minima of $\varphi_{\text{eff}}(r)$).

The numbers between the brackets of (3), namely $5/4, 9/4, 13/4, 17/4, \cdots$, are taken as the fingerprint of spherical periodicity. Hard-sphere models give similar sequences but are unable to explain the systematic variation of nearest-neighbour distances with the changing composition and hence changing $2k_{\text{F}}$, as observed experimentally.

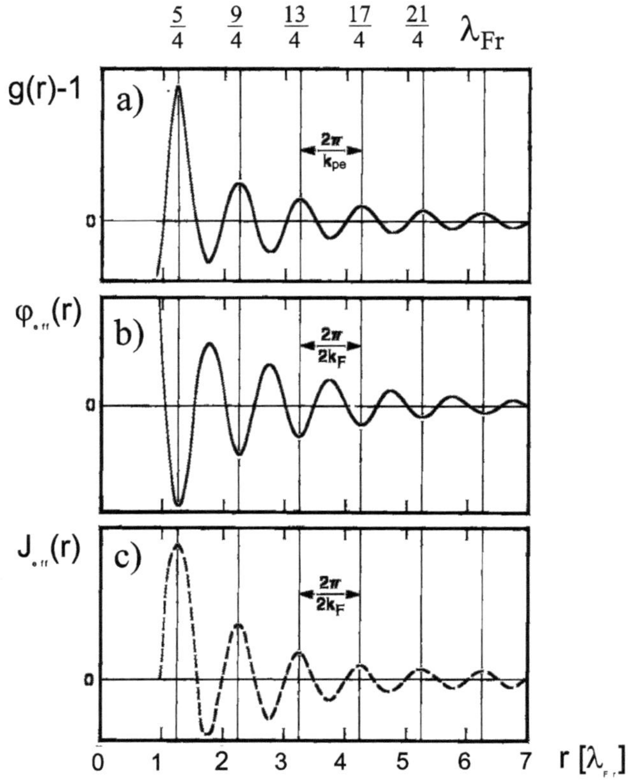

Fig. 1. (**a**) Schematic pair distribution function $g(r)$ in resonance with the pair potential $\varphi_{\mathrm{eff}}(r)$ [5] or an oscillating effective magnetic interaction $J_{\mathrm{eff}}(\mathrm{r})$. (**b**) Schematic view of an effective pair potential with Friedel oscillations at medium distances. (**c**) Schematic view of an effective magnetic interaction with RKKY oscillations at medium distances. All the curves are drawn in a scaled version, with $\lambda_{\mathrm{Fr}} = 2\pi/2k_{\mathrm{F}}$ as an internal length scale

The momentum-based resonance can be seen as an interference effect between a once-scattered plane electron wave, spherically outspreading from a chosen adatom, and all the partial waves back-scattered coherently from all the equidistant nearest-neighbour shells (Fig. 2a).

Any atom on a shell serves itself as an adatom in the center of its own consecutive nearest-neighbour shells. Accordingly, the spherical-periodic regions are strongly interpenetrating each other. The grade of the SPO dominates the grade of the interference and hence the depth of the pseudogap in the electronic density of states (DOS) at E_{F} [5, 29]. The pseudogap itself becomes quite broad due to the spatial limitation of the SPO. The more optimal the resonance becomes (the better the nearest neighbours are occupying the spherical-periodic nearest-neighbour shells, and the more the SPO extends),

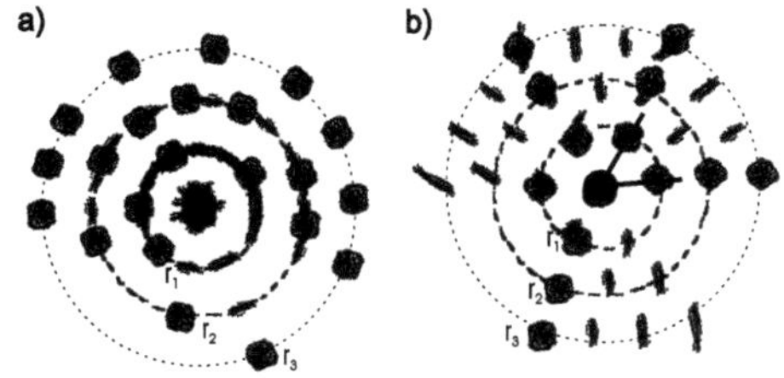

Fig. 2. (**a**) Simplified view of a 2D momentum-based resonance causing spatial order or spherical periodicity. (**b**) Simplified view of an angular-momentum-based resonance causing angular order. The broken circles indicate the optimal location of nearest-neighbour shells according to (3). The broad shadowed parts of circles indicate electron waves coherently back-scattered by the nearest-neighbour shells. Between the shells they would form standing spherical electron waves (not shown here). In (a) only those parts of circles are shown which are directed to the adatom in the center, in (b) only those which are directed around the adatom

the deeper becomes the pseudogap and the more stable and less conducting becomes the phase [29] (quite similar to the crystalline case where plane electron waves with the wavelength λ_F may be in resonance with the mirror planes of a lattice, causing standing plane electron waves or 'covalency'). As a consequence, transport properties and in particular their anomalies contain additional information on resonance stabilisation. Subsequently, electronic transport properties, such as the resistivity, the Hall coefficient, as well as the thermopower, are additional tools to study resonance stabilisation effects [5, 6].

Whereas the SPO, discussed so far, goes back to the screening of the local *charge* of an ion and hence uses the charge of the electrons, years ago indications were reported [18] that similar features exist whenever *local moments* (e.g. at Fe) are involved. In this case the electrons try to screen with their spin the local moments. Spin-density oscillations occur, causing an indirect interaction between local moments. Spherical-periodic nearest-neighbour shells may arise which are occupied by those atoms carrying local moments (RKKY interaction [30]) (Fig. 1c).

2.2 Description of Resonance Stabilisation in *k*-Space

In momentum-space (k-space) the momentum-based resonance can be explained with the help of the dispersion of the conduction electrons, shown in Fig. 3 for the crystalline (a) and disordered (b) cases, both drawn schematically together with the corresponding electronic DOS and the gap in the former and the pseudogap at E_F in the latter. The optimal state is given whenever the lifting of the degenerate states, the formation of a resonance gap/pseudogap, occurs at $\pm k_F$ or E_F, respectively, since only those states

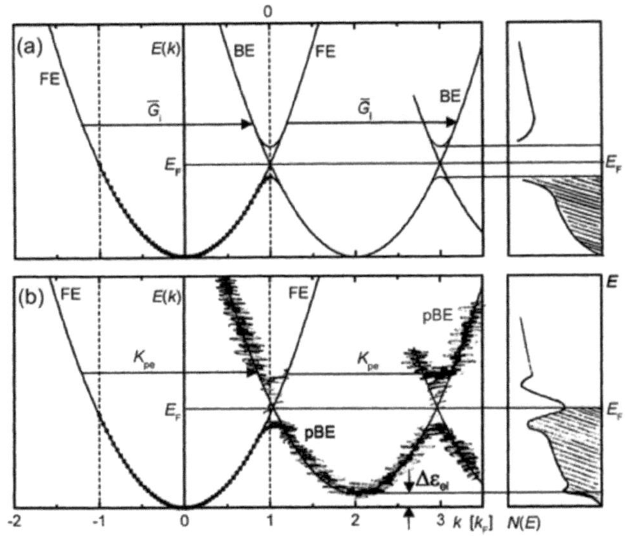

Fig. 3. (**a**) Schematic dispersion of electrons in the crystalline case. (**b**) Schematic dispersion of electrons in the disordered state, in both cases together with the electronic DOS. The k-scale is normalized by k_F

get shifted to higher energies that are empty, whereas those states shifted to lower energies are occupied, bringing the largest net effect.

In a crystal the resonance is expressed by

$$\boldsymbol{G}_{\mathrm{i}} = 2k_{\mathrm{F}} \quad , \tag{5}$$

with $\boldsymbol{G}_{\mathrm{i}}$ a reciprocal lattice vector in a particular direction i (Fig. 2a). In the disordered state it is expressed by

$$K_{\mathrm{pe}} = 2k_{\mathrm{F}} \quad , \tag{6}$$

in both cases by the equality of the position of a characteristic peak in $S(K)$ at $\boldsymbol{G}_{\mathrm{i}}$ or K_{pe}, respectively, and a characteristic momentum of the electronic system, namely $2k_{\mathrm{F}}$.

Without any order, only the free-electron parabola (FE) would exist, with the states occupied between $-k_{\mathrm{F}} \leq k \leq +k_{\mathrm{F}}$ and $0 \leq E \leq E_{\mathrm{F}}$ (far left side of Fig. 3a, b). In the crystalline case (a), where (5) is fulfilled, there are further electron parabolae, describing Bloch electrons (BE). Due to their elastic scattering with the static structure, they are no more describing free electrons alone. Instead, they are shifted by $n \times \boldsymbol{G}_{\mathrm{i}}$ along the k-scale ($n = 1,2,3,\cdots$) and, therefore, contain structural features. Along the energy scale there is no shift due to the quasi-infinite mass of the static structure and, therefore, complete elastic scattering.

Under the optimal situation the FE and the BE are in a degenerate state at $k = \pm k_{\mathrm{F}}$ or $E = E_{\mathrm{F}}$ (in Fig. 3a only the situation at $k = +k_{\mathrm{F}}$ is shown).

Due to the coupling between the FE and the BE, the degeneracy gets lifted and a gap opens (Fig. 3a; middle, far right). Due to the decrease of the electronic states at E_F, this particular structure is more stable than those with the gap somewhere else. Subsequently, through self-organization of its static structure, the total system attempts to reach an optimal configuration. In crystalline systems the optimum may be fulfilled in a few directions only; in other directions there will be no gap. Accordingly, the mean decrease of the total energy has to be calculated as the complex sum over all directions [12].

In the liquid and amorphous cases (Fig. 3b), with both phases having spatially limited spherical-periodic medium-range order, similar features arise. But, instead of a well defined shift of the FE-parabola by $n \times \boldsymbol{G}_i$, now there is a diffuse shift by $n \times K_{pe}$, since K_{pe}, describing the location of the pseudo-Bragg peak in the disordered state, itself is broad. Accordingly, the so-called pseudo-Bloch electronic states (pBE) are broadened. Instead of a gap, now there is a pseudogap at E_F only [29]. Its depth will depend on the degree of the SPO [5].

But, as will become important in Sects. 8/9, now there is a slight shift of the pBE-parabola by $\Delta\varepsilon_{el}$ along the energy scale since the spherical-periodic order involves a finite number of atoms and with any Umklapp scattering there is an energy loss involved. This has consequences for the DOS. There is, measured on the scale of the bandwidth (eV), an extremely narrow second pseudogap at the lower band edge (Figs. 3b; middle, far right). Its width depends on the size (mass) of the SPO regions. The shift will be small, in the range of meV (see also Sect. 9). The larger the SPO regions become, the smaller becomes the corresponding energy shift [31]. The energy shift itself is broadened due to the different extensions of the SPO regions.

In Fig. 4 some earlier measurements of the valence-band region of amorphous binary alloys are shown. Close to or at the lower band edge there is a tiny peak at E_{sp} which could not be explained at the time of its first publication [5]. Today we may interpret this peak as induced by the tiny energy shift of the electronic states due to the finite mass of the SPO.

In Sect. 9 we show that, together with similar effects on dynamic excitations, this energy shift will lead to the so-called spherons, a resonance state between three subsystems, namely the electrons, the static structure, and the dynamic excitations. The spherons have additional effects on transport properties.

Due to the electronic effects on the static structure, below in this contribution λ_{Fr} will be taken as an internal length scale in r-space to measure distances, and $2k_F$ as an internal scale in k-space to measure momentum transfers.

It should be emphasized that the SPO and its optimization via a self-organization process gives only part of the total energy. Similar to crystalline matter where, besides the global Bloch theorem, the systems have to fulfill all the local quantum-chemical requirements in a unit cell, in disordered systems

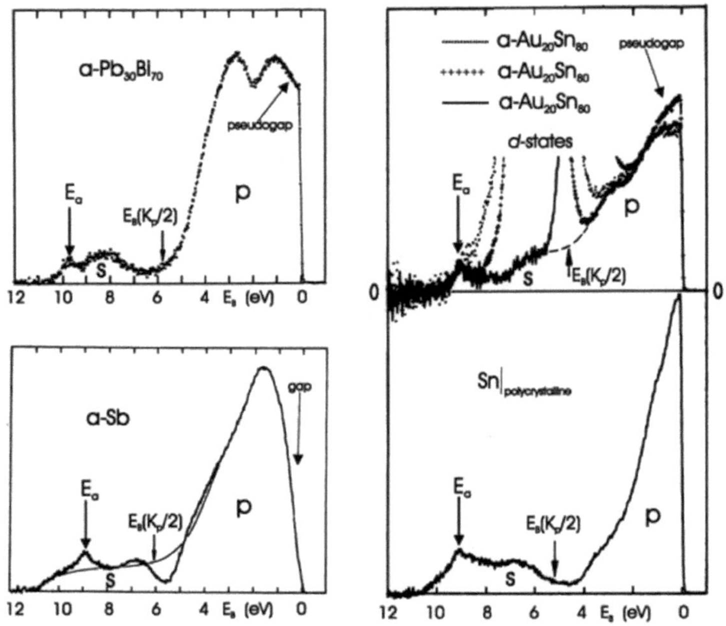

Fig. 4. Photoelectron data (UPS) of some binary amorphous systems and strongly disordered Sn [5, 11] showing indications of a peak of electronic states at the lower band edge (vertical arrows at E_{sp}). The pseudogaps indicated by $E_{\mathrm{B}}(K_{\mathrm{P}}/2)$ are related to the first structural peak in $S(K)$ (see Sect. 6) and the pseudogaps (or gap, in Sb) at E_{F} to the SPO

the total system has also to serve the local chemical needs of the individual atoms. The local *and* the global effects both stabilize the total system, dependent on their composition, the charge or local moment of the atoms, their difference in electronegativity, as well as the fact that empty *d*- or *f*-states may exist.

3 Adjustments of the Static Structure Alone

In this section we first present the simple amorphous glasses where the main effect is the adjustment of the static structure to the electronic constraints.

3.1 Amorphous Metals Showing Spherical-Periodic Order Only

The best known examples of disordered resonance-stabilized systems are binary metallic glasses of the type *l*/*a*-AuSn. They show two clear indications of electronic influence. First of all, there is a structural peak in $S(K)$ close to $2k_{\mathrm{F}}$ which shifts parallel to the changing $2k_{\mathrm{F}}$ by composition changes [5]. Accordingly, all the structural data known so far can be scaled by $2k_{\mathrm{F}}$ in k-space (vertical axis in Fig. 5c) and by λ_{Fr} in r-space (vertical axis in Fig. 5d) as expected for a resonance-stabilized system. Secondly, most of the properties of the metallic glasses under consideration, including the structural data, can, as for Hume-Rothery alloys [22], be scaled versus the mean effective valency $\bar{Z}$ (horizontal axis of Fig. 5c, d) [5].

In the noble-metal-poor range, the momentum-based resonance in these alloys is relatively weak (Figs. 5a1, b1), although they show radial order of the spherical-periodic type mentioned above, seen by the nearest-neighbour shells at $r_{\mathrm{n+1}}^{\mathrm{Fr}}$ (Fig. 5d), as given by (3). The corresponding peak in k-space is slightly below $2k_{\mathrm{F}}$ (Fig. 5c) and quite low (Fig. 5a1). A large first peak is located far below $2k_{\mathrm{F}}$ (Fig. 5a1). The pseudogap at E_{F} is weak, the amorphous phase is rather unstable (below room temperature), the resistivity is relatively low, and its temperature coefficient is positive [5]. Both the Hall coefficient as well as the thermopower are close to the free-electron values (FEM values) [5].

The resonance improves whenever the mean valency, $\bar{Z} = 1.8\,\mathrm{e/a}$, is approached (Figs. 5a2, b2, c, d) [5]. But, even now, when only one dominant peak exists precisely at $2k_{\mathrm{F}}$ and all the ions are occupying the equidistant shells, from the first up to relatively far nearest-neighbour shells, the systems are still not under optimal resonance. The occupation of the shells is nearly perfect but, *at* the shells, still random. The systems stay metallic with a still

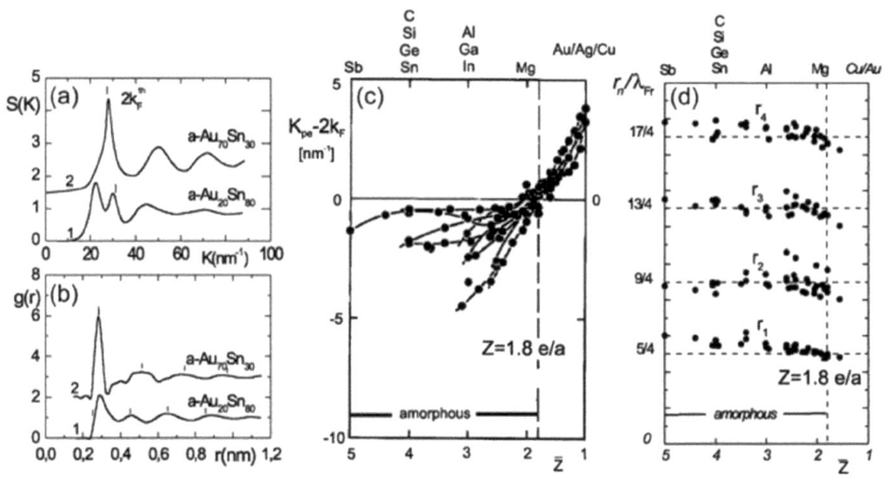

Fig. 5. Structural data of some simple metallic glasses. (**a**) Two characteristic structure factors of a-AuSn [32] compared with $2k_{\mathrm{F}}$ (*solid vertical lines*). (**b**) Corresponding pair distributions compared with $5/4$, $9/4$, $13/4$, $\cdots \times \lambda_{\mathrm{Fr}}$ (*vertical lines*), the fingerprint of SPO. (**c**) Distance of the resonance-induced structural peak in $S(K)$ to $2k_{\mathrm{F}}$ scaled at the horizontal axis by $\bar{Z}$ [5]. (**d**) Corresponding positions of nearest-neighbour shells measured in units of λ_{Fr} scaled again by $\bar{Z}$ [19]. Some amorphous semiconductors (Sb, C, Si, Ge) are included for comparison with the metallic glasses (for references to the individual alloys see [5])

weak pseudogap at E_F, although it goes down to 50% of the free-electron value (Fig. 14) [5, 29, 33]. The resistivity as well as the thermal stability both become high, transport anomalies appear as negative temperature coefficients of the resistivity and positive thermopowers [5]. The Hall coefficient deviates by roughly 40% from the FEM but stays negative [5]. Particle-density anomalies are found to be small [32], in contrast to a-NaSn [7] (see below), and the valencies are as expected (from the position of the elements in the periodic table), in contrast to what we found for TM-Al alloys (TM: Mn, Fe, Co, Ni) (see below) [6].

The l/a-AuSn alloys, at a mean valency of $\bar{Z} = 1.8$ e/a (e.g. a-$Au_{73}Sn_{27}$, a-$Au_{80}Sb_{20}$), are completely dominated by the momentum-based resonance. At these particular compositions only one first peak in $S(K)$ exists exactly at $2k_F$ and in r-space all the nearest neighbours are occupying their Friedel minima. Subsequently, all their properties, known to us, are indistinguishable from one system to the other [5, 6]. Although far from compositions with $\bar{Z} = 1.8$ e/a the individual systems may differ by orders of magnitude in their electronic transport properties, at this particular composition different systems cannot be distinguished. More details on the simple binary metallic glasses of the type a/l-AuSn have been reported elsewhere [5].

4 Additional Adjustments of the Electronic System

Besides the adjustment of the ionic positions to the electronic constraints, which has been observed in all the systems under consideration, additional optimization scenarios may occur due to the additional adjustment of $2k_F$ (the adjustment of the electronic subsystem to the structural constraints) by varying $\bar{n}_o$ or $\bar{Z}$ in (2). Variations of $\bar{n}_o$ [7] as well as $\bar{Z}$ [8, 16] both have been reported.

4.1 Amorphous Zintl Systems with the Additional Adjustment of $2k_F$ via the Density

Zintl systems, alloys of the type a-NaSn, containing alkalides instead of the noble metals, belong also to the resonance-stabilized systems but now with the additional adjustment of the electronic system via $\bar{n}_o$ [7]. Due to the large difference of the electronegativity ($\varepsilon_{Na} = 0.9$, $\varepsilon_{Sn} = 1.8$), by charge transfer and the stripping off of its single outer electron, Na atoms can reduce their effective volume. The particle density, $\bar{n}_o$, increases (up to 50%) (Fig. 6a), causing an adjustment of the Fermi sphere diameter $2k_F^n > 2k_F^{th}$ to the structural peak at K_{pe} versus composition (Fig. 6b). The index n in $2k_F^n$ has been introduced to express the electronic adjustment of $2k_F$ to the static structure by the density anomaly. $2k_F^{th}$ represents the FEM value without density anomalies. The resonance condition $K_{pe} = 2k_F^n$ is well fulfilled over the complete composition range, with small differences close to pure Sn (Fig. 6b).

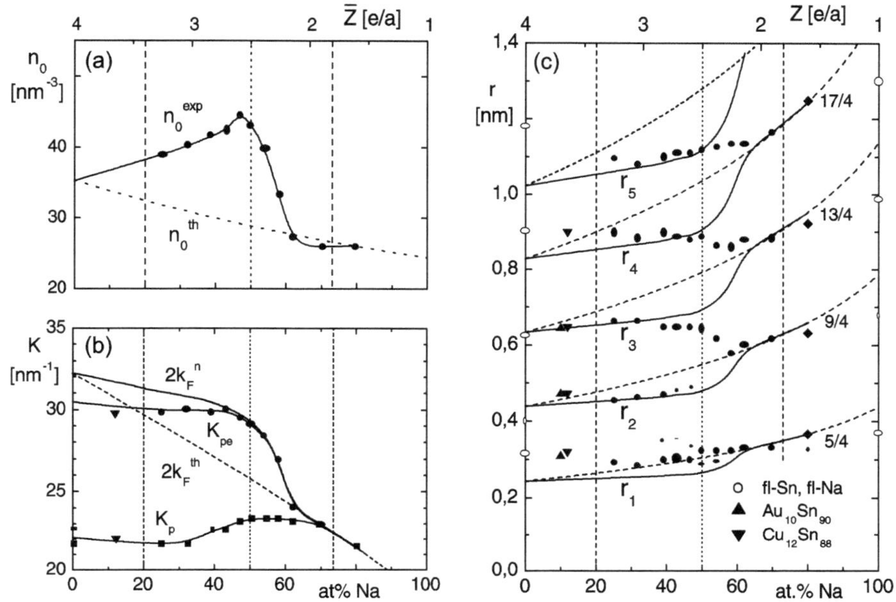

Fig. 6. Structure data of a-NaSn [7]. (**a**) Particle-density, (**b**) peak positions in k-space, and (**c**) peak positions in r-space. The broken lines indicate $\bar{n}_o^{th}$ (**a**), $2k_F^{th}$ (**b**), and the Friedel minima (**c**) without density anomalies. Close to pure Sn where a-NaSn has not been prepared, a-$Au_{10}Sn_{90}$ and a-$Cu_{12}Sn_{88}$ [32] have been added. The vertical lines at 20 at%Na and 73 at%Na include the amorphous range, the dotted vertical line at 50 at%Na the range of strongest ionic bonding [34]

There is no free parameter in this comparison. The particle density has been measured [7], the valency of Na has been taken as $Z_{Na} = 1\,e/a$, and the valency of Sn as $Z_{Sn} = 4\,e/a$, respectively, according to their position in the periodic table of the elements. Above 50-60 at%Na, in the range where the density anomaly disappears, the peak at K_{pe} shifts back to $2k_F^{th}$, to a new resonance, but now without density anomalies.

In r-space the nearest-neighbour shells follow the adjusted Friedel oscillations (Fig. 6c) with the exception of a range from 50-60 at%Na. In this range, medium positions – positions between the Friedel minima, those preferred by negative ions – are occupied. And indeed, strong ionic-bonding features and the formation of so-called Zintl clusters have been reported especially for this region [34].

Subsequently, a-NaSn shows three resonance-stabilized regions. One is at the Na-poor side, where the density anomaly helps to optimize the resonance. One is at the Na-rich side, where the resonance has adjusted again, but without any density anomaly. And one is between both, where ionic-bonding features arise andZintl clusters exist.

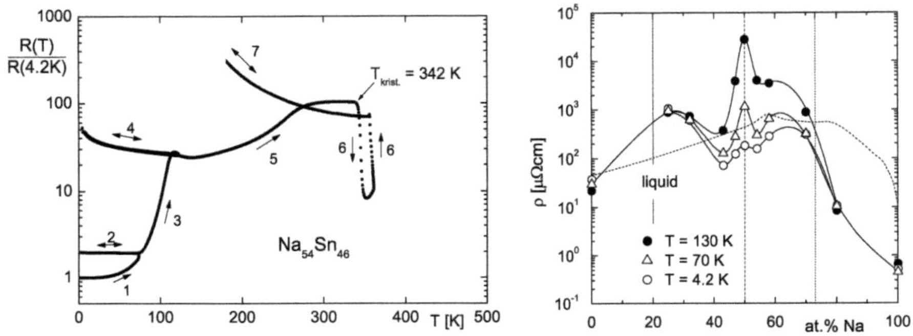

Fig. 7. Resistivity of a-NaSn [7]. (*left*) a-$Na_{54}Sn_{46}$ versus temperature, (*right*) versus composition

Due to the improved resonance and, subsequently, the ionic-bonding features, the resistivity is extremely high in the amorphous state and increases tremendously by annealing [7]. The resistivity is even anomalously high in the liquid state, especially around $Na_{60}Sn_{40}$ [35]. In Fig. 7a the evolution of the resistivity of one representative amorphous alloy is shown versus temperature. Upon annealing, the resistivity increases by two orders of magnitude. A more detailed research of the related structural changes has still to be performed. But, first measurements show indeed an improved SPO by annealing [7]. In Fig. 7b the variation of the resistivity, measured immediately after deposition at $T = 4$ K and at two annealing steps, is drawn versus composition, showing the very high resistivity around 50-60 at%Na, where ionic-bonding features exist. The resistivity of the liquid alloys is shown for comparison. They show the highest resistivities around 60 at%Na.

Over the complete concentration range the density and subsequently the charge transfer seem to optimize the resonance quite flexibly. The charge transfer itself and the formation of Zintl clusters both seem to be triggered by the optimization process of the resonance. Ionic-bonding features, the density anomaly, as well as the formation of Zintl clusters are caused by the optimization of the resonance.

4.2 Amorphous TM-Al Alloys with the Additional Adjustment of $2k_F$ via Hybridization

Aluminum is an excellent metal in its crystalline state. In binary amorphous alloys with the noble metals, of the type described above, aluminum shows the spherical resonance but stays metallic [5] (see Fig. 5c, d and Fig. 9a5), whereas in alloys with late-transition elements containing empty *d*-states at E_F (*a*-TM-Al; TM: Mn, Fe, Co, Ni, CuPd, PdMn, PdRe), semiconducting features arise [36]. The electronegativity of Al is $\varepsilon_{Al} = 1.5$, that of Mn, Fe, Co, Ni, Cu is 1.5, 1.8, 1.8, 1.8, 1.9, respectively. With the exception of Mn, all are larger than for the Al. Accordingly, electrons would preferentially stay

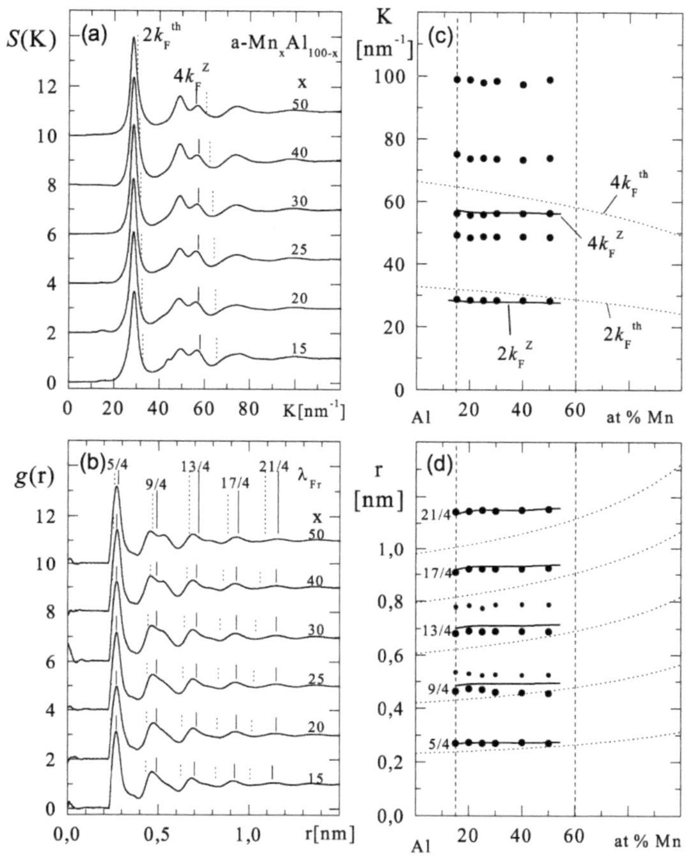

Fig. 8. Structural properties of a-MnAl after annealing to 300 K showing clear indications of SPO [6, 16]. (**a**) Structure factor and (**b**) pair-distribution function. (**c**) Peak positions in $S(K)$ and (**d**) in $g(r)$. The dotted lines indicate $2k_F$ without any hybridization effect. The thick lines assume $2k_F^Z$ in exact agreement with K_{pe}

closer to the TM, improving hybridization effects which become important in these alloys, as we show below. Hybridization effects between the *s,p*-states and the TM *d*-states have already been proposed in earlier publications on the basis of DOS measurements [37].

In a-MnAl, a-FeAl, and a-CoAl, at a first glance, the resonance condition ($K_{pe} \simeq 2k_F^{th}$) seems to be fulfilled at about 60-70 at%Mn, Fe, Co (compare e.g. in Fig. 8c the position of the first peak in $S(K)$ for a-MnAl with the corresponding dotted line). In this range the resistivity should be highest [38] and the system most stable, which is contrasted experimentally (Fig. 9a, b). Instead, surprisingly, the highest resistivity and stability are at the Al-rich side. The simple presence of *d*-states at E_F, as strong scatterers, can not be made responsible for this anomaly, as their content becomes low. Accordingly,

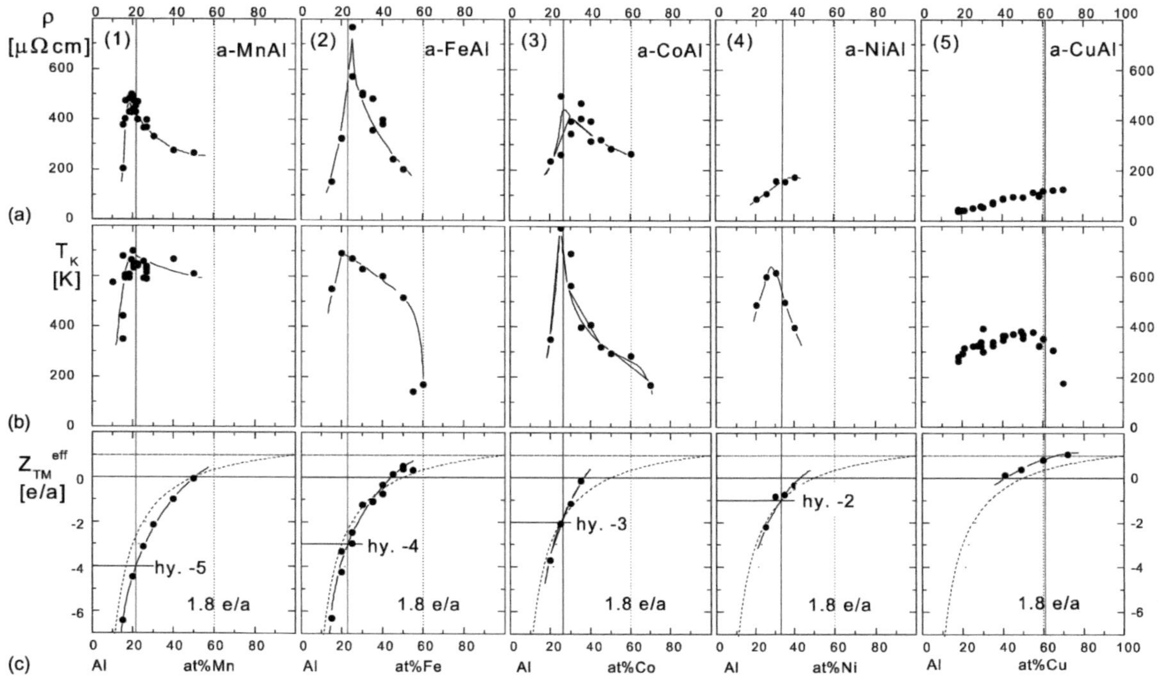

Fig. 9. For the binary TM-Al alloys, in (**a**) the resistivities, in (**b**) the thermal stability, and in (**c**) the effective valencies, $\bar{Z}_{\mathrm{TM}}^{\mathrm{eff}}$, due to hybridization, are drawn versus composition [6]. The broken lines in (**c**) indicate, as a rough estimate, an effective valency by hybridizing all the Al p-states with the TM d-states. The full dots in (**c**) are calculated by the assumption that $2k_{\mathrm{F}}^{\mathrm{Z}}$ equals K_{pe} as described in the text

we assume that again the electronic system has found a way to adjust. But now, at the Al-rich side, the effective $2k_F$ has to be smaller than $2k_F^{th}$ (Figs. 8a, c), in contrast to a-NaSn above. Because density anomalies are small and the $\bar{n}_o^{exp}$ are even larger, instead of smaller, as needed [6, 16], we claim that now it is $\bar{Z}$ which is adjusting.

Since only a single first peak exists and, in addition, the resistivity is high at the Al-rich side as mentioned, we assume that the mutual adjustment of the static and the electronic structure is quite strong. $2k_F$ may be as close to K_{pe} as in semiconducting systems (Subsect. 5) [9, 16, 19, 39, 40]. The first strong peak is quite narrow, indicating spherical-periodic correlations over large distances. This is supported by the existence of a peak at $2\times K_{pe}$ (solid lines in Fig. 8a, c) and, as has been observed after annealing, at $3\times K_{pe}$ [5].

As a trial we assume an exact agreement of $2k_F^Z$ with K_{pe} over the complete amorphous range (the index Z in $2k_F^Z$ has now been introduced to express the electronic adjustment of $2k_F$ to the static structure via Z). This assumption also determines the positions of the Friedel minima, r_{n+1}^{Fr} (solid lines in Fig. 8b, d), and allows us to calculate the mean valency, $\bar{Z}^{eff}$, by (2) and from this the effective valency, Z_{Mn}^{eff}, of the transition elements, by fixing the valency of Al as $Z_{Al} = 3\,e/a$. The adjustment of $2k_F^Z$ to K_{pe} in k-space causes in r-space an excellent matching of the Friedel minima with all the strong maxima, indicating again spherical periodicity.

The effective valency of Mn itself varies strongly with composition (full dots in Fig. 9c1). In order to answer the question of how this may happen, we assume that only the Al p-states but not the s-states become hybridized with the Mn d-states and that the hybridization is *complete* in the sense that no Al p-states are left unhybridized in the Fermi sea. The knowledge of the Al content then automatically determines how many Al p-electrons have to be taken by the individual Mn atoms. The effective valency of Mn is then given by the valency of pure Mn minus the number of p-electrons which are hybridized with the Mn d-electrons. Assuming a valency of all the pure late-transition elements from Mn to Ni as $Z_{TM}^{eff} = 1$ - $1.1\,e/a$, as has been proposed by Waseda for Fe [41], and as has been successfully applied to those amorphous TM alloys without Al [18], yields the dashed lines in Fig. 9c1, in quite good agreement with the measured effective valency of Mn (full dots). This confirms our assumption that all the Al p-states are hybridized with the Mn d-states.

We still have to explain the high resistivity and high thermal stability at the Al-rich side. This may be caused by a *complete* filling-up of *all* the empty d-states with the Al p-states by hybridization. Whenever, e.g. Mn has gone into hybridization with 5 Al p-electrons (short horizontal line in Fig. 9c1), its d-states are completely occupied, which happens at 21 at%Mn. Subsequently, the highest resistivity and thermal stability should occur at this composition, which actually is the case and supports our interpretation of an enhanced resonance stabilisation triggered by hybridization effects. Under the given conditions the s- and d-states are completely full and the p-states completely empty. The d-states have been shifted below E_F, the p-states above. A deep pseudogap exists at E_F. Under these conditions, the resistivity as well as the thermal stability both should be high.

Hybridization effects in quasicrystals and, as we show later, their angular correlations, both seem to be triggered by the optimized resonance. The empty d-states at E_F act as a 'sink' for conduction electrons. The diameter of the Fermi sphere is able to adjust to the given constraints and in this case shrinks. Versus composition the Fermi sea gets partially 'drained' by the exact amount that is necessary to optimize the resonance.

But, the complete hybridization of Mn, Fe, Co, Ni, Cu would need different numbers of Al p-states. The composition where the maximum resistivities and the highest stabilities exist, therefore, should shift systematically on the concentration scale. In order to show that for different alloys similar features indeed occur at different compositions, we present resistivity data (Fig. 9a), data of the crystallisation temperature (Fig. 9b), and data of the effective valency, $Z_{\mathrm{TM}}^{\mathrm{eff}}$ (Fig. 9c). $Z_{\mathrm{TM}}^{\mathrm{eff}}$ has been calculated via $2k_{\mathrm{F}}^{\mathrm{Z}} = K_{\mathrm{pe}}$, as described above. And, indeed, the maxima shift from Mn-Al to Cu-Al systematically. The short horizontal lines in Fig. 9c indicate the individual effective valencies of the TM, when all the empty d-states would be hybridized completely. Comparing the related compositions with the maxima in Figs. 9a, b (vertical solid lines) it becomes obvious that the systematic variation of the resistivity and stability maxima shows the strong correlation to the complete hybridization of the TM d-states. This correlation is even better seen in Fig. 10.

CuAl shows no hybridization effect since all the Cu d-states are already occupied. A 'sink' for Al p-states is not available. Accordingly, its resistivity as well as thermal stability are both quite low [42, 43]. The highest resistivity and highest stability are both around 60 at%Cu, where the mean effective

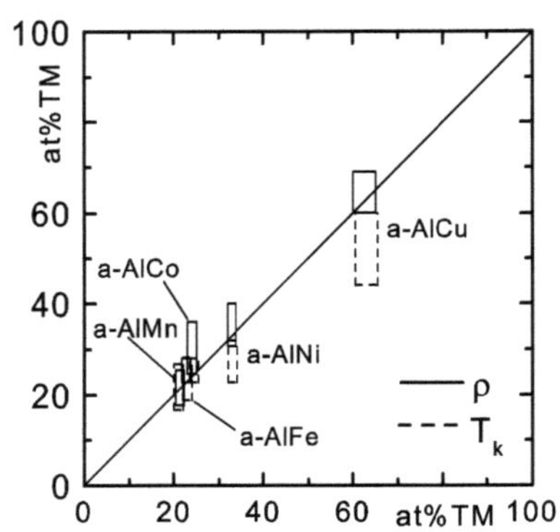

Fig. 10. Location of the resistivity and crystallisation maxima on the concentration scale versus the approximate concentration where the Al p- TM d-hybridization is completed [6]. The size of the squares indicates error bars

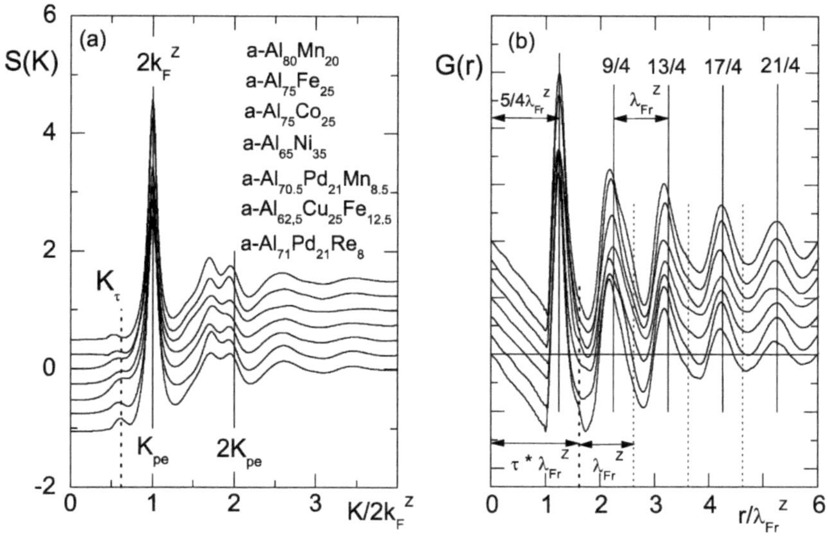

Fig. 11. (**a,b**) Structural data of Al-rich alloys taken at 350 K showing clearly the indications of spherical resonance in k- and r-space [6]. The ternary systems become quasicrystalline by heat treatments. In addition to the SPO there are τ-related structural features (*broken vertical lines*)

valency is 1.8 e/a and the system belongs to the simple glassy metals with SPO, but with no hybridization effects.

In Figs. 11a, b, structural data of some binary TM-Al alloys are shown for comparison. Their compositions were chosen from the optimal composition ranges just discussed. Some ternary alloys are included [6]. They become quasicrystalline by heat treating their amorphous phase [44, 45, 46, 47]. The K- and r-scales are normalized by $2k_F^Z$ or λ_{Fr}^Z, respectively. Surprisingly, there are no major differences among all the alloys, either in k- or in r-space. There is one dominant peak at $K_{pe} = 2k_F^Z$ and multiples of it at $2K_{pe} \approx 4k_F^Z$ and, at this stage of annealing less expressed, at $3K_{pe} \approx 6k_F^Z$ [6], indicating a high structural order.

It is obvious that the SPO is well pronounced and exists to very large distances. For each alloy the position of the first peak in r-space is in excellent agreement with $5/4 \times \lambda_{Fr}^Z$. Angular correlations as well as the spherical periodicity do not need any compromise at short distances, as in amorphous semiconductors (see below). Or, the compromise is weak since both are going hand in hand if the systems are close to becoming quasicrystalline.

4.3 Long-range Angular Correlations by Annealing the TM-Al

By annealing the TM-Al alloys we observed, for the first time, direct indications of angular correlations in the amorphous state. Taking a closer look at the structural details of $S(K)$ in those alloys at the optimum (Fig. 11), they all show a tiny pre-peak located, within experimental resolution, at $K_\tau = K_{\rm pe}/\tau = 2k_{\rm F}^{\rm Z}/\tau$, with $\tau = 1.618...$, the irrational number characterizing the icosahedral order and its angular correlations.

In r-space (Fig. 11b), in addition to the SPO, further nearest-neighbour distances occur at τ-related positions, with [6, 16, 19]

$$r^\tau_{\rm n+1} \approx (\tau + n) \times \lambda_{\rm Fr}, \qquad n = 0, 1, 2, 3, \cdots \quad . \tag{7}$$

At 350 K these structural features are strongly covered by the SPO but become more pronounced with heat treatments (Fig. 9 in [6]). Whereas the SPO starts its periodic sequence of spherical nearest-neighbour shells at $r_1^{\rm Fr} = 5/4 \times \lambda_{\rm Fr}$, here we have a second sequence of $\lambda_{\rm Fr}$-distances starting at $r_1^\tau = \tau \times \lambda_{\rm Fr}$. So, all together the system is dominated by three characteristic distances

$$\begin{aligned} r_1^{\rm Fr} &= 5/4 \times \lambda_{\rm Fr} \ , \\ r_1^\tau &= \tau \times \lambda_{\rm Fr} \ , \quad \text{and} \\ \Delta r^\tau = \Delta r^{\rm Fr} &= \lambda_{\rm Fr} \ , \end{aligned} \tag{8}$$

all related to the Fermi wavelength $\lambda_{\rm F} = 2 \times \lambda_{\rm Fr}$. Two of them, namely $r_1^\tau = \tau \times \lambda_{\rm Fr}$ and $\Delta r^\tau = \Delta r^{\rm Fr} = \lambda_{\rm Fr}$ give a short and a long distance, distinguished by the irrational number τ, and therefore are characteristic for an icosahedral structure, although the systems are still clearly within the amorphous state. The first nearest-neighbour distance $r_1^{\rm Fr}$ itself is a distance which, surprisingly, nowhere appears in an icosahedron, if we assume $\lambda_{\rm Fr}$ as its edge and $\tau \times \lambda_{\rm Fr}$ as its diagonal.

The peak in r-space at $r_1^\tau = \tau \times \lambda_{\rm Fr}$ is related to the pre-peak at K_τ in k-space, but the distance between the higher τ-related nearest neighbours, namely $\Delta r^\tau = \lambda_{\rm Fr}$, contributes to the strong peak at $K_{\rm pe} = 2k_{\rm F}^{\rm Z}$. Accordingly, from the τ-related nearest-neighbour distances, those at medium-range distances contribute additionally to the depth of the DOS at $E_{\rm F}$ and hence to the transport properties as well as the thermal stability.

Since the τ-related peaks are based on three-body correlations and hence on angular correlations, we claim them to be angular-momentum based. As has been shown elsewhere [5, 6, 48, 49], there are indeed strong indications of two types of pseudogaps at the Fermi energy (see Fig. 15): a broad one, which seems to be related to the SPO [5, 29], and a sharp one. Here we claim, as has been supported theoretically [50], that the sharp pseudogap in quasicrystals is related to long-range angular-correlated order or planarity.

5 Angular-Momentum-Based Resonances

As just shown for the TM-Al alloys, there are indications for angular-momentum-based effects on structure formation. They may additionally improve the stability. In those cases either three-body correlations (Fig. 2b) are important, or coherent interference effects by electron waves sent entirely around the primarily scattering adatom by consecutive scattering events. And in fact angular order, long-range as in the quasicrystals or short-range as in amorphous semiconductors [51], is found to be effective and to represent the first step from spherical-periodic to planar order in a crystal. In the optimized a-TM-Al alloys (Fig. 11), in r-space the first SPO-shell distance exactly matches $r_1^{\mathrm{Fr}} = 5/4\ \lambda_{\mathrm{Fr}}$, whereas in semiconductors, as a compromise between both resonances, often the first nearest-neighbour distance gets shifted to larger distances than expected for the SPO alone, although the higher coordination shells indicate improved SPO [9, 16, 40]. If the corresponding pseudogap becomes, once again, located at the Fermi energy, the total pseudogap may evolve to a real gap, with consequences up to a metal-insulator transition and enhanced stability.

Here we review the semiconductors. Further results have been published elsewhere [9, 16, 19, 20].

5.1 Disordered Semiconductors Showing Additional Angular Correlations at Short Distances

Amorphous and liquid semiconductors – elements as well as alloys – show, under the simple systems far from $\bar{Z} = 1.8\,\mathrm{e/a}$, structural peaks in $S(K)$ at $K = K_{\mathrm{pe}} = 2k_{\mathrm{F}}$ which are outstandingly high (Fig. 12a). In the amorphous semiconductor state even peaks at $2\times 2k_{\mathrm{F}}$ exist, indicating a much higher degree of SPO than in the amorphous metals. After normalizing the K-scale by the internal momentum $2k_{\mathrm{F}}$, they all show similar structural features (Fig. 12a). Figure 12b shows the high degree of SPO by the high occupation of the nearest-neighbour shells around 5/4, 9/4, 13/4, 17/4,$\cdots \times\lambda_{\mathrm{Fr}}$. $2k_{\mathrm{F}}$ is calculated by the same formula as above (2), with $\bar{n}_{\mathrm{o}}$ from experiments and $\bar{Z}$ from the position of the elements within the periodic table (e.g. $Z_{\mathrm{C}} = 4\,\mathrm{e/a}$, $Z_{\mathrm{Te}} = 6\,\mathrm{e/a}$).

In order to justify our estimation of $2k_{\mathrm{F}}$ and to extract the electron density, $\bar{n}_{\mathrm{o}}\times\bar{Z}$, independently, plasmon energies, $E_{\mathrm{p}}^{\mathrm{EELS}}$, have been measured and compared with the FEM value

$$E_{\mathrm{p}} = \hbar\sqrt[2]{\frac{\bar{n}_{\mathrm{o}}\bar{Z}e^2}{\varepsilon_{\mathrm{o}}m_{\mathrm{e}}}}\ , \tag{9}$$

calculated along two lines (see the table attached to Fig. 13). One of the FEM values, $E_{\mathrm{p}}^{\mathrm{n}}$, has been calculated by using the particle density as measured by the scattering data. These theoretical data are shown as solid vertical lines in Fig. 13. The plasmon loss energies, $E_{\mathrm{p}}^{\mathrm{liq}}$, have been calculated by assuming the particle density of the liquid elements, since their structure is closer to the amorphous state than the crystalline one. There is again a quite good agreement to the measured plasmon energy. Obviously, even in semiconductors $2k_{\mathrm{F}}$ is a well defined quantity, although itinerant electrons

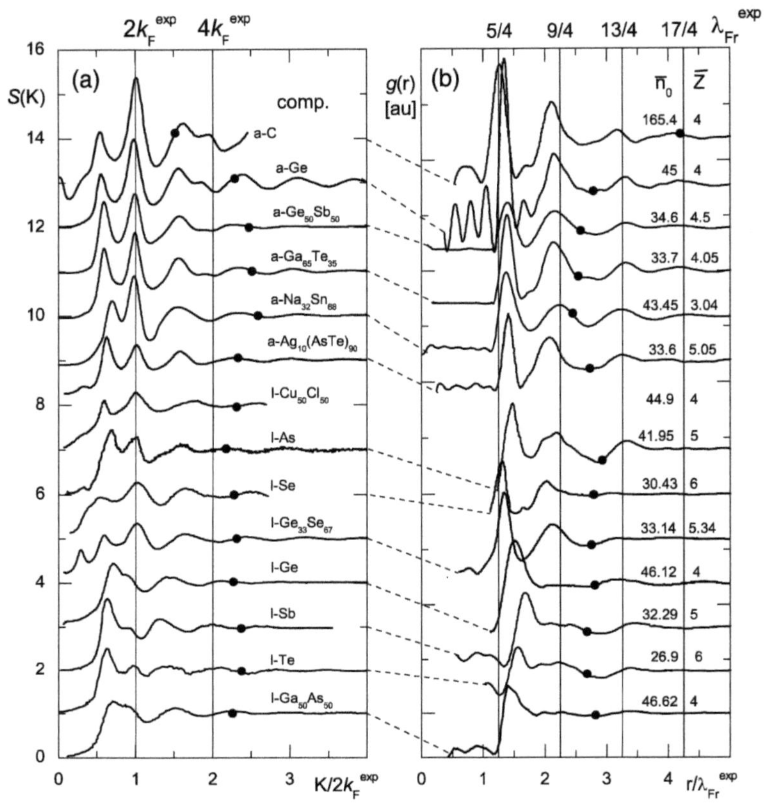

Fig. 12. (**a**) Structure factors of amorphous and liquid semiconductors after normalizing the K-scale by $2k_F$ [9]. (**b**) Pair distribution after scaling with the internal length λ_{Fr}. For comparison with absolute scales, in (**a**) and (**b**) the dots on each curve represent $120\,nm^{-1}$ in k-space or $0.5\,nm$ in r-space, respectively [9]. In (**b**) the particle density and the valency are indicated at each system

are absent. If plasmon data are not available, free-electron data can be taken as a quite good estimation.

The SPO of semiconducting materials (Fig. 12b) is enhanced compared to the amorphous metals (Fig. 5b). On the other hand, these systems are network-forming glasses with strong AC at short distances. This additional order may improve the spherical resonance. But, although K_{pe} is exactly at $2k_F$ and the first nearest-neighbour distance is sharper than in the glassy metals above, the latter is shifted to larger distances compared to $5/4\times\lambda_{Fr}$ (with the exception of a-C). This shift may indicate a compromise of the momentum-based to the angular-momentum-based resonance at short distances. The exception of a-C can be understood by the position of its first peak in $S(K)$ close to k_F, half the value for the peak at $2k_F$. One is a harmonic of the other, both supporting each other. The compromise at short distances

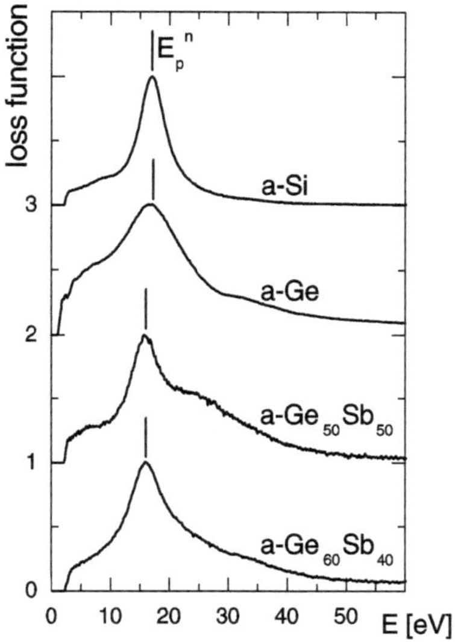

	E_p^{EELS}	E_p^n	E_p^{liq}
a-Si	17,00	17	18,38
a-Ge	16,65	17,2	16,88
a-$Ge_{50}Sb_{50}$	16,00	16,01	16,15
a-$Ge_{60}Sb_{40}$	16,10	16,02	16,05

Fig. 13. Energy loss spectra of amorphous Si, Ge, and Ge_xSb_{100-x} alloys at annealing temperatures $T_{ann} = 300\,K$, showing plasmon resonance peaks [20]. The intensities are normalized to their maxima. The table compares (from left to right): peak positions, E_p^{EELS}, as measured; the free-electron value, E_p^n, calculated by the application of the measured particle density; and the free-electron value, E_p^{liq}, calculated under the assumption of particle densities of the liquid state and for the alloys' constant atomic volumes of the individual atoms within the alloys

is no longer necessary. The medium-range shells, on the other hand, are more pronounced and range up to larger distances, improving the depth of the DOS at E_F exceptionally. We assume that, enhanced by the angular correlations, the atoms *at* the shells are less randomly located (that is, they order at the shells).

At the end of the discussion of the structural features of disordered systems, and under the impression of the importance of resonance effects at early stages of structure formation, we propose that the diffraction data may be analyzed differently than it was in the past. We may ask, how much SPO is present in a disordered system in comparison to AC, and how are both related to each other. So, we may get three partials, one describing the SPO, one the AC, and a mixed one describing the correlation between both.

6 Pseudogaps Induced by Resonance Stabilisation

As mentioned above, the resonance effects naturally cause pseudogaps or gaps in the electronic density of states, more favourable at E_F than elsewhere.

6.1 Pseudogaps at E_F Due to SPO

It was shown years ago by photoelectron spectroscopy (UPS-region) that, indeed, in the simple glassy metals a pseudogap exists at E_F [29]. It becomes deepest for those alloys with $\bar{Z}=1.8\,e/a$, where the SPO is most pronounced [5]. In Fig. 14 for different glassy alloys spectroscopy data close to the Fermi energy are redrawn, clearly showing the occupied part of the broad pseudogap at E_F in each case. Away from these compositions, the pseudogap becomes weaker [6, 29, 43]; with increasing temperature it becomes weaker too [52], due to the weakening of the SPO with changing composition or increasing temperature. The pseudogap at E_F has been confirmed by measurements of the specific heat of the electrons at low T [33], as well as by the electronic susceptibility [54], and they all agree with each other. For alloys with $\bar{Z}=1.8\,e/a$ the deviation from the free-electron value is approximately 50%. Consequences for electronic transport properties have been reported [5], and for the Au alloys of Fig. 14 they are nearly identical due to the identical DOS at E_F.

In the simple glassy alloys only one pseudogap has been resolved at E_F. This is different from the a-TM-Al alloys, where we have shown above that due to the additional angular correlations we expect two pseudogaps at E_F. Direct measurements of even the broad pseudogap by photoelectron spectroscopy are strongly hindered by the large contribution of the TM *d*-states close to or at the Fermi energy. But, together with X-ray emission and photoabsorption spectroscopy, it was convincingly shown that there is indeed

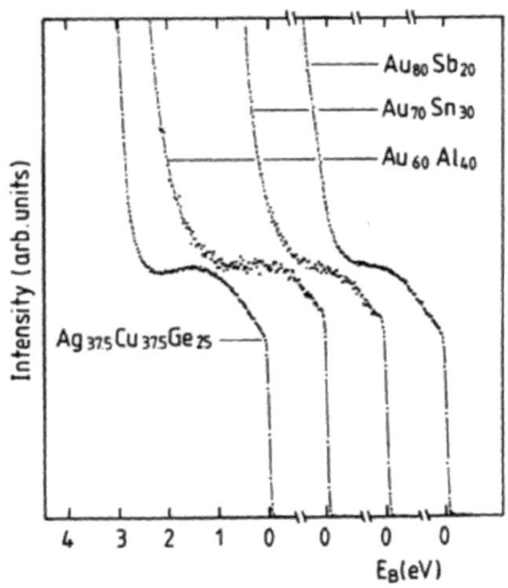

Fig. 14. Electronic DOS close to E_F, measured by photoelectron spectroscopy, for different amorphous simple glassy metals with $\bar{Z}\approx 1.8\,e/a$ [53]. The spectra are normalized to the intensity at $E_B=1.5\,eV$. The strong increase at large binding energies is related to the noble-metal d-states

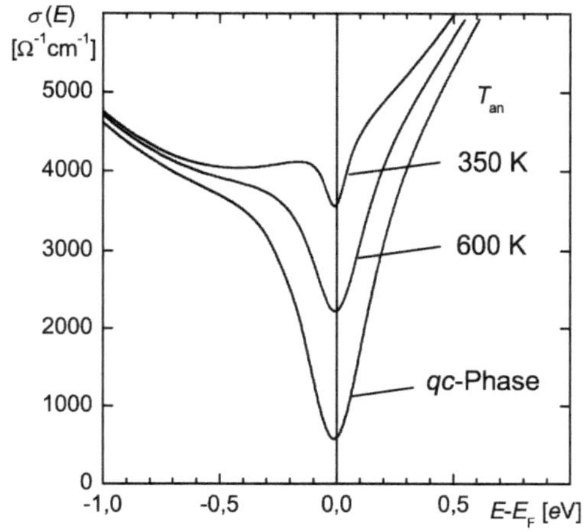

Fig. 15. The spectral function, $\sigma(E)$, of amorphous $Al_{62,23}Cu_{25,20}Fe_{12,57}$ at different annealing states [6, 49]. The spectral function can be interpreted as a representation of the electronic DOS

such a broad pseudogap at E_F [37]. The distinction between a broad and a sharp one, on the other hand, was impossible. By tunnelling experiments, however, the sharp one was later resolved [55].

In Fig. 15 indirect indications of both pseudogaps are presented by the so-called spectral conductivity, $\sigma(E)$, of a-$Al_{62,23}Cu_{25,20}Fe_{12,57}$, an amorphous ternary alloy which becomes quasicrystalline by heat treatments [6, 49]. $\sigma(E)$ is the result of a coupled fit of the measured resistivity, thermopower, and Hall coefficient, and their T-dependences, to the theoretical expressions containing $\sigma(E)$ [49, 50]. Under certain conditions, $\sigma(E)$ is proportional to the electronic DOS [50]. Accordingly, we interpret Fig. 15 as indicating two pseudogaps at E_F: a broad one which comes from the SPO, and a sharp one which comes from the relatively long-range angular correlations [6, 49]. By heat treatments, in particular the sharp one gets more pronounced, already within the amorphous state (Fig. 15).

6.2 Pseudogaps Far Below E_F Due to Angular Correlations and the First Peak in $S(K)$

As shown above, together with the structural peak in $S(K)$ at $2k_F$, very often there are further peaks well below $2k_F$ (Fig. 5). It was recognized years ago that they also have an influence on the electronic DOS, although they may have no influence on electronic transport or phase stability if no further effects occur at E_F. First indications of those effects have been found in liquid elements of the main group of the periodic table [56]. It was observed that a well pronounced gap exists far below the Fermi energy, somewhere in the middle of the valence band. Using synchrotron radiation it was recognized that this gap separates the s- from the p-states. Later it was recognized that it is related to the *first* peak in the static structure factor at K_p well below $2k_F$ [11]. It was later also observed in glassy metals (Fig. 4) [5, 11]. At the time these observations were reported, there was no convincing explanation [11, 56, 57].

Figure 16a shows the valence band of some of the liquid elements mentioned [56], and Fig. 16b shows the strict correspondence between the binding energies of those electrons that have a momentum $K_p/2$ and the position of the pseudogaps measured relative to the Fermi energy (their binding energies). Whenever structural features are correlated with s-, p-, or d-states, as just described, we take this as an indication of angular-momentum-based effects on the static structure, causing angular-correlations. So, the experimental finding of this additional pseudogap well below E_F, its separation of

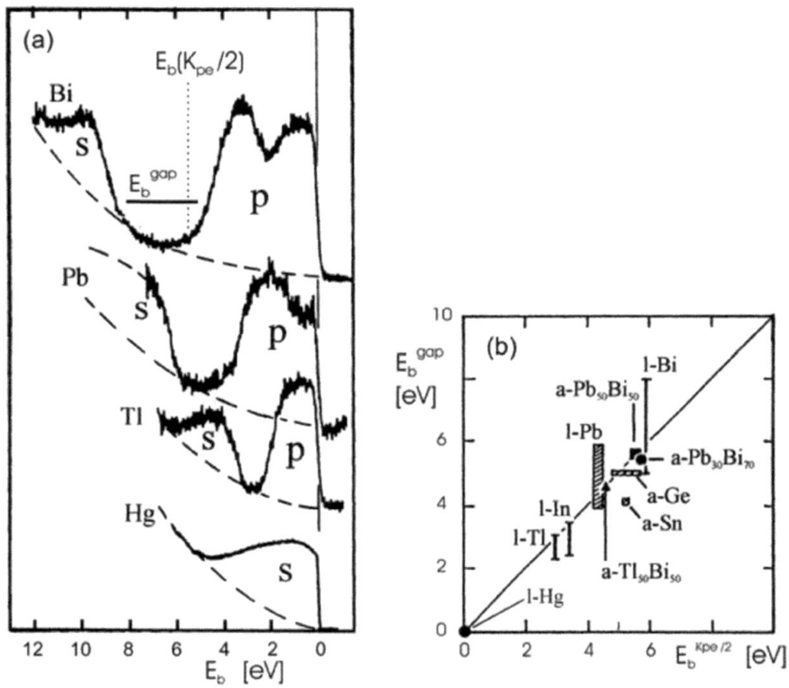

Fig. 16. (**a**) Photoelectron spectra (UPS) of liquid elements showing a gap within the valence band separating the s- from the p-states [56]. The dashed curves in (**a**) indicate qualitatively the background from secondary electrons. The vertical broken line indicates exemplarily for liquid Bi those electronic states which are related to the first peak in $S(K)$ at $K_p \ll 2k_F$. The horizontal bar for Bi shows the position and width of the pseudogap. (**b**) Shows this strict relation for some liquid as well as amorphous systems [5, 11]. In liquid Hg the pseudogap, separating the s- from the p-states, is located at the Fermi energy and explains many of its properties [58]

the s- from the p-states, and its relation to the first peak in the static structure factor support our assumption that the first peak is telling us something about angular correlations, as has already been claimed in the amorphous quasicrystals by the peak at K_τ.

Whether the angular correlation helps to stabilize the system depends on the position of the related pseudogap relative to E_F. Whenever the first peak shifts closer to $2k_\mathrm{F}$, is located at a rational fraction of the peak at K_pe (as in glassy carbon (Fig. 12)), or is related to τ (as in the quasicrystals), then there are additional stabilizing effects due to additional pseudogaps at E_F. Therefore, optimizing the structural stability occasionally may require, together with the momentum-based resonance, the optimization of the angular-momentum-based resonance.

7 Consequences of Spherical Periodicity on Electronic Transport

Whenever due to resonance effects there is the formation of a pseudogap at E_F, electronic transport properties are affected. This has been shown for the glassy metals [5, 43, 53, 59], as well as for the a-TM-Al alloys, which have the potential to become quasicrystalline [6, 19, 47, 48]. For the former, the resistivity increases, the Hall coefficient deviates more and more from the free-electron value, the thermopower may even change its sign, and the stability of the phase against crystallisation increases, with the increasing depth of the pseudogap, when alloys with $\bar{Z} = 1.8\,\mathrm{e/a}$ and strongest SPO are approached. All these features could be explained by the broad pseudogap at E_F alone. For the a-TM-Al alloys it became clear that the transport anomalies could only be understood by two pseudogaps at E_F [49, 50]: the broad one which is related to the SPO, and the sharp one which is related to the relatively long-range angular correlations. Subsequently, electronic transport properties versus composition and temperature could well be explained. Details have been reported elsewhere and will not be part of the present contribution [6]. Anomalies at low T for both types of disordered systems are briefly discussed below (9).

8 Consequences of Spherical-Periodicity on Dynamics

From crystals we know that whenever there is periodicity there are dramatic effects on sound propagation. The corresponding dispersion of their dynamic excitations (the phonons) shows the same periodicity, but is, by no means, further more linear as the one of Debye phonons due to the opening of a gap.

Quite similar to the electrons in Sect. 2 (Fig. 3) in their k-space description, particular effects of the SPO on the dynamic behaviour can be explained

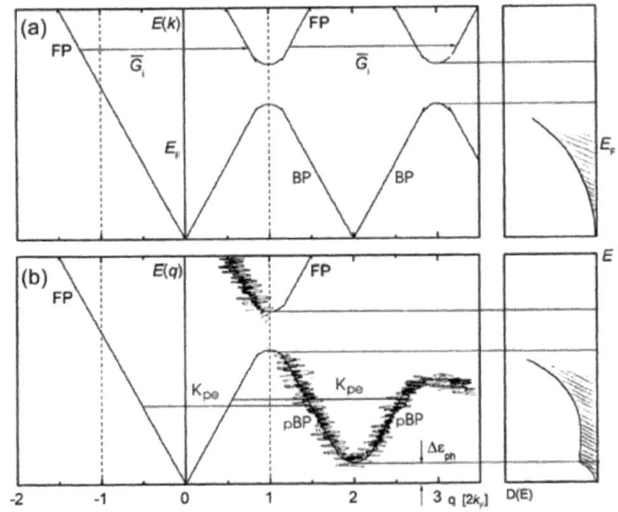

Fig. 17. (**a**) Schematic dispersion of dynamic excitations in the crystalline case. (**b**) Schematic dispersion of dynamic excitations in the disordered state. The q-scale is normalized by k_F

with the help of the dispersion of dynamic excitations [14]. The corresponding wave number in this case is called q. Figure 17 shows schematically in (a) the situation in the crystalline case and in (b) the disordered case, together with the dynamic DOS for both. Analogous to the case of the electrons, we may start with 'free-sound' in a system without periodicity. The corresponding dispersion may, somewhat incorrectly[3], be called free phonons (FP) or Debye phonons. They have a linear dispersion (far left side of Fig. 17a). In the crystalline case (a) with (5) fulfilled, there are further linear branches, shifted by $n \times \boldsymbol{G}_i$, describing Bloch phonons (BP). Analogous to the electronic case, they no longer describe FP; instead they contain structural features. Due to the quasi-infinite mass of the static structure, once again, there is no shift along the energy scale.

The FP and the BP are in a degenerate state at $q = \pm k_F$. Due to the coupling between both, the degeneracy is lifted and a gap opens (Fig. 17a; middle, far right). Resonance effects dominate the dynamic features. But, contrary to the electronic system, this gap now has, to our belief, no influence on the stability of the system.

In the liquid and amorphous cases, both with SPO (Fig. 17b), analogous to the crystalline case, similar situations exist for the FP and the pseudo-Bloch phonons (pBP); there is an opening of a (pseudo) gap at $q = \pm k_F$.

[3] The phrase phonon is reserved for quantized dynamic excitations in single crystals.

Disregarding the fact that the pBP are broadened, thus far there are no major differences from the crystalline case. But there is one important difference, namely the shift of the pBP by $\Delta\varepsilon_{\mathrm{ph}}$ at $q = 2k_{\mathrm{F}}$ along the energy scale, since the spherical-periodic order involves only a finite number of atoms. Accordingly, with any Umklapp-process there is an energy loss involved. In the literature these states are called *phonon-roton states*, due to their similarities to dynamic excitations in liquid helium. Approximately, we assume a parabolic shape of the dispersion in this range [14]. The shift has consequences for the dynamic DOS, namely, a small broadened gap at very low energies. In this case, the width of the gap (meV) becomes comparable to the width of the energy band, which itself is, for dynamic excitations, in the range of a few dozen meV. The larger the SPO regions become, the smaller becomes the corresponding shift. Dispersions of the type just described have been observed in many liquid [60] and amorphous materials [61].

What will become important for the following is the fact that the shift of the electronic states along the energy scale (Fig. 3b) and the analogous shift of the dynamic excitations (Fig. 17b) are identical, since the reason for these shifts – the finite mass of the atoms involved in the spherically periodic regions – is the same for both.

9 New Dynamic Modes by Spherical-Periodic Order

Whereas in ideal crystals electrons and acoustic phonons can only resonate with each other in the first Brillouin zone at small k-values, in disordered systems this limitation is no longer valid [62]. Accordingly, a resonance between the pBEs and the pBPs can be expected at larger k whenever at the same wave vectors they have identical energies. This is true especially around $\boldsymbol{G}_{\mathrm{i}} = K_{\mathrm{pe}} = 2k_{\mathrm{F}}$, because both show the energy shift by the finite mass of the spherical-periodic regions.

Figure 18 shows schematically the dispersions of the pBE and pBP (the phonon-roton states) in an enlarged version around this particular point (their blurring has been neglected). Due to a coupling between both, there will be again a splitting of the states. Because the curvature of the pBE parabola at K_{pe} is much greater than that of the pBP parabola, the resonance and therefore the splitting is strongly localized around K_{pe}. We may define a new quasiparticle for these states with a density of its k, q states of the form given by the broken curve in Fig. 18. Due to the importance of the spherical-periodic regions we call the new quasiparticles *spherons*. A spheron, accordingly, is a state which combines spherical electron waves with the spherical-periodic static structure and the spherical dynamic excitations of the disordered systems. Due to the broadening of the pBE as well as the pBP, the additional forming of a gap between the bonding state and the anti-bonding state may not be resolved. Whereas at ground state, spheron states are unoccupied

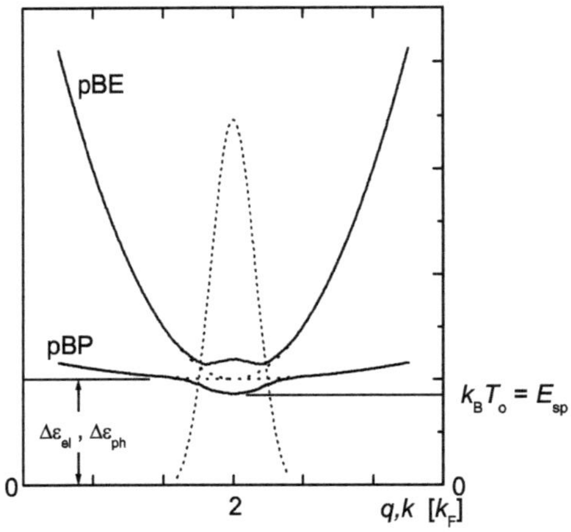

Fig. 18. Schematic representation of the resonance splitting between the pBE and pBP states around $k, q = 2k_F$ causing a new state, the so-called *spheron* [14]. The broken curve shows, schematically again, the density of the k, q states of the spheron. The blurring of the pBE and pBP states is disregarded

because the dynamic part is unoccupied, for $T > T_o$ one would note their occupation.

Experimental evidence of the spherons may be seen in the electronic density of states at E_{sp}, measured by photoelectron spectroscopy close to the lower band edge, already presented above in Fig. 4. The spherons may also explain the so-called Boson peak in the phonon density of states [61], as well as the unusually strong scattering spots at low energies in the dynamic structure factor at those q values where a large peak in the static structure factor exists [63].

Besides the Debye phonons, the spherons were found to be necessary to explain anomalous electronic transport properties. In previous papers their influence at low T was reported [14, 15, 64, 65]. Spherons are claimed to be responsible for resistivity anomalies such as the transition from a positive to a negative temperature coefficient with rising temperature, as well as the 'knee' in the thermopower, both around $T = 15$-$30\,K$. Although the spheron influences mentioned here are all seen in electronic transport properties, they cannot be excluded in dielectric systems, since, as has been shown above with the semiconductors, spherical-periodic regions and resonances between the electrons, the static structure, and the dynamic excitations may also exist in those systems without itinerant electrons.

In r-space the spherons can be seen as localized spherical electron-structure-phonon states with the wavelength λ_{Fr}. It can be considered as a shear mode or a breathing mode [66] within the spherically periodic regions. The wavelength of the spherons exactly fits the distances between the consecutive spherical shells.

Spherons, or less specifically the SPO, have ingredients of the so-called mode-coupling theory[4] which is especially important to describe the glass temperature. The so-called cage effect may be given by the consecutive spherical shells around any adatom; α- and β-processes may be related to these spherons, localized within the spherical-periodic arrangements and diffusing from one spherical arrangement to the next. For the future, there are still many open questions to be answered.

10 Conclusions

We reviewed that structure formation at an early stage is a quite general effect of optimized resonances between the static structure and the electronic system. Instead of mirror planes in a crystal, at early stages there is a well defined spherical-periodic order (mirror spheres) at short and medium distances, in the mean around any adatom. In order to find the energetically most preferred phase, both the static structure as well as the electronic system may have to adjust mutually. Resonance gaps at E_{F} are formed. Short- as well long-range angular correlations improve the sphericity and, under certain conditions, cause additional pseudogaps at E_{F}. The spherical-periodic ordering principle serves as a global, autonomous concept of structure formation, and it may help to reduce the huge number of independent parameters in, e.g. band-structure calculations or may contribute to the solutions of some of the still unsolved severe problems in disordered condensed matter. The model has a large universality. Up to now amorphous metals, ionic glasses, amorphous quasicrystals and quasicrystals themselves, as well as semiconductors, could well be described. All follow the same fundamental resonance principle, stronger at low temperatures and weaker at high temperatures. Accordingly, metallicity, covalency, and ionicity are extreme cases of optimized resonances.

One of the most important consequences of the spherical-periodic order is the existence of resonant spherical electron-static structure-phonon states: the spherons. They will not exist in crystals. They may have a high importance for transport anomalies at low T. Their role and, in particular, their correct mathematical description are still undetermined.

Acknowledgement

Many thanks to H. Solbrig, H. Nowak, and J. Barzola-Quiquia for many stimulating discussions, and the Deutsche Forschungsgemeinschaft for the partial financial help during the years.

[4] See contributions in this book.

References

1. W. Götze, L. Sjögren: Rep. Prog. Phys. **55**, 241 (1992); Phys. Rev. A **43**, 5442 (1991)
2. W.A. Phillips (ed.): *Amorphous Solids: Low-Temperature Properties* (Springer, Berlin 1981)
3. U. Buchenau, M. Prager, N. Nuker, A.J. Dianoux, N. Ahmad, W.A. Phillips, Phys. Rev. **34**, 5665 (1986); A.P. Sokolov, U. Buchenau, W. Steffen, B. Frick, A. Wischnewski: Phys. Rev. B **52**, R9815 (1995); C. Masciovecchio, V. Mazzacurati, G. Monaco, G. Ruocco, T. Scopigno, F. Sette, P. Benassi, A. Cunsolo, A. Fontana, M. Krisch, A. Mermet, M. Montagna, F. Rossi, M. Sampoli G. Signorelli, and R. Verbeni: Phil. Mag. B **79(11/12)**, 2013 (1999)
4. Y. Waseda: *The Structure of Non-Crystalline Materials*, (McGraw Hill Inc. 1980); M. Shimoji: *Liquid Metals*, (Academic Press, London - New York - San Francisco 1977)
5. P. Häussler: Phys. Rep. **222(2)**, 65 (1992)
6. P. Häussler, J. Barzola-Quiquia, R. Haberkern, C. Madel, M. Lang, K. Khedhri, and D. Decker: On fundamental structure forming processes. *Quasicrystals-Structure and Physical Properties*, ed H.-R. Trebin: (WILEY-VCH 2003) pp 289
7. O. Madel: Atomare Struktur und elektrischer Widerstand amorpher Na_xSn_{100-x}-Legierungen. PhD. Thesis, Technical University, Chemnitz (1998) O. Madel, C. Lauinger, and P. Häussler: J. Non-Cryst. Solids **250-252**, 267 (1999)
8. M. Lang, G. Schwalbe, C. Madel, R. Haberkern, P. Häussler: Ferroelectrics **250**, 257 (2001)
9. P. Häussler J. Barzola-Quiquia: J. Non-Cryst. Solids **312-314**, 498 (2002)
10. P. Häussler: J. Physique (Paris) **C8**, 361 (1985)
11. P. Häussler, G. Indlekofer, H.-G. Boyen, P. Oelhafen, H.-J. Güntherodt: Europhys. Lett. **15(7)**, 759 (1991);
    P. Häussler, G. Indlekofer, H.-G. Boyen, P. Oelhafen and H.-J. Güntherodt: Mat. Sci. Eng. A **133**, 120 (1991)
12. V. Heine and D. Weaire: Pseudopotential theory of cohesion and structure. In: *Solid State Physics, Advances in Research and Application*, vol 24, ed H. Ehrenreich, F. Seitz and D. Turnbull (Academic Press, New York 1970) pp 247
13. N.W. Ashcroft, N.D. Mermin: *Solid State Physics* (Saunders, London 1976).
14. H. Nowak P. Häussler: J. Non-Cryst. Solids **250-252**, 389 (1999)
15. P. Häussler: Physica B **219&220**, 299 (1996)
16. J. Barzola-Quiquia: Szenarien der Strukturbildung in $Al_{100-x}\ddot{U}M_x$-Legierungen und Halbleitern sowie Konsequenzen daraus für elektronischen Transport. PhD Thesis, Technical University, Chemnitz (2003)
17. J. Barzola-Quiquia, M. Lang, D. Decker, P. Häussler: J. Non-Cryst. Solids, to be published
18. P. Häussler, D. Moorhead, R. Hauert, H. Poppa, E. Kay: J. Non-Cryst. Solids **117/118**, 293 (1990)
19. J. Barzola-Quiquia, M. Lang, R. Haberkern, P. Häussler: Proc. *Aperiodic Structures 2001*, Krynica, Poland; ISBN 83-914795-3-6

20. J. Barzola-Quiquia, M. Spindler, P. Häussler: Proc. XIII. Int. Conf. on *Non-Oxide Glasses and New Optical Glasses*, 9.-13.9.2002, Pardubice, Czech Rep.
21. J. Barzola-Quiquia, P. Häussler: J. Non-Cryst. Solids **299-302**, 269 (2002)
22. W. Hume–Rothery: J. Inst. Met. **35**, 295 (1926);
W. Hume–Rothery, G.-V. Raynor, *The Structure of Metals and Alloys*, 4th edn (Institute of Metals, London 1962)
23. R.E. Peierls: *Quantum Theory of Solids* (Oxford Univ. Press, Oxford, 1955) pp 108
24. A.P. Blandin: Theoretical Considerations of the Hume–Rothery Rules. In: *Phase Stability in Metals and Alloys*, ed P.S. Rudman, J. Stringer, R.I. Jaffee: (McGraw-Hill, New York 1967) pp 115
25. J. Friedel: Adv. Phys. **3**, 446 (1954)
26. R. Oberle, H. Beck: Solid State Commun. **32**, 959 (1979)
27. J. Kroha, A. Huck, T. Kopp: Czech. J. Phys. **46**, 2275 (1996);
J. Kroha, D. Walther, R. v. Baltz: Interplay between the geometrical and the electronic structure in quasicrystals. In: *Quasicrystals- Structure and Physical Properties*, ed H.-R. Trebin (WILEY-VCH 2003) pp 236
28. T. Kaneyoshi: J. Phys. Jap. **45**, 94 (1978); J. Phys. F **5**, 1014 (1975);
T. Kaneyoshi, S. Iwabuchi: J. Mag. Mag. Mat. **15-18**, 1421 (1980)
29. P. Häussler, F. Baumann, J. Krieg, G. Indlekofer, P. Oelhafen, H.-J. Güntherodt: Phys. Rev. Lett. **51**, 714 (1983)
30. J. Kondo: Theory of dilute magnetic alloys. In: *Solid State Physics*, vol 23, ed H. Ehrenreich, F. Seitz, D. Turnbull (Academic Press, New York 1969) pp 183;
M.A. Ruderman, C. Kittel: Phys. Rev. 96, 99 (1954)
31. K. Handrich, J. Resch: phys. stat. sol. (b) **108**, K57 (1981); phys. stat. sol. (b) **152**, 377 (1989)
32. H. Leitz: Z. Phys. B**40**, 65 (1980);
H. Leitz, W. Buckel: Z. Phys. B **35**, 73 (1979)
33. M. Sohn, F. Baumann: J. Phys.: Condens. Matter **8**, 6857 (1996)
34. H.T.J. Reijers, W. van der Lugt, M.-L. Saboungi, Phys. Rev. B **42**, 3395 (1990)
35. C. van der Marel, A.B. van Oosten, W. Geertsma, W. van der Lugt: J. Phys. F: Metal Phys. **12**, 2349 (1982)
36. F.S. Pierce, S.J. Poon: Science **261**, 737 (1993)
37. E. Belin-Ferré: Electron density of states in quasicrystals and approximants. *Quasicrystals - An introduction to Structure, Physical Properties, and Applications* Materials Science, vol 55, ed J.-B. Suck, M. Schreiber, P. Häussler (Springer, 2003) pp 338
38. J.M. Ziman: Adv. Phys. **16**, 551 (1967)
39. P. Häussler, H. Nowak, R. Haberkern: Mat. Sci. Eng. **294-296**, 283 (2000)
40. P. Häussler, J. Barzola-Quiquia: J. Non-Cryst. Solids **299-302**, 269 (2001)
41. Y. Waseda, M. Ohtan: Z. Naturforsch. **30a**, 485 (1975)
42. A. Lambrecht: Zur Struktur abschreckend kondensierter Al-Cu Schichten. Diploma, University Karlsruhe, Germany, 1980; A. Lambrecht, H. Leitz, J. Hasse: Z. Phys. B: Condens. Matter **38**, 35 (1980)
43. P. Häussler, F. Baumann, U. Gubler, P. Oelhafen, H.-J.Güntherodt: *Proc. 5th Int. Conf. on Rapidly Quenched Metals*, Würzburg, Germany, 797 (1984)
44. C. Roth, G. Schwalbe, R. Knöfler, F. Zavaliche, O. Madel, R. Haberkern, P. Häussler: J. Non-Cryst. Solids **250-252**, 869 (1999)
45. R. Haberkern, J. Barzola-Quiquia, C. Madel, P. Häussler: Mat. Res. Soc. Symp. **643**, K8.3.1 (2001)

46. R. Haberkern, C. Roth, R. Knöfler, L. Schulze, and P. Häussler: Mat. Res. Soc. Symp. **553**, 13 (1999)
47. R. Haberkern: Electronic transport properties of quasicrystalline thin films. In: *Quasicrystals - An Introduction to Structure, Physical Properties, and Applications*, Materials Science, vol 55, ed J.-B. Suck, M. Schreiber, P. Häussler, (Springer 2003) pp 364
48. P. Häussler, R. Haberkern, C. Madel, J. Barzola-Quiquia, M. Lang: J. Alloys and Compounds **342**, 228 (2002)
49. C. Madel: Elektronische Transporteigenschaften von amorphem und quasikristallinem Al-Cu-Fe. PhD Thesis, Technical University, Chemnitz (2000)
50. C.V. Landauro and H. Solbrig: Physica B **301**, 267 (2001)
51. D. Weaire: Contemp. Phys. **17(2)**, 173 (1976)
52. P. Häussler: J. Non-Cryst. Solids **156-158**, 332 (1993)
53. H.-G. Boyen, P. Rieger, P. Häussler, F. Baumann, G. Indlekofer, P. Oelhafen, H.-J. Güntherodt: J. Phys: Condens. Matter **2**, 7115 (1990); H.-G. Boyen, P. Häussler, F. Baumann, U. Mizutani, R. Zehringer, P. Oelhafen, H.-J. Güntherodt, and V.L. Moruzzi: J. Phys. Cond. Matter **2**, 7699 (1990)
54. U. Mizutani, F. Nakamura: J. Phys. F **13**, 2685 (1983)
55. D.N. Davidov, D. Mayou, C. Berger, C. Gignoux, A.G.M. Jansen: Mat. Sci. Eng. A **226-228**, 972 (1997); D.N. Davidov, D. Mayou, C. Berger, C. Gignoux, A. Neumann, A.G.M. Jansen, P. Wyder: Phys. Rev. Lett. **77(15)**, 3173 (1996); R. Escudero, J.C. Lasjaunias, Y. Cavayrac, M. Boudard: J. Phys.: Condens. Matter **11**, 383 (1996)
56. G. Indlekofer: Systematics in the electronic structure of polyvalent liquid metals determined by electron spectroscopy. PhD Thesis, University, Basel (1987); G. Indlekofer, P. Oelhafen, and H.-J. Güntherodt: New preparation method for liquid alloy surfaces and their characterisation by electron spectroscopy. In: *Physical and Chemical Properties of Thin Metal Overlayers and Alloy Surfaces*, ed D.M. Zehner and D.W. Goodman
57. W. Jank, J. Hafner: Phys. Rev. B **41**, 1497 (1990); ibid, 11530
58. N.F. Mott: Phil. Mag. **13**, 989 (1966)
59. P. Häussler: Mat. Sci. Eng. A **133**, 10 (1991)
60. K. Tankeshwar, G.S. Dubey, K.N. Pathak: J. Phys. C **21**, L811 (1988); E. Nassif, P. Lamparter, and S. Steeb: Z. Naturforsch. A **38**, 1206 (1983); T. Bodensteiner, C. Morkel, P. Müller, and W. Gläser: J. Non-Cryst. Sol. **117/118**, 116 (1990)
61. J.-B. Suck, H. Rudin: Vibrational dynamics of metallic glasses studied by neutron inelastic scattering. In: *Glassy Metals II*, vol 53, ed H. Beck, H.-J. Güntherodt: (Springer, Berlin, Heidelberg, New York 1983) pp 217; J.-B. Suck, H. Rudin, H.-J. Güntherodt, H. Beck: Phys. Rev. Lett. **50**, 49 (1983); J. Phys. C **13**, L1045 (1980); J. Phys. C **14**, 2305 (1981); J.-B. Suck: Mat. Sci. Eng. A **133**, 40 (1991)
62. S. Takeno, M. Goda: Progr. Theor. Phys. **45(2)**, 331 (1971)
63. T. Nakayama: Physica B **263-264**, 243 (1999); M. Arai, Y. Inamura, T. Otoma, N. Kitamura, S.M. Bennington, A.C. Hannon: Physica B **263-264**, 268 (1999)
64. P. Häussler, H. Nowak, M. Bhuiyan, J. Barzola-Quiqia: Physica **316-317**, 489 (2002)

65. C. Lauinger, J. Feld, J. Rimmelspacher, P. Häussler: Mat. Sci. Eng. A **181/182**, 916 (1994)
66. M. Braden, W. Reichardt, A.S. Ivanov, A.Yu. Rumiantsev: Europhys. Lett. **34(7)**, 531 (1996);
P. Vashishta, R. Kalia, I. Ebbsjo: Phys. Rev. B **39**,6034 (1989);
R.L. Cappelletti, M. Cobb, D.A. Drabold, W.A. Kamitakahara: Phys. Rev. B **52**, 9133 (1995)

Part II

Dynamics of Disordered Systems

Introduction to Part II

Disordered systems provide a continuing challenge for the scientific community. Such systems often exhibit anomalous long-time behavior, which may be hard to access experimentally and difficult to explain theoretically. Such a behavior is found, for example, at the glass transition from undercooled liquids to the amorphous solid state. In addition non-equilibrium phenomena such as aging and anomalous transport are typically found. The difficulty of this subject lies in the appearance of non-trivial cooperative phenomena, which may result from competing interactions, e.g. between coupled spins, or from complex dynamic modifications of structural properties as in glasses.

The first two contributions to this part review experimental aspects of the dynamics of disordered systems. The first one by Reiner Zorn and Ulrich Buchenau is devoted to the structural glass transition. They show that suspensions of colloidal particles, such as polymer spheres, provide ideal systems for investigations of the glass transition. They are ideal in two respects: firstly experiments can be simply conducted with light (dynamic light scattering, DLS) instead of neutrons, which have to be used to probe the dynamics on a molecular scale. And secondly, these systems appear to be perfectly described by mode-coupling theory (MCT). The latter is only briefly sketched here, but treated in detail in the contribution by Rolf Schilling. In addition the differences between colloidal suspensions and undercooled liquids and dynamical mechanical measurements at the glass transition are discussed.

Jens-Boie Suck in his article reviews the experimental aspects of neutron inelastic scattering (NIS) for investigations of the dynamics of disordered systems. After introducing the density correlation functions measured in these experiments, the specific aspects for the observation of collective excitations in amorphous solids and in fluids are presented. In particular the successful applications of this technique e.g. to liquid metals or metallic glasses are reviewed, but also its current limitations are discussed.

In the following two contributions theories for the glass transition and glassy dynamics are presented and reviewed. The paper of Rolf Schilling deals mostly with the structural glass transition, where, in contrast to the spin-glass transition, static disorder is self-generated as result of this transition. After a concise review of several phenomenological theories, the currently most prominent microscopic theory for the structural glass transition, the mode-coupling theory, is derived and its consequences are explained in detail. In the framework of MCT the glass transition appears as a dynamical phase transition as it is identified as an ergodic-to-nonergodic transition, which can be measured via the intermediate scattering function obtainable from DLS or NIS experiments. Subsequently another prominent microscopic theory, replica theory, is treated in detail and it is shown that the latter yields a static glass transition. The possibly complimentary nature of these theories is discussed.

Heinz Horner's contribution is mainly concerned with mean-field like models and theories for glassy dynamics and in addition puts more emphasis on non-equilibrium phenomena such as aging. He first gives an overview over disordered systems, for which glassy dynamics is relevant. These range from spin-glasses to structural glasses, but also from systems such as neural networks to combinatorial optimization problems. The main part is devoted to dynamic mean-field theory and its consequences for the dynamics of the p-spin interaction spin-glass. It is shown how this theory yields results analogous to MCT for temperatures above the glass transition. In addition the non-equilibrium dynamics in the regime below the glass transition temperature is captured, which allows e.g. the description of aging phenomena. A comparison with replica theory is made and common as well as distinct features are worked out. Finally it is shown how properties of cage relaxation in supercooled liquids can be captured in such spin-glass models by including the possibility of relaxing bonds.

The last paper in this part by Ted Janssen presents models and theories for the nonlinear dynamics of aperiodic crystals. The term 'aperiodic'refers to a state, which is intermediate between random and periodic structures. After providing some typical examples in the class of quasi-crystals and from incommensurate structures, their description in superspace is explained. The main part of the contribution deals with models for these structures, such as the generalized Frenkel-Kontorova model or the Double Chain Model, and the dynamical collective excitations that can occur. Of special interest are the stability properties of phasons, characteristic low frequency excitations in incommensurate structures, and of solitons, both of which may decay, e.g. due to anharmonic interactions with phonons. Nonlinearity plays a two-fold role insofar as it is responsible for the occurrence of so-called phason gaps, but also because the ground state of these model systems is related to low-dimensional nonlinear mappings. The advancement in the understanding of the latter provides deeper insights into the ground state properties of these structural models and vice versa.

Glass Transition in Colloids and Undercooled Liquids

Reiner Zorn and Ulrich Buchenau

Institut für Festkörperforschung, Forschungszentrum Jülich,
Postfach 1913, 52425 Jülich, Germany
r.zorn@fz-juelich.de; u.buchenau@fz-juelich.de

1 Introduction

In current solid state theory the understanding of the crystalline state of matter is much more developed than that of the amorphous state. Amorphous materials constitute a large part of condensed matter. A theoretical understanding is of immense importance for basic science as well as application purposes: silicate glasses in windows, bottles etc. glassy polymers, e.g. polymethylmethacrylate (Plexiglas), polycarbonates as substrate for compact discs, polystyrene in various forms as Styrofoam and metallic glasses with their special magnetic properties.

Amorphous materials comprise a large number of chemically strongly different materials as polymers, silicate glasses, salt melts, molecular and metallic glasses. The study of these materials involves numerous experimental methods as neutron scattering, dielectric spectroscopy, nuclear magnetic resonance, rheology, and computer simulation.

If one compares results from different materials or different methods, one finds striking universalities, encouraging the hope for a single theoretical concept able to describe the experimental observations. If the glass transition is indeed universal, one should study the simplest possible glass former, containing only one sort of particles, and with a simple interparticle potential. On the atomic level, such systems are noble gases. The problem which completely obstructs the study of the glass transition in these systems is the required speed of cooling. Calculations show that only by quenching at rates of the order 10^{13} K/s they can be transformed into a glass. For realistic cooling rates crystallisation is inevitable.

A way to avoid this problem in experimental studies is to use model glasses where the atoms are replaced by colloidal particles which are more than 1000 times bigger. These particles are suspended in a solvent which makes viscous friction forces more important than inertial forces. Altogether, the use of colloids leads to a much slower dynamics enabling experiments on the time scale ranging from hours to months.

The present contribution discusses first this colloid case, where both experiment and theory nowadays present a beautifully closed and consistent picture. Thereafter, dynamical mechanical experiments are described to demonstrate the remaining problems for the understanding of undercooled liquids, to provide at least some experimental background for the theoretical overview of R. Schilling in this book.

2 Colloids

The "standard material" for the study of the glass transition in colloids are suspensions of polymer spheres in a suitable solvent. By using microemulsion polymerisation it is possible to synthesize nearly monodispersely distributed rigid spheres [1]. The size chosen for the spheres is usually about 200 nm with a coating of flexible polymer chains of about 10 nm thickness.

Besides the advantage of extremely slow crystallisation these systems have the advantage that the interparticle potential is close to that of hard spheres (i.e. $U(r) = \infty$ if r becomes smaller than the diameter of a sphere). This enables the calculation of the structure factor and dynamical properties. The application of mode coupling theory concepts (see Sect. 5.1) is facilitated by the absence of thermally activated processes ("hopping" over energy barriers) which dominate the dynamics in molecular glass forming liquids at low temperatures.

From the experimental point of view the systems are easy to study because the characteristic dimensions are close to the wavelength of light. Therefore, dynamic light scattering can be applied to observe times ranging from microseconds to hours (10 decades) (the same studies on molecular systems require neutron scattering studies with a dynamical range from $10^{-13} \ldots 10^{-8}$ s, 5 decades.)

As we will see later from the experimental results in Sect. 4 the analogy to atomic or molecular systems is far-reaching. Indeed, dependent on the particle density of the colloids, one finds liquid-like and crystalline order and also a dynamically frozen glass-like structure.

In addition to the classical system of polymer spheres, monodisperse amorphous silica spheres have been used recently [2]. Also star branched polymers with a high number of branches turn out to be effectively hard sphere colloids [3].

3 Dynamic Light Scattering

3.1 General Principles

The most important experimental method to investigate the dynamics of colloidal glasses is *dynamic light scattering* (DLS). In this chapter the basics of this method will be explained. For a detailed derivation of the statements and equations the reader is referred to standard textbooks [5, 6, 7, 8].

Figure 1 shows the schematic experimental set-up of the DLS experiment. Monochromatic laser light of wavelength λ is scattered from the sample. The

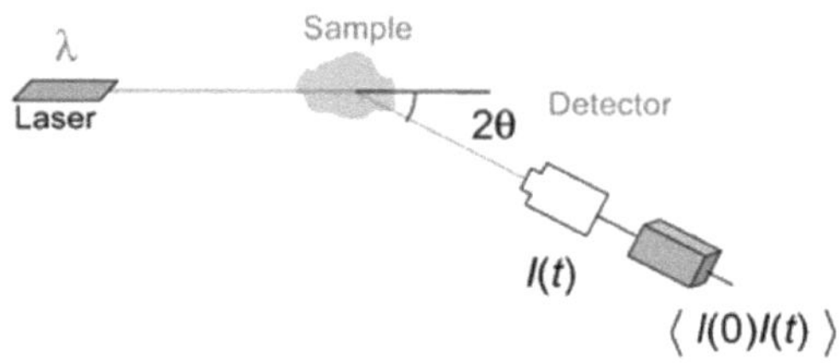

Fig. 1. Schematic set-up of a dynamic light scattering experiment

scattered photons are registered in a detector giving a time dependent intensity signal $I(t)$. From this signal a special purpose computer ("correlator") constructs the correlation function $\langle I(0)I(\tau)\rangle$ as explained later in this section.

The experiment involves two parameters: the scattering vector $\mathbf{Q}$ and the correlation time τ. The former is defined as the difference of incident and scattered wave vectors:

$$\mathbf{Q} = \mathbf{k}_1 - \mathbf{k}_0 \,. \tag{1}$$

Since $k_1 \approx k_0 = 2\pi n/\lambda$, the scattering vector is determined by the scattering angle, the wavelength and the index of refraction n:

$$\mathbf{Q} = \frac{4\pi n}{\lambda} \sin\theta \,. \tag{2}$$

The dependence of the scattered intensity on $\mathbf{Q}$ contains the *structural* information obtained by the experiment

The *dynamical* information is contained in the intensity correlation function

$$\langle I(0)I(\tau)\rangle = \lim_{T\to\infty} \frac{1}{T} \int_{t_0}^{t_0+T} I(t)I(t+\tau)\mathrm{d}t \,. \tag{3}$$

(Of course $I(t)$ and also the correlation functions depend on the scattering angle and therefore on $\mathbf{Q}$. To simplify the expressions this dependence will only be shown where necessary.) The practical meaning of this equation is the following: The integral with the factor $1/T$ is the average over the time interval of length T. In order to obtain a stable value independent of the short-time fluctuations one has to take the limit of infinite time interval length T. For comparison the average intensity is defined by

$$\langle I\rangle = \lim_{T\to\infty} \frac{1}{T} \int_{t_0}^{t_0+T} I(t)\mathrm{d}t \,. \tag{4}$$

In contrast to this definition equation (3) contains the product of two intensity measurements separated by the time τ. Therefore, $\langle I(0)I(\tau)\rangle$ expresses in how far intensities measured in this temporal distance are correlated. It is easy to see that if there is no temporal correlation (3) reduces to the square of (4), $\langle I\rangle^2$. On the other hand if there is complete correlation (especially for $\tau = 0$) it assumes the value $\langle I^2\rangle$ which is always greater than or equal to the former value.

In order to express the degree of correlation independently of the intensity itself a normalised quantity is defined

$$g_2(t) = \frac{\langle I(0)I(t)\rangle}{\langle I\rangle^2} \,. \tag{5}$$

Under certain ideal experimental conditions this function decays from 2 to 1. In order to obtain a connection to the microscopic properties of a colloid one needs the correlation function of the electric field instead of that of the intensities, namely

$$g_1(t) = \frac{\langle E(0)E(t)\rangle}{\langle E^2\rangle} \,. \tag{6}$$

With certain assumptions, the most important being that the electrical field is a Gaussian-distributed quantity, one can derive the *Siegert relation* between the two correlation functions:

$$g_1(t) = \sqrt{g_2(t) - 1} \,. \tag{7}$$

Scattering theory gives a simple relation between the field correlation function (6) and the time dependent particle positions of a liquid-like colloid assuming particles which are identical (monodisperse) with respect to their shape, size, and optical properties:

$$g_1(t) = \frac{S(\mathbf{Q}, t)}{S(\mathbf{Q})} \,. \tag{8}$$

Here, $S(\mathbf{Q})$ and $S(\mathbf{Q}, t)$ are the static and dynamic structure factor. The former is defined as usual in the theory of diffraction:

$$S(\mathbf{Q}) = \left\langle \sum_{i,j} \exp\left(\mathrm{i}\mathbf{Q} \cdot (\mathbf{r}_i - \mathbf{r}_j)\right) \right\rangle \,. \tag{9}$$

From this definition it can be seen that it is the spatial Fourier transform of the two-particle correlation function. The particle positions $\mathbf{r}_i$ entering here are those at the same instant. Thus, the static structure factor represents the Fourier transform of a "snapshot picture" of the structure. The dynamic structure factor is a generalisation of (9) involving the correlation of particles *at different times*:

$$S(\mathbf{Q}, t) = \left\langle \sum_{i,j} \exp\left(\mathrm{i}\mathbf{Q} \cdot (\mathbf{r}_i(0) - \mathbf{r}_j(t))\right) \right\rangle \,. \tag{10}$$

The definition implies that the value at $t = 0$ is identical to $S(\mathbf{Q})$ and therefore $g_1(t)$ starts at 1. From this value it decays indicating that the spatial correlation of the particles gets lost due to their motion. If the material is liquid-like, i.e. particles can reach any position in the sample volume after sufficient time, decay is complete: $\lim_{t\to\infty} g_1(t) = 0$. If it is solid like (crystalline or glassy) there is a finite value, called the non-ergodicity parameter f_Q.

3.2 Non-Ergodicity

This naming of f_Q points to a problem arising in the study of the glass-like phases of colloids: If the particle density is so high that the particles block each other they cannot reach any point in the sample within the experimental time of some hours. Then, the system's trajectory in phase space does not reach all its regions any more. In consequence the ergodic hypothesis cannot be employed. But it is clear that the average in the intensity correlation function (3) is a time average while that of the structure factors (9) and (10) are ensemble averages. So the simple relation between dynamic light scattering and microscopic dynamics is invalid in this case.

This is immediately clear already from static light scattering: $S(\mathbf{Q})$ is a smooth function of Q in a disordered sample. It shows a maximum corresponding to a preferred distance of scattering centres, $Q_{\max} = 2\pi/d$ but no Bragg peaks as in a crystalline material. But if one puts a piece of scratched or smoked glass into a laser beam the light is scattered into small patches, so-called speckles. The speckles are more dense close to $Q_{\max}$ but the exact shape of $I(\mathbf{Q})$ is clearly different from that of $S(\mathbf{Q})$. There are several simple ways to obtain the correct $S(\mathbf{Q})$ in this situation[1]: (1) Because a glassy sample is isotropic, only the absolute value $Q = |\mathbf{Q}|$ is relevant. Therefore one can average over detector positions with different azimuthal angle but identical Q. This "radial averaging" involves many speckles and therefore creates a kind of ensemble average. (2) Because the sample is macroscopically homogeneous it can be shifted through the beam during the experiment. Then, different parts of the sample are illuminated creating an ensemble average in an obvious way. (3) Because of the isotropy of the sample the same effect can be achieved by rotating the sample.

The situation for DLS is somewhat more complicated. Figure 2 shows schematically the experimental result for three situations: liquid-like ergodic, completely arrested (non-ergodic), and partially non-ergodic, i.e. the particles can perform a local motion on the experimental time scale but are confined to certain volumes. In the liquid situation all intensities measured by a variation of detector or sample position as described before are different. Nevertheless,

[1] Actually one will obtain the product of structure factor and form factor of the particle, $S(Q)P(Q)$. Here $P(Q)$ is derived from the shape of a single particle by Fourier transform. But since $P(Q)$ is a smooth function of Q this does not affect the argument.

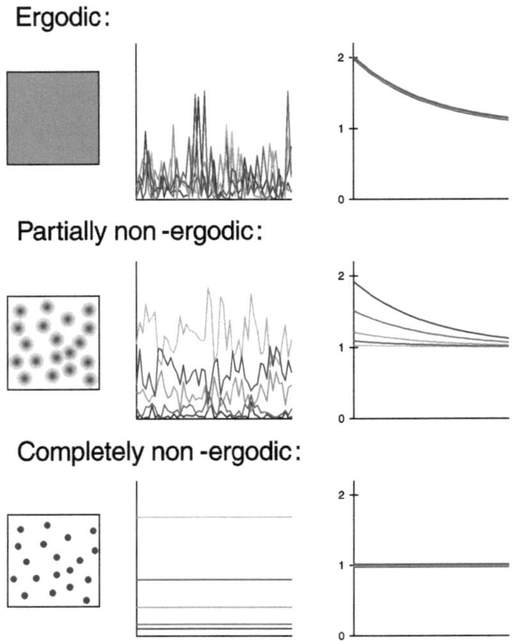

Fig. 2. Three schematic scenarios of dynamic light scattering on a colloid. Left column: microscopic picture of the sample, averaged over the typical experiment duration. Centre column: intensities $I(t)$ measured for different detector/sample positions corresponding to a single absolute scattering vector Q. Right column: intensity correlation functions at the same positions, the colours correspond to those of the $I(t)$ curves

the correlation functions are identical, decay from 2 to 1, and can be converted to $g_2(t)$ by the Siegert relation (7). For the completely arrested particles it is clear that the intensities are constant because the scattering centres do not move. They are different because the detector may be placed on a speckle or in between. $g_2(t) = 1$ but from the definition for an arrested sample $S(Q,t)/S(Q) = 1$ so the Siegert relation is clearly inapplicable.

3.3 Incoherent Light Scattering

A recent extension to the methods of light scattering is that of incoherent DLS [2]. In contrast to the techniques described above which give information about the pair correlation function this methods yields the self-correlation of individual particles.

Incoherent scattering is a standard technique for neutron scattering. There, it is caused by different scattering lengths of chemically identical nuclei (either as different isotopes or by different spin orientation). The analogous technique for light scattering requires the use of particles with identical shape and surface properties but different index of refraction. This has recently been achieved by using polymer spheres and silica spheres of identical size with identical polymer coating grafted on the surface.

As in neutron scattering the total scattering signal can be decomposed into a coherent part and an incoherent part

$$g_1(t) \propto \left(\overline{b^2} - \overline{b}^2\right) S_{\text{inc}}(Q,t) + \overline{b}^2 S_{\text{coh}}(Q,t) \tag{11}$$

where $S_{\text{coh}}(Q,t)$ is the dynamic structure factor as defined in equation (10) called "coherent" her for distinction. $S_{\text{inc}}(Q,t)$ is its incoherent counterpart

$$S_{\text{inc}}(\mathbf{Q},t) = \left\langle \sum_i \exp\left(\mathrm{i}\mathbf{Q} \cdot (\mathbf{r}_i(0) - \mathbf{r}_i(t))\right) \right\rangle . \tag{12}$$

Here, only one particle index i occurs. So each particle's position is linked with its own position at a later time but not that of another particle as in (10).

The prefactors in (11) refer to the distribution of scattering lengths of the particles. They depend on size, shape, and refractive index difference with respect to the solvent of a particle. Because of the form factors they are also Q-dependent unless the particles are much smaller than the wavelength of light. It can be seen that if the particles are monodisperse also concerning their optical properties the first prefactor (the variance of scattering lengths) vanishes. In this situation the light scattering is solely coherent. If there is a polydispersity there is in general a mixture of coherent and incoherent scattering. Interestingly, also the situation of pure incoherent scattering can be achieved: Because b may be positive or negative depending on whether the index of refraction of a particle is larger or smaller than that of the solvent the average $\overline{b}$ can be put to zero by a suitable choice of the solvent.

4 Experimental Results

Figure 3 shows cuvettes with suspensions of poly(methyl methacrylate) latex spheres ($\approx$ 200 nm radius) stabilised by a surface coating of poly(12-hydroxystearic acid) of about 10 nm thickness. As solvent a mixture of a hydrocarbon (e.g. decalin) with carbon disulfide is used to closely match the refractive index and thus avoid excessive turbidity. The cuvettes are illuminated by white light from behind. One can see different colourful structures in the scattered light. For high as well as low concentrations (right- and leftmost cuvette, respectively) only diffuse scattering is visible. This corresponds to a random distribution of colloid particles in the solvent. For intermediate concentrations multicoloured patches are visible which originate from "crystals", i.e. volumes where the latex spheres assume a regular repetitive structure. Often, these coexist with amorphous regions.

Fig. 3. Suspensions of spherical latex particles four hours after randomisation by tumbling. Right to left in the order of increasing concentration: $\phi_\mathrm{E} = 0.48 \ldots 0.64$. The cuvettes are illuminated by white light from behind. From [10]

Depending on the concentration the following phenomenology is found:

$\phi_\mathrm{E} < 0.49$	liquid-like structure, no tendency of crystallisation
$0.49 < \phi_\mathrm{E} < 0.54$	coexistence region liquid-crystal, fast crystallisation, small crystals, homogeneously nucleated
$0.54 < \phi_\mathrm{E} < 0.58$	only crystalline phase, fast crystallisation, small crystals, homogeneously nucleated
$0.58 < \phi_\mathrm{E} < 0.61$	slow, often only partial crystallisation, large needle-shaped crystals, heterogeneous nucleation
$0.61 < \phi_\mathrm{E}$	amorphous structure, only traces of crystals at surface and walls

Up to a concentration $\phi_\mathrm{E} \approx 0.58$ this behaviour is what one expects from theoretical considerations and computer simulations on the hard-sphere system[2] [12]. In the concentration range above $0.574 \ldots 0.581$ an obvious change takes place [10]: The crystallisation becomes slower and the crystals have different shape because of the different crystallisation mechanism. This is interpreted as a hindrance of crystallisation due to tight packing of the spheres, a mechanism analogous to the forming of a glass by quenching a molecular liquid to low temperatures.

The same behaviour is visible in the static light scattering patterns, Fig. 4. Immediately after "randomisation" of the samples, i.e. tumbling the sample cell in order to destroy long-range spatial correlations, all samples show a

[2] Indeed, the effective concentrations ϕ_E given here and in the literature are calculated by assuming that the volume of a latex sphere is such that the crystals start to occur at a volume fraction of 0.494 which is the theoretical hard-sphere value. (This is done because the "volume" of a coated polymer sphere is subject to the definition in how far the coating has to be included.)

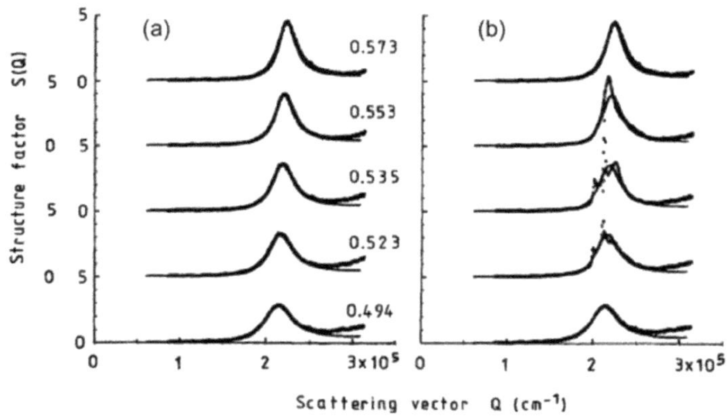

Fig. 4. Structure factor $S(Q)$ from static light scattering from latex suspensions. The numbers at the curve indicate the effective volume fraction ϕ_E. The continuous curves show the expected $S(Q)$ from the Percus-Yevick theory of hard-sphere liquids. (**a**) Measurements made immediately after randomisation by tumbling the samples. (**b**) Measurements made several hours later. From [13]

liquid-like structure factor $S(Q)$ with a broad peak indicating the short-range correlation of the particles. The structure factor agrees reasonably well with that expected from the Percus-Yevick approximation for hard-spheres [14]. After four hours rest, Bragg peaks arise in the samples of intermediate concentrations while for large and small concentrations the samples remain amorphous. This is the typical behaviour found in molecular glasses, only that the role of temperature is taken by the concentration here. For high temperatures (low concentrations) the liquid phase is thermodynamically stable. For low temperatures (high concentrations) the dynamics are so slow that crystallisation is inhibited and the amorphous phase is kinetically stable, which is called the glassy phase.

In order to explore the dynamics of the colloid particles dynamic light scattering experiments have been done on the same samples. Figure 5 shows some of the results. It can be seen that the curves show an exponential decay in the beginning, as expected from diffusion: $S(Q,t)/S(Q) = \exp(-Q^2Dt)$. Then a slowing down takes places giving rise for a plateau. The plateau becomes longer for higher concentrations and is higher close to the maximum of $S(Q)$. At very high concentration no complete decay of $S(Q,t)$ is visible.

The usual interpretation of these curves is that at short time the particles move freely without hitting their neighbours. This leads to initial decay of correlation obeying the laws of Brownian motion. As soon as the average displacement of the particles reaches the nearest-neighbour distance their motion becomes impeded like they were confined in a cage formed by the neighbour particles. At that stage the correlation does not decay any further (plateau in $S(Q,t)$). The extent of this caging effect increases with concentration. If the concentration is low fluctuations of the particles forming the cage wall

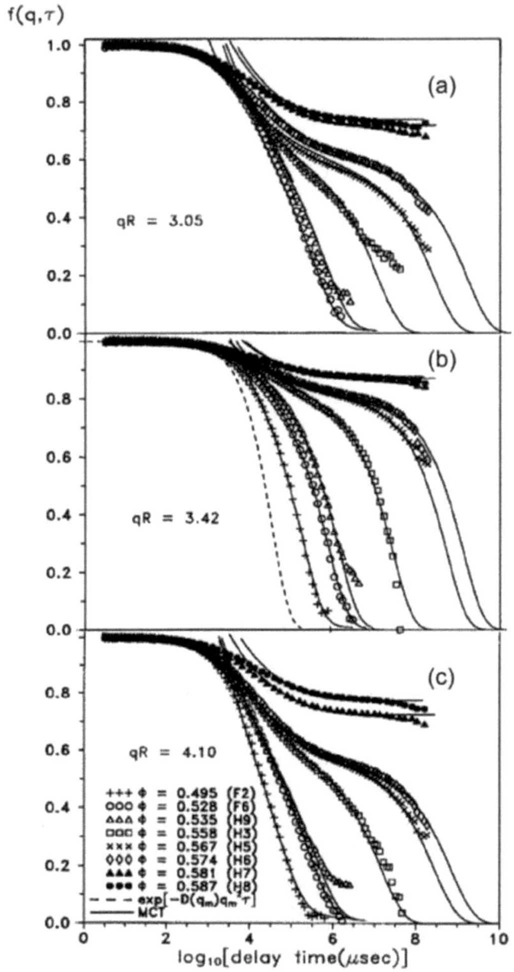

Fig. 5. Normalised dynamic structure factors $S(Q,t)/S(Q)$ for three scattering vectors Q around the structure factor peak. The solid curves are mode-coupling theory fits to the data. The dashed curve fits the initial exponential decay. From [9]

create openings from time to time allowing the central particle to escape. In this situation the correlation will decay completely for infinite time. For high concentration this may become impossible so that the plateau in $S(Q,t)$ exceeds to infinite times. This situation would be that of an ideal glass.

The $S(Q,t)/S(Q)$ curves from DLS of Fig. 5 are very similar to those from neutron scattering on molecular glasses, shown exemplarily in figure 6 for the glass forming salt melt $Ca_{0.6}K_{0.6}(NO_3)_{1.4}$ [15]. Apart from the differences by several orders of magnitude in the length and time scales (1 s $\leftrightarrow$ 1 ps,

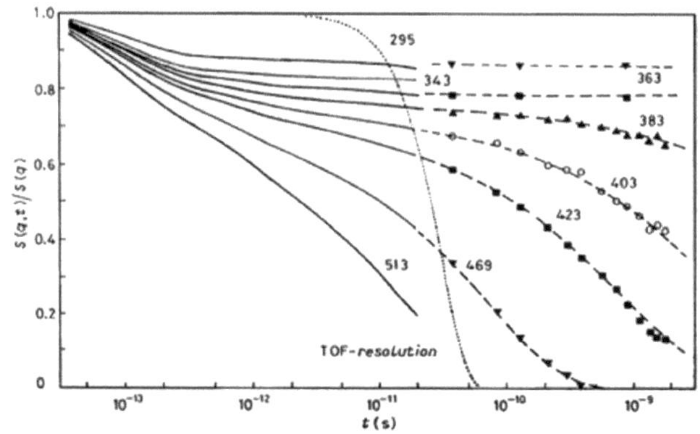

Fig. 6. Normalised dynamic structure factors $S(Q,t)/S(Q)$ of $Ca_{0.6}K_{0.6}(NO_3)_{1.4}$ for $Q = 19\,\mathrm{nm}^{-1}$, the position of structure factor peak. The solid curves are obtained by Fourier transform from a time-of-flight spectrometer. The individual points at long time are results from neutron spin echo. From [15]

$200\,\mathrm{nm} \leftrightarrow 0.1\,\mathrm{nm}$) the curves show the same qualitative behaviour with temperature instead of concentration as control parameter.

Qualitatively similar curves are obtained by incoherent light scattering. But for this method the data is more directly interpretable. In the Gaussian approximation the incoherent scattering function is given by

$$S_{\mathrm{inc}}(Q,t) = \exp\left(-Q^2 \left\langle \Delta r^2 \right\rangle /6\right) . \tag{13}$$

This means one can calculate the average mean squared displacement of the colloid particles, see figure 7. Here the three regimes can be seen evidently: For short times $\left\langle \Delta r^2 \right\rangle$ increases linearly as expected for Brownian motion. Then there is a plateau at the value corresponding to the distance at which collisions with the neighbour particles take place. Finally, there is an increase of $\left\langle \Delta r^2 \right\rangle$ beyond this length which shifts to longer times when the concentration is increased.

5 Mode Coupling Theory

5.1 Outline of the Theory

The theory is treated in some detail in the contribution of R. Schilling in this book. Here, we show only some basic equations and results which are relevant for the interpretation of the data on colloidal glasses.

The mode coupling theory (MCT) has been developed in the 80s and has since then made a big contribution to the understanding of structural glasses [16]. The central quantity of this theory is the density-density correlation function:

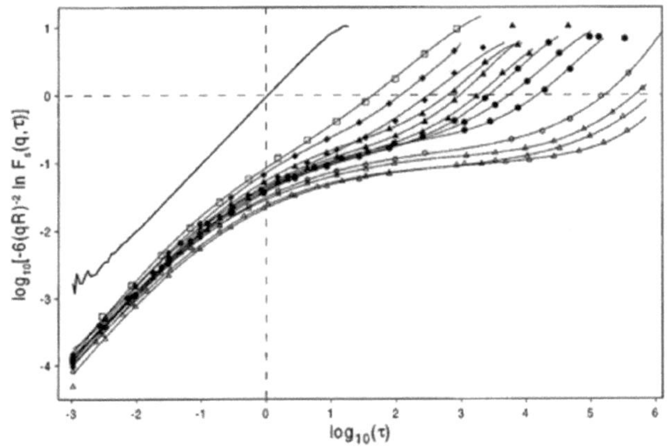

Fig. 7. Mean squared displacements of latex/silica spheres in the suspensions calculated using the Gaussian approximation (13). The line without symbols corresponds to a very dilute suspension where the particles can diffuse freely. From [2]

$$\Phi_Q(t) = \frac{\langle \delta\rho_Q^*(t)\delta_Q(0)\rangle}{\left\langle \delta\rho_Q^*(0)\delta_Q(0)\right\rangle} . \tag{14}$$

If all particles are identical this quantity is equal to the normalised dynamic structure factor $S(Q,t)/S(Q)$. Therefore, dynamical light scattering experiments should be immediately comparable to the results of the theory.

The main results of MCT are (1) the prediction of a kinetic glass transition, (2) explicit solutions for $\Phi_Q(t)$ for simple systems, e.g. hard spheres, (3) scaling laws, (4) relations between the exponents of scaling laws.

The fundamental equations of MCT have a form like

$$\ddot{\Phi} + \gamma\dot{\Phi} + \Omega^2\phi + \Omega^2 \int_0^t M(t-t')\dot{\Phi(t')}\mathrm{d}t' = 0 . \tag{15}$$

(For a real structural glass forming system the variables Φ and M are also Q dependent and instead of a single equation (15) one has an infinite set. But many fundamental results can already be obtained from this extremely reduced version of the MCT.) The first three terms are those also found in the description of a damped harmonic oscillator. For colloidal systems the second and third term dominate the first, which can be omitted (Brownian motion instead of Newtonian). γ/Ω^2 is a friction coefficient including the viscosity of the solvent and the size of the particles. The additional integral term contains a memory function (passed times $t - t'$ enter through the integration) acting as an additional damping term (time derivative of Φ).

The addition of a memory function in order to describe complex viscoelastic behaviour has already been done before MCT [7]. The crucial step forward by MCT was to relate the memory function itself to the density correlation

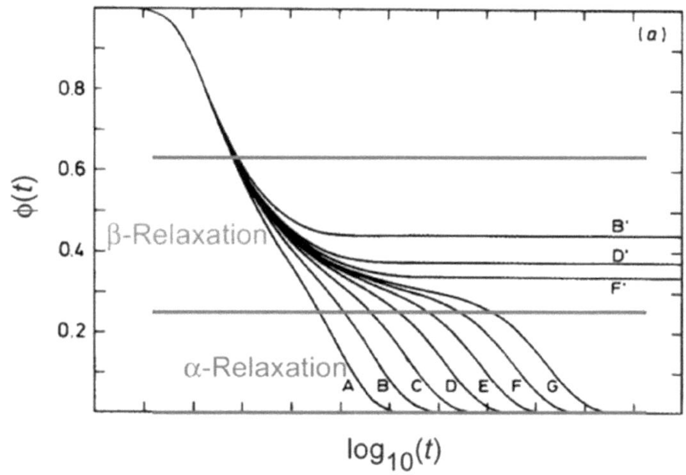

Fig. 8. Numerical solution of the mode coupling equation (15) with a second order polynomial (16). Curves A…G denote coupling parameters in the liquid range (with decreasing distance to the critical line), B'…F' those in the glass range. For exact parameter values see ref. [17]. The grey lines denote the approximate boundaries of the validity of the scaling laws (17) and (18)

function. In this simple model this would be done by a polynomial

$$M(t) = \sum_k v_k \Phi^k(t)\,. \tag{16}$$

In the full description of a structural glass also mixed products of $\Phi_Q(t)$ with different Q occur implying that modes of different Q are coupled. The coefficients which then have the form $v_{Q;q_1,q_2\ldots}$ can be calculated from the static structure factor $S(Q)$ alone. For the full as well as the simplified model the equations (15) and (16) form a closed set which can be solved numerically.

The MCT equation systems invariably show two kinds of solutions: (1) For small values of the coupling parameters v_k the solution $\Phi(t)$ decays to zero for long times. As already stated in Sect. 3.1 this corresponds to the liquid state. Therefore, small values of v_k are associated with high temperatures in molecular liquids or small densities in colloids. (2) For high values of v_k a finite value $\Phi(t) = f_Q$ remains, the non-ergodicity parameter. At a certain surface in the space of v_k a discontinuous transition between the two solution types takes place. This is connected with a critical value of the physical control parameter (temperature or density). The distance from the critical value is expressed by a separation parameter $\sigma \propto (T_c - T)/T_c$ or $\sigma \propto (\phi - \phi_c)/\phi_c$, respectively.

Figure 8 shows numerical solutions of (15) with a second order polynomial as (16). The curves show two scaling regimes: The final decay (α relaxation) has always the same shape in the semilogarithmic representation. This implies

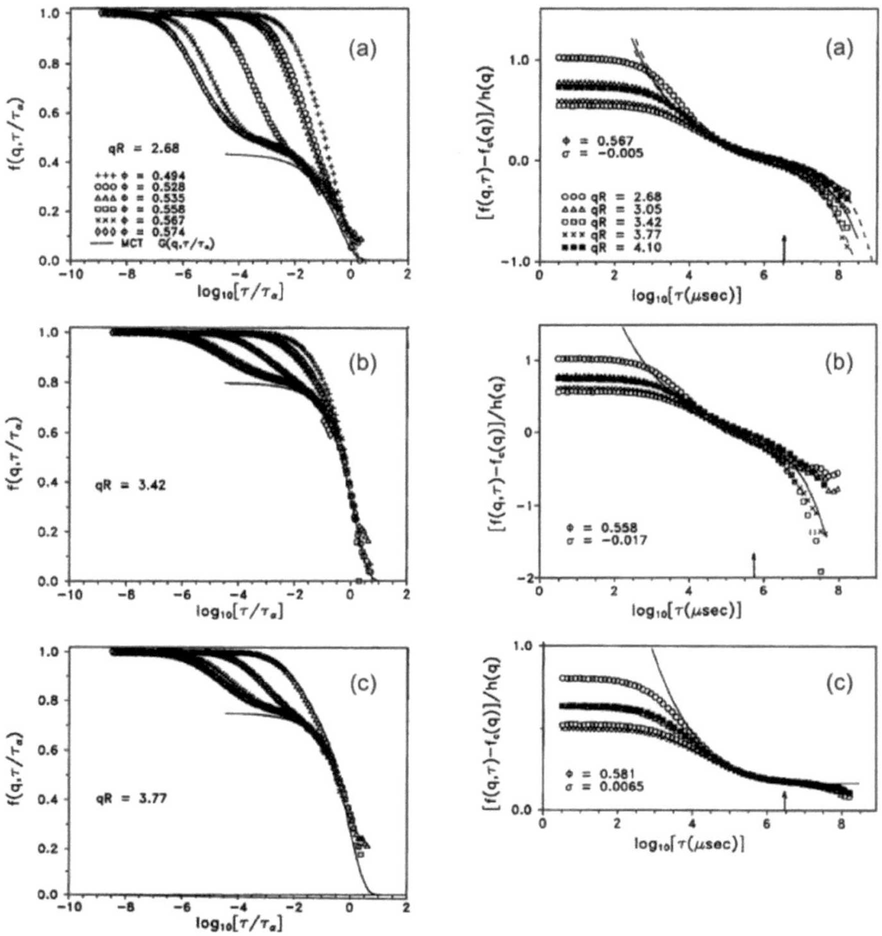

Fig. 9. Test of the scaling properties of $\Phi_Q(t)$ predicted by MCT. The left column shows the test of the α relaxation scaling (17) of data from different concentrations at identical Q values. The right column shows the β relaxation scaling (18) of curves for different Q. From [9]

that by a single scaling time τ_α all curves can be rescaled to a single master curve (time-temperature superposition principle):

$$\Phi_Q(t) = f_Q^c G_Q(t/\tau_\alpha)\,. \tag{17}$$

For shorter times there is another region where scaling is possible using a linear transformation of the ordinate in addition to a scaling time τ_β:

$$\Phi_Q(t) = f_Q^c + |\sigma|^{-1/2} h_Q g_\pm(t/\tau_\beta)\,. \tag{18}$$

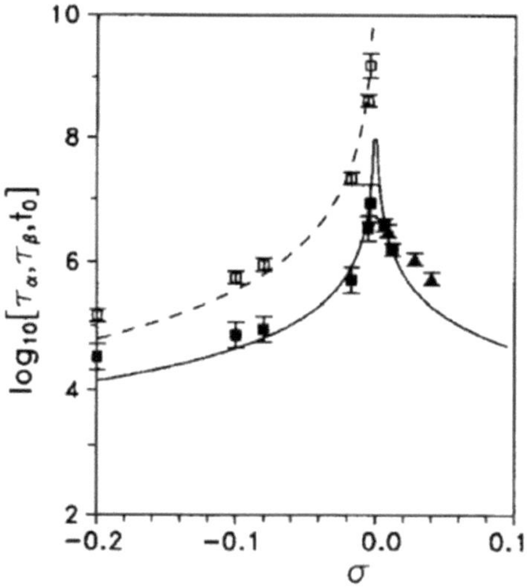

Fig. 10. Scaling times from DLS data. The empty symbols represent the α relaxation scaling time τ_α, the filled symbols that of the β relaxation τ_β. The dashed and continuous curve show the MCT predictions for α and β relaxation scaling, respectively. From [9]

It can be seen that the α relaxation expression requires a Q dependent master function G_Q while the β relaxation has only two, g_+ in the glass and g_- in the liquid.

The scaling times depend on the separation from the critical point by power laws

$$\tau_\beta \propto |\sigma|^{-\frac{1}{2a}} \tag{19}$$

$$\tau_\alpha \propto |\sigma|^{-\left(\frac{1}{2a}+\frac{1}{2b}\right)} \tag{20}$$

where a and b can be calculated from the v_k and are universally related by

$$\frac{\Gamma^2(1-a)}{\Gamma(1-2a)} = \frac{\Gamma^2(1+b)}{\Gamma(1+2b)} . \tag{21}$$

The exponent parameters a, b and scaling amplitudes f_Q^c, h_Q can be calculated if $S(Q)$ is known. Especially, the Percus-Yevick approximation for hard spheres yields $1/2a = 1.66$ and $1/2a + 1/2b = 2.58$.

5.2 Comparison to Experiments

Figure 9 shows a test of the scaling properties using the DLS data of Fig. 5. One can see that both scaling laws are reasonably well fulfilled. As predicted by MCT, different curves Q values can be collapsed in the β relaxation region

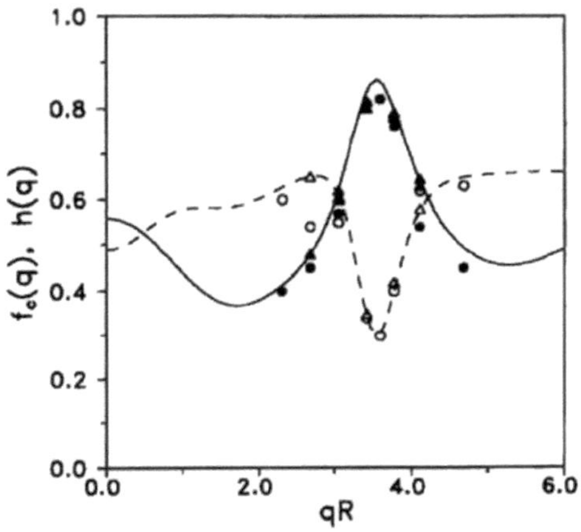

Fig. 11. β relaxation ordinate scaling parameters of equation (18). The filled and empty symbols represent the values from DLS experiments, the continuous and dashed curves the MCT predictions for f_Q^c and h_Q, respectively. From [9]

while in α relaxation region each Q needs an individual master function but different concentrations can be scaled together.

Because there are also other theories predicting e.g. time-temperature superposition it is important that the scaling times used in Fig. 9 also obey the expected power laws. This is demonstrated in Fig. 10. The lines in this figure follow power laws with the calculated values of the exponents. Thus, they contain no free parameters.

Finally, the scaling of the β relaxation in Fig. 9 yields the parameters f_Q^c and h_Q which, as Fig. 11 shows, exactly coincide with those calculated from the Percus-Yevick $S(Q)$ for hard spheres.

6 Undercooled Liquids

6.1 Similarities and Differences to Colloids

As shown above, colloidal suspensions do indeed have a glass transition. It differs from the classical glass transition of undercooled liquids in an important aspect: thermal activation plays no role for the structural rearrangements. The equilibration occurs exclusively by the quasicontinuous Brownian motion of the particles. Keeping the temperature constant, the Brownian motion keeps its time scale and the structural arrest of the glass transition occurs with increasing concentration of the colloidal particles, in close analogy to the theoretically heavily studied hard-sphere system. Dynamic light scattering data from colloids provide a very impressive and detailed proof of the mode coupling theory.

If a liquid is a good glass former, the interatomic potentials are usually too complicated to allow the calculation of the parameters of the mode coupling theory. So one has to fit the parameters to measured data. This has been done for a large number of undercooled liquids, combining the available experimental data. Above T_c, the separation of the α-process (the elementary process of the flow) from the microscopic picosecond motion in an undercooled liquid seems to be reasonably well described by the mode coupling theory [18], though there is no general agreement on this statement [19]. As a general rule [20], one finds $\tau_\alpha(T_c) \approx 10^{-7}$ s, a bit longer than the originally considered value 10^{-9} s [21]. The question whether T_c marks the transition to an energy-landscape-dominated dynamics is at present heavily studied by simulations [22].

6.2 Soft and Hard Matter

At lower temperatures, one seems to enter a completely different world, dominated by thermally activated hopping between different valleys of the energy landscape [21]. There is no generally agreed theory of the so-called calorimetric glass transition at the glass temperature T_g, where τ_α approaches the time scale of 100 s. This glass transition is the borderline between hard and soft matter; if one speaks about soft matter, one implicitly assumes a temperature above the glass temperature of the material in question. Soft matter implies a shear modulus of a few MPa, while hard matter has a shear modulus of the order of GPa, three orders of magnitude higher.

Now the glass transition in the colloids is not a transition from soft to hard matter, because there is practically no change in the shear modulus; colloid liquid, glass and crystal belong all three to the soft matter class. So the question arises whether the mechanism of the glass transition in undercooled liquids is different from the one in the colloids. That would imply that the mode coupling theory describes the dynamics of the undercooled liquid at higher temperatures, down to the critical temperature T_c, but not the final structural arrest at a still lower glass temperature T_g, where the long time shear modulus rises to a value of the order of GPa. We cannot answer this question here, but we will describe some rheological experiments characterizing the calorimetric glass transition.

7 The Breakdown of the Shear Modulus at the Glass Transition

7.1 Comparison to a Single Debye Process

The quantitative treatment of the mechanical properties introduces the shear strain ϵ. The shear stress σ is related to the shear strain via the shear modulus G

$$\sigma = G\epsilon. \tag{22}$$

In a viscoelastic medium, the shear stress after a given initial shear strain at time zero will decay. If one keeps the initial strain $\epsilon(0)$ constant, one can describe the decay of the shear stress with a time-dependent shear modulus $G(t)$

$$\sigma(t) = G(t)\epsilon(0). \tag{23}$$

Instead of measuring the decay in time, one can also measure with a periodic strain at the frequency ω. One will find a periodic stress with the same frequency, but with a phase difference. This phase difference is a measure for the loss at the given frequency. One characterizes the response of such a dynamic experiment by a complex shear modulus, with a real part $G'(\omega)$ and an imaginary part $G''(\omega)$. The phase angle δ is given by the relation $\tan\delta = G''/G'$. Thus $G'(\omega)$ measures the restoring force constant at the frequency ω and $G''(\omega)$ the loss.

Here, we focus on torsion pendulum measurements of polystyrene, done at a frequency of 0.64 Hz, i.e. $\omega = 4\ s^{-1}$. One measures the resonance frequency and the damping of the pendulum. The filament consists of the polymer being studied. One follows the development of the resonance frequency and of the damping as a function of temperature, from very low temperatures up to temperatures above the glass transition. Thus one obtains the data points shown in Fig. 12. The data are taken from reference [23]. One finds the breakdown of the shear modulus at this frequency around 360 K in the undercooled liquid. At this temperature, there is a peak in $G''(\omega)$, called the α-peak (glassy relaxations are numbered in greek letters with decreasing temperature, starting with the breakdown of the shear modulus).

These data should be describable in terms of an energy landscape picture, where the system makes thermally activated jumps from one valley to the other.

Let us see whether this idea describes the breakdown of the shear modulus in Fig. 12. We start with the simplest possibility, assuming a single thermally activated process with a relaxation time τ_α. One can calculate $G'(\omega)$ and $G''(\omega)$ of such a single-exponential decay (usually denoted as Debye process) from the Debye equations

$$G'(\omega) = G\frac{\omega^2\tau_\alpha^2}{1+\omega^2\tau_\alpha^2} \tag{24}$$

$$G''(\omega) = G\frac{\omega\tau_\alpha}{1+\omega^2\tau_\alpha^2}. \tag{25}$$

The equations imply that the height of the peak in the imaginary part at $\omega\tau = 1$ is precisely one half of the high frequency modulus G.

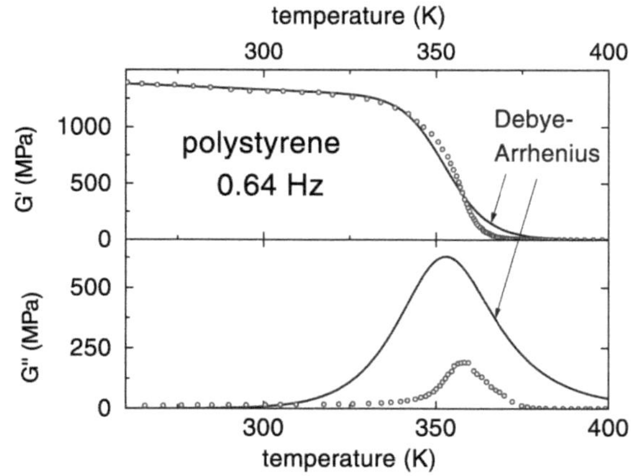

Fig. 12. Fit of torsion pendulum measurements [23] of $G'(\omega)$ (upper part) and $G''(\omega)$ (lower part) of polystyrene at the glass transition in terms of the Debye-Arrhenius model

The temperature dependence of a thermally activated process is given by the Arrhenius relation

$$\tau_\alpha = \tau_0 \mathrm{e}^{V/kT} \tag{26}$$

where τ_0 is a microscopic vibrational time of the order of 10^{-13} s and V is the barrier height which has to be overcome to get from one valley of the energy landscape to a neighbouring one.

Combining eqs. (24-26), one can calculate the temperature dependence of the real and the imaginary part of the shear modulus. Figure 12 shows an attempt to fit the glass transition of polystyrene in this simplest possible picture. The barrier height V and the high frequency shear modulus G were adapted to the real part of the modulus. In addition, a small temperature dependence of G was allowed for.

The fit of $G'(\omega)$, though far from perfect, is not too bad, especially when compared the very bad fit of $G''(\omega)$ in the lower part of Fig. 12.

One learns a lot from the failures of this fit. The main failure is the height of the peak in the imaginary part, which is a factor of three too high. There is an obvious explanation for this failure: One does not have a single relaxation time. Instead, one should superimpose a large number of processes with different relaxation times. This would indeed bring the peak height down. But: If we do that, we should find a strong broadening of the peak. The measured peak, however, is three to four times narrower than the single Debye peak. This seems to contradict the Kramers-Kronig relation between real and imaginary parts, which states that a step in the real part requires a proportional peak in the imaginary part. Is Kramers-Kronig wrong here?

The answer is obviously that the Kramers-Kronig relation is not wrong, but does indeed hold. The decay cannot be described by a single relaxation time, but must be described by an integral over many relaxation times (this is called the stretching of the glass transition process). The reason why the measured peak is so narrow is not its narrow width in frequency, but lies in its temperature dependence: it just runs through the frequency window of the measurement with a much higher speed than the one expected from the Arrhenius relation eq.(26) (this is called the fragility of the glass transition process). From the comparison with the Arrhenius-Debye picture in Fig. 12, one can immediately quantify the stretching and the fragility: The stretching is a factor of three (this is the ratio of the amplitudes in G'' of single Debye process and experiment), the temperature dependence of the relaxation time must be a factor of seven faster than the expected Arrhenius relation (the ratio of the areas of the peak in G'').

7.2 Fragility and Angell Plot

In fact, this fragile behaviour is what one finds if one changes the temperature and measures the corresponding change of time scale of the breakdown of the shear modulus. To compare with the expected Arrhenius behaviour, one usually plots the logarithm of the shift factor $a_T = \tau_\alpha(T)/\tau_\alpha(T_g)$ versus T_g/T, the so-called *Angell plot*. Such an Angell plot is shown in Fig. 13 for vitreous silica and polystyrene, defining the glass temperature T_g by a τ_α of 100 s. The data for silica are long time dynamical mechanical measurements [24] and photon correlation work [25], for polystyrene torsional creep data [26]. In polystyrene, the slope at T_g is even larger than the factor of seven estimated in the previous subsection; it is a factor of nine larger than the one expected for the Arrhenius law. This is a consequence of the curvature; the data of the previous subsection correspond to $a_T = 1/400$, two and a half decade away from T_g.

The two examples in Fig. 13 correspond to two extremes on the fragility scale, polystyrene being one of the most fragile glass formers and silica being the strongest known glass former (strong as opposed to fragile; a possible measure of the fragility is the slope at T_g). Even in the silica case, the slope at T_g is clearly larger than the Arrhenius expectation, and there is a clear curvature. Experimentally, one observes a correlation [27] between fragility and stretching (the deviation of the time decay of $G(t)$ from a single exponential): the more fragile the glass former, the stronger the stretching.

Though there seems to be general agreement that the glass transition of the undercooled liquid is due to thermally activated processes in its energy landscape, the simple Arrhenius picture fails completely.

8 Summary

In the colloid case, one finds a structural glass transition, which occurs as a function of concentration if there is not enough time for crystallisation.

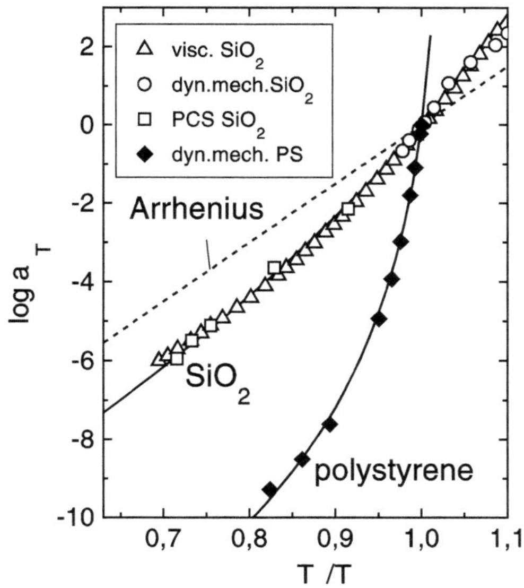

Fig. 13. Angell plot of the temperature dependence of the breakdown of the shear modulus in polystyrene and vitreous silica

One can adapt the potential parameters to mimic the theoretically accessible hard-sphere system. Beautiful light scattering data illustrate the structural arrest and prove the validity of the mode coupling theory.

The undercooled liquid case is much less clear. Unlike the colloid case, there is a transition from the liquid to a true solid with a solid-like shear resistance, again competing with crystallisation. The mode coupling theory seems to be valid at higher temperatures. Below its critical temperature T_c the energy landscape dominates the dynamics. Whether this implies a new theory is not clear. At present, one is limited to phenomenological concepts like fragility and stretching.

References

1. L. Antl, J.W. Goodwin, R.D. Hill, R.H. Ottewill, S.M. Owens, S. Papworth, J.A. Waters, Coll. Surf. **17**, 67 (1986)
2. W. van Megen, T.C. Mortensen, S.R. Williams, J. Müller, Phys. Rev. E **58**, 6073 (1998)
3. J. Stellbrink, J. Allgaier, D. Richter, Phys. Rev. E. **56**, R3772 (1997)
4. E.R. Weeks, J.C. Crocker, A.C. Levitt, A. Schofield, D.A. Weitz, Science **287**, 627 (2000)
5. B. Chu, *Laser Light Scattering* (Academic Press, New York, 1974).
6. H.Z. Cummins, E.R. Pike eds., *Photon Correlation Spectroscopy and Velocimetry* (Plenum Press, New York, 1977).

7. B.J. Berne, R. Pecora *Dynamic Light Scattering* (Wiley-Interscience, New York, 1976).
8. R. Pecora, *Dynamic Light Scattering* (Plenum Press, New York, 1985).
9. W. van Megen, S.M. Underwood, Phys. Rev. E **49**, 4206 (1994).
10. P.N. Pusey, W. van Megen, Nature **320**, 340 (1986).
11. W. van Megen, S.M. Underwood, Nature **362**, 616 (1993).
12. W. G. Hoover, F.H. Ree, J. Chem. Phys. **49**, 3609 (1968).
13. W. van Megen, P. Pusey, Phys. Rev. A **43**, 5429 (1991).
14. J.K. Percus, G.J. Yevick, Phys. Rev. **110**, 1 (1958).
15. W. Knaak, F. Mezei, B. Farago, Europhys. Lett. **7** 529 (1988).
16. W. Götze, *Aspects of Structural Glass Transition* in *Les Houches. Session LI. Liquids, Freezing and the Glass Transition*, eds.: J.P. Hansen, D. Levesque, J. Zinn-Justin (North Holland, Amsterdam, 1991).
17. W. Götze, L. Sjögren, J. Phys. C: Solid State Phys. **21** 3407 (1988).
18. W. Götze, J. Phys: Condens. Matter **11**, A1-A45 (1999)
19. F. Stickel, E.W. Fischer, R. Richert, J. Chem. Phys. **102**, 6251 (1995); F. Stickel, E.W. Fischer, R. Richert, J. Chem. Phys. **104**, 2043 (1996); C. Hansen, F. Stickel, T. Berger, R. Richert, E.W. Fischer, J. Chem. Phys. **107**, 1086 (1997)
20. A.P. Sokolov, J. Non-Crystalline Solids **235-237**, 190 (1998)
21. M. Goldstein, J. Chem. Phys. **51**, 3728 (1968)
22. T.B. Schrøder, S. Sastry, J.C. Dyre, S.C. Glotzer, J. Chem. Phys. **112**, 9834 (2000); K. Broderix, K.K. Bhattacharya, A. Cavagna, A. Zippelius, I. Giardina, Phys. Rev. Lett. **85**, 5360 (2000); L. Angelani, R. Di Leonardo, G. Ruocco, A. Scala, F. Sciortino, Phys. Rev. Lett. **85**, 5356 (2000); but see also B. Doliwa and A. Heuer, cond-mat/0209139 (2002)
23. F.R. Schwarzl, *Polymermechanik* (Springer, New York 1990), Abb. 5.16
24. J.J. Mills, J. Non-Crystalline Solids **14**, 255 (1974)
25. J.A. Bucaro, H.D. Dardy, J. Non-Crystalline Solids **24**, 121 (1977)
26. D.J. Plazek, V.M. O'Rourke, J. Polym. Sci: Part A-2, **9**, 209 (1971)
27. R. Böhmer, K.L. Ngai, C.A. Angell, D.J. Plazek, J. Chem. Phys. **99**, 4201 (1993)

Experimental Investigations of Collective Excitations in Disordered Matter

Jens–Boie Suck

Materials Research and Liquids, Institute of Physics, Chemnitz University of Technology, 09107 Chemnitz, Germany
suck@physik.tu-chemnitz.de

1 Introduction

The topic of this chapter fits very well in the framework of the main heading of this volume "Collective Excitations in Disordered and Nonlinear Systems", as in liquids the vibrational atomic dynamics is generally nonlinear and in amorphous solids some of them are also and certainly the atomic positions in both systems are disordered. First some basic quantities used in connection with experimental investigations of the dynamics in disordered systems will be introduced and then some typical more recent results obtained so far for dense fluids (liquids and compressed gases) and some of the many types of different glasses will be given. Because of the introductory character of this volume, neither rigor nor completeness, but simplicity is aimed for in this description. References to more complete reviews on either of the two system classes or on both are given in Ref. [1].

2 Neutron Inelastic Scattering and Correlation Functions

Even though all results discussed below will have been obtained using neutron inelastic scattering (NIS), and we shall therefore only discuss this method more explicitly, the following considerations are not related to a special probe. The special character of the radiation used enters the coupling of the probe to the scatterer and with this the corrections, which have to be applied to arrive at a more or less radiation independent result. With this, they also influence to some extend the interpretation of the data, depending on the degree to which one is able to determine the dynamical information on the sample independent of the radiation used, and because different probes may be scattered from different parts of the atoms: electromagnetic radiation from the electronic shell, implying already for one single atom a distribution of scattering entities and with this a form factor. The same argument applies to magnetic neutron scattering, which however will not be taken into account here, as we shall concentrate on vibrational excitations. With this omission, neutrons are scattered only from the nuclei, which have negligible extension compared to the wavelength of thermal and cold neutrons ($> 0.05\,nm$). This allows us to use very successfully an effective interaction between probe and nuclei, the Fermi pseudopotential, which effectively reduces to an isotope (and spin) dependent scattering length. Within approximations, this enables to represent the double differential scattering cross-section, which is obtained from the measured intensity after corrections and normalization, as a *product* between the properties of the probe and the incident and scattered flux and the dynamic structure factor $S(Q,\omega)$, which contains all the information on the dynamics of the sample [2]

$$\frac{d^2\sigma}{d\Omega dE} = \left(b^2 \frac{k}{k_0}\right) S(Q,\omega) \,. \tag{1}$$

Here σ is the scattering cross-section ($\sigma = 4\pi b^2$), Ω the solid angle, under which the detector surface is viewed from the centre of the sample and dE is the actual energy bin in the measured spectrum. Q is the modulus of the wave-vector transfer

$$\boldsymbol{Q} = \boldsymbol{k_0} - \boldsymbol{k} \tag{2}$$

where $\boldsymbol{k_0}$ and $\boldsymbol{k}$ are the wave-vectors of the probe before and after the single scattering event, with moduli k_0 and k. In the case of disordered systems discussed here one can safely assume an isotropic sample and omit the vector properties of the momentum transfer and just retain its modulus. In eq.1 ω is the energy transfer divided by $\hbar$

$$\hbar\omega = E_0 - E = \frac{\hbar^2}{2m}(k_0^2 - k^2) \,. \tag{3}$$

Here E_0 and E is the energy of the probe before and after the single scattering event and m is the mass of the scatterer. In an inelastic scattering experiment one therefore measures the scattered intensity as a function of the incident and scattered wave vector. The fact that one can only conclude (via energy- and momentum conservation) in *single* scattering events from the energy- and momentum transfer of the probe on the same quantities of the excitation in the sample, shows the restriction of $S(Q,\omega)$ to the time-dependent *two* particle correlations and the necessity to correct for intensities due to multiple scattering. This latter correction is of special importance in the investigation of the collective excitations in *disordered* systems using NIS, as these investigations are best conducted at smallest momentum transfers (near to the hydrodynamic limit of small Q and ω), where the *relative* amount of multiply scattered intensity often reaches 40% of the total measured intensity.

$S(Q,\omega)$ is the spectrum of the van Hove time dependent pair correlation function, G(r,t), where r is the modulus of the distance vector between the correlated pair of particles at $\boldsymbol{r'}$ and $\boldsymbol{r''}$. In the classical limit this function gives the probability to find a particle on a shell with radius r and thickness dr at time t, when there was a particle (the same or another) at $r = 0$ at $t = 0$. A two-step Fourier transformation relates the dynamical $(Q - \omega)$ space of the experiments with the space-time $(r - t)$ space of the correlation functions. In the first step from real- to reciprocal-space (more exactly: Fourier modul) the *intermediate scattering function* is obtained

$$F(Q,t) = \int d^3r\, e^{-i\,Qr}\, G(r,t) = N^{-1} \sum_1^N \sum_1^N \left\langle e^{-i\,Qr_i(0)} e^{i\,Qr_j(t)} \right\rangle \tag{4}$$

which in the classical approximation describes the time decay of the correlations, reflected at t=0 in the static structure factor $S(Q)$, i.e.$F(Q,t=0) = S(Q)$ and the acute brackets represent the ensemble average. The second transformation from time to energy leads to the dynamic structure factor

$$S(Q,\omega) = \frac{1}{2\pi\hbar N} \int dt\, e^{-i\omega t} F(Q,t)\,. \tag{5}$$

The exponentials of the type $e^{i\,Qr''(t)}$ are *time dependent local number densities*, which in this case are the *dynamical variables.* Thus the dynamic structure factor is the spectrum of a time dependent density-density correlation function. As these correlation functions contain auto- and distinct correlations, i.e. the correlation of a particle with itself in time and with other particles, the measured dynamic structure factor reflects the *single particle motion* as well as the *collective* atomic dynamics. The Fourier transformation of the van Hove auto-correlation function and the distinct correlations translates these into the self- part of the dynamic structure factor and the dynamic structure factor, respectively, which are proportional to the *incoherent* and the *coherent* scattered intensity in the neutron spectra. If one investigates collective excitations, only the *in-phase motion* of different particles is of interest, and thus one would have to remove the incoherent scattering before interpreting the results. As this is very difficult in most cases, one has to interpret the coherent scattering "on top" of the incoherent one, which is often very difficult to do unequivocally. One therefore chooses the sample for the investigation of collective excitations in a way that the dominant part of the scattering is coherent and that the incoherent contribution can be estimated or neglected. In the dynamic structure factors shown below the single particle motion is alway still included.

In the interpretation of the results, one has to keep in mind that different elements, - in the case of neutrons even different isotopes of the same element -, scatter differently. A straight forward interpretation of the measured intensities is therefore only possible if the sample contains just one scattering element, where in the case of neutrons a mean value over the contributions from different isotopes (and spins) is taken. As soon as more than one element is involved, a further complication arises, because in that case the *total* dynamic structure factor is determined from the measured intensity, which is the weighted sum over the *partial* dynamic structure factors, of which there are as many as different time-dependent correlations exist in the van Hove correlation function, i.e. $n(n+1)/2$, if n is the number of different elements in the sample.

$$\sigma S(Q,\omega) = 4\pi \sum_i \sum_j b_i b_j c_i c_j S_{i,j}(Q,\omega) + \sum_i \sigma_i^{inc} c_i S_i^s(Q,\omega) \quad (i,j = 1...n) \tag{6}$$

where σ is the total scattering cross-section, b and c are the bound scattering lengths and the atomic concentration of the actual scattering unit, respectively. The incoherent scattering due to the incoherent scattering cross-section σ^{inc} was added to the total dynamic structure factor (S_i^s is the self part of the dynamic structure factor of the element i under consideration) in order to show the quantity which can optimally be determined from the measured intensities. For samples with different scatterers another incoherent part comes in due to the disturbance of the interference of the neutron waves coming from different scatterers (often called the monotonous Laue scattering). With two exceptions [3, 4] in all experimentally investigations of the collective dynamics of fluids and amorphous solids done so far, the total dynamic structure factor has been obtained. The weighting factors in front of the partial dynamic structure factors $S_{i,j}(Q,\omega)$ can render the interpretation of the spectra more difficult and in the end only the partial dynamic structure factors will give a final answer.

As liquids are flowing, one can also use in place of the time dependent local number densities the current density $\boldsymbol{j}$ as dynamical variable [5],

$$\boldsymbol{j}(\boldsymbol{r},t) = \frac{1}{\sqrt{N}} \sum_{l=1}^{N} \boldsymbol{v}_l(t)\, \delta(\boldsymbol{r} - \boldsymbol{R}_l(t)) \tag{7}$$

defining thus the current correlation function

$$J_{\alpha\beta}(r,t) = V \left\langle j_\alpha(r',0) j_\beta(r'',t) \right\rangle \quad (\alpha,\beta = x,y,z, \text{ and } V \text{ the volume}). \tag{8}$$

As in the case of the dynamic structure factor, one obtains its spectrum by a double Fourier transform via the corresponding intermediate scattering function

$$J_{\alpha\beta}(Q,t) = \left(\frac{Q_\alpha Q_\beta}{Q^2}\right) J_l(Q,t) + \left(\delta_{\alpha\beta} - \left(\frac{Q_\alpha Q_\beta}{Q^2}\right)\right) J_t(Q,t) \quad (\alpha,\beta = x,y,z) \tag{9}$$

$$J_l(Q,\omega) = \int_{-\infty}^{\infty} e^{-i\omega t}\, J_l(Q,t)\, dt\,. \tag{10}$$

As neutrons and X-rays couple directly only to longitudinal excitations (to transverse only via Umklapp scattering), only the *longitudinal* part of its spectrum is measurable. Therefore the function has been split up in its longitudinal and transverse part, of which the latter one can be determined directly in computer simulations. Even though a different dynamical variable has been chosen in the case of the longitudinal current-current correlation function, its spectrum does not give new information, as it is linked to the dynamic structure factor [5]

$$J_l(Q,\omega) = \left(\frac{\omega}{Q}\right)^2 S(Q,\omega) \; . \tag{11}$$

In spite of this, J_l plays a *central* role in the *experimental* determination of dispersion relations of collective excitations in disordered matter: due to the multiplication of $S(Q,\omega)$ with ω^2 and due to the fact that $S(Q,\omega)$ falls off at larger ω more rapidly than any power of ω, J_l always has one or more maxima, which emphasize the weak and broad maxima in the dynamic structure factor. Compared to its usefulness, the danger of a misinterpretation (e.g. in case of $S(Q,\omega)$ being represented by a Lorentzian) seems acceptable.

3 Collective Excitations in Amorphous Solids

As we concentrate here on the collective excitations in amorphous solids, - omitting other dynamical properties characteristic for amorphous solids like the two-level-systems at energies mainly below 250 GHz (0.1 meV) and the low energy modes (so called Boson peak) at about 1 THz (4.186 meV) -, we assume that the results have been obtained at temperatures well below the Debye temperature Θ_D and that the harmonic approximation therefore is applicable. The most simple analogue to what we want to look at (noninteracting vibrations) is the one-dimensional undamped harmonic oscillator with its equation of motion

$$m\,\ddot{z} = -Kz \qquad \ddot{z} + \frac{K}{m}z = 0 \tag{12}$$

with the restoring force- or spring-constant K and m the oscillating mass. The Ansatz for the solution of this differential equation is

$$z(t) = a\ \sin(\omega t), \tag{13}$$

which transforms the equation to

$$-\omega_o^2 + \frac{K}{m} = 0, \tag{14}$$

which delivers the solution

$$\omega_0 = \sqrt{\frac{K}{m}}\,. \tag{15}$$

This result is fundamental for the interpretation of the measured spectra, *if* the assumption of rather similar interaction potentials between the different constituents of the sample investigated can be justified: then one can attribute the higher frequencies predominantly to the scatterer with the smaller mass and vice versa. This most simple argument will be used below.

The equations above describe the dynamics of one oscillator. We have to deal in principle with 10^{24}. We therefore start with the simplest case, the periodic lattice, where the calculations can be limited to the atoms in one unit cell. Following closely ref. [6] the calculation of harmonic vibrations in solids can be described in essentially the same way as in equ.12 to 15 above. The expansion of the interatomic interaction potential in the small displacements $\boldsymbol{u}(R)$ is limited to the second derivative within the harmonic approximation (corresponding to the force constant in the equations above).

$$\begin{aligned} U^{\text{harm}} &= \frac{1}{4}\sum_{R}\sum_{R'}\sum_{\mu,\nu}(u_\mu(\boldsymbol{R}) - u_\mu(\boldsymbol{R'}))\Phi''_{\mu\nu}(\boldsymbol{R}-\boldsymbol{R'})(u_\nu(\boldsymbol{R}) - u_\nu(\boldsymbol{R'})) \\ &= \frac{1}{2}\sum_{R}\sum_{R'}\sum_{\mu,\nu} u_\mu(\boldsymbol{R})D_{\mu\nu}(\boldsymbol{R}-\boldsymbol{R'})u_\nu(R') \qquad (\nu,\mu = x,y,z) \end{aligned} \tag{16}$$

$$\Phi''_{\mu\nu}(\boldsymbol{r}) = \frac{d^2\,\Phi(\boldsymbol{r})}{dr_\mu dr_\nu} \tag{17}$$

The equation of motion now reads

$$m\ddot{\boldsymbol{u}}(\boldsymbol{R}) = -\sum_{R'}\boldsymbol{D}(\boldsymbol{R}-\boldsymbol{R'})\boldsymbol{u}(\boldsymbol{R'}) \tag{18}$$

and can be solved with the Ansatz

$$\boldsymbol{u}(\boldsymbol{R},t) = \varepsilon e^{i(\boldsymbol{qR}-\omega t)} \tag{19}$$

where $\boldsymbol{q}$ is the wave-vector of the vibrational excitation. This leads to

$$m\omega^2\varepsilon = \mathbf{D}(\mathbf{q})\varepsilon \tag{20}$$

where $\boldsymbol{D}(\boldsymbol{q})$ is the *dynamical matrix*, the Fourier transform of $\boldsymbol{D}(\boldsymbol{r})$, and ε are the eigenvectors and ω the corresponding frequencies. The fact that the determinant of this set of equation has to be zero, if the equations are zero for all values of $\boldsymbol{u}(\boldsymbol{R},t)$, leads to the eigenvalues $\lambda_j(\mathbf{q})$, i.e. $\omega_j(\mathbf{q}) = \sqrt{(\lambda_j(\mathbf{q}))/\mathbf{m}}$, which are the dispersion branches we are looking for (comp. eq. (15)).

The equations used above are general within the harmonic approximation. If the sample has its own system of coordinates, like e.g. a single crystal does, the direction of the eigenvectors can be determined with respect to these coordinates. In addition, for a periodic crystal the calculation of vibrational frequencies can be restricted to the atoms in the unit cell and the irreducible part of the Brillouin zone. For amorphous solids, a structural model of the system has to be defined (within the same interaction potentials as used for the calculation of the vibrational excitations) and the dynamical matrix has to be diagonalized for the system as a *whole*. For not too complicated model potentials this can be done for models up to about 350 to 500 vibrating entities. Within *a priori* techniques, involving density functional theory (DFT), the models have to be about a factor of 5 smaller, but no predefined interactions are required. For larger systems than 500 vibrating entities (atoms, rigid molecules etc.) one can get very detailed information from the calculation of the local density-of-states (LDOS) using model potentials and the recursion method [7]. When speaking of 'dispersions' in disordered matter, one has to keep in mind that in contrast to what is done in single crystals, the excitations cannot be determined as a function of their respective wave vector $\boldsymbol{q}$ but only as a function of the modulus of the *experimental* wave-vector transfer Q!

From the eigenvectors and eigenfrequencies obtained via the diagonalization of the dynamical matrix, Bloch's spectral function $f(Q,\omega)$ can be calculated [7]

$$f(Q,\omega) = 2\omega \sum_{R} \sum_{R'} \sum_{j} (\varepsilon \cdot \mathbf{u}_j) e^{-i\mathbf{QR}} (\varepsilon \cdot \mathbf{u}'_j) e^{-i\mathbf{QR}'} \delta(\omega^2 - \omega_j^2)\,, \tag{21}$$

which is directly proportional to the one-phonon, more generally the one-excitation part of $S(Q,\omega)$

$$S^{1\text{phon}}(Q,\omega) = \frac{\hbar}{m} [\frac{(n(\omega)+1)}{\omega}] e^{-2W(\mathbf{Q})} (\varepsilon \cdot \mathbf{Q})^2 f(Q,\omega) \tag{22}$$

where $e^{-2W(\mathbf{Q})}$ is the Debye-Waller factor. For the dynamics of a solid treated in the harmonic approximation, one can distinguish (within approximations) between single and multiple excitations, and thus $f(Q,\omega)$ can be obtained from the measured dynamic structure factor after this has been reduced to its one-excitation part by multiple-excitation corrections. At high enough temperature (classical limit) the exponential in the Bose-Einstein occupation factor, $n(\omega)$, can be developed. In this case $f(Q,\omega)$ has a very simple connection with $S^{1\text{phon}}(Q,\omega)$, which is essentially the same as has the spectrum of the longitudinal current correlation function (see eq.11), often used in the case of fluids, - apart from the fact that here the *one-phonon* part of $S(Q,\omega)$ is used.

$$f(Q,\omega) = A\,(\frac{\omega}{Q})^2\, S^{1\text{phon}}(Q,\omega) \tag{23}$$

In Fig. 1 four different representations of the same cut at $Q = 21.5\,nm^{-1}$ through the measured $S(Q,\omega)$ are given: the data as determined from the measurement, the symmetrized version of the same data (where half of the

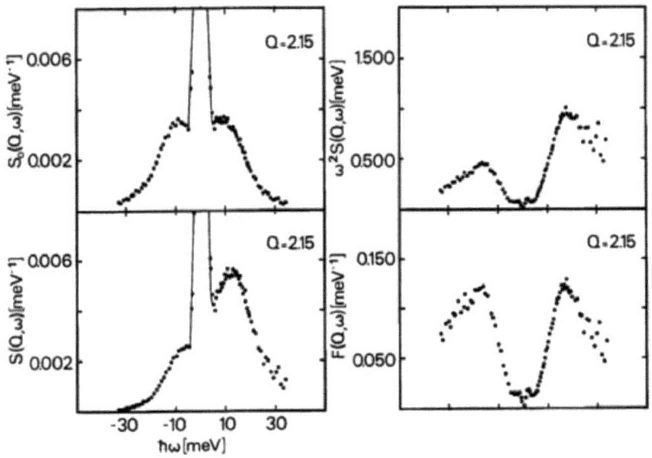

Fig. 1. Cut through the dynamic structure factor of amorphous MgZn at $Q = 21.5\,\mathrm{nm}^{-1}$ in four different representations: as determined (S), symmetrized (S_0), as spectrum of the longitudinal current correlation function $J_l(Q,\omega)$ and as spectral function $f(Q,\omega)$

detailed balance factor, $\exp(-\hbar\omega/2k_BT)$ has been attributed to $S(Q,\omega)$, enhancing the energy gain side ($\omega < 0$) and reducing the energy loss side ($\omega > 0$) of the spectra, a very common representation of $S(Q,\omega)$), the spectrum of the longitudinal current correlation function and the spectral function. In the latter case it was assumed that the spectrum is still dominated at this momentum transfer by the one-phonon processes.

For a multi-component system $f(Q,\omega)$ is again the *total* spectral function, if obtained from the total dynamic structure factor, as this is normally the case, and it represents again a weighted sum over the *partial* spectral function similar to $S(Q,\omega)$. In spite of this, $f(Q,\omega)$ is *extremely helpful in the determination of the dispersion of collective excitations in amorphous solids.*

However it should be clear that the way from measured data to the spectral function is a difficult one, as is roughly shown in Fig. 2. The measured intensity, already normalized and corrected for absorption and detector efficiency (every correction is a threedimensional function) still contains the intensity from multiple scattering, M, and the effect of the resolution function, $R(\omega)$. If one succeeds in correcting for these, one still has to separate the incoherent from the coherent scattering and the one-excitation contributions from the multi-excitations before one can compare the experimental results with the *total* spectral function.

In using eq. (23) one has to keep in mind, that only for phonons, which are presented by δ functions, the dispersions obtained from (23) are the same as obtained from the phonon peaks in the dynamic structure factor. *For disordered systems this is not the case*, because their excitation spectra are

$$\left[\tilde{S}_{exp}(Q,\omega) + \widetilde{MS}(Q,\omega)\right] * R(\omega)$$

$$\downarrow$$

$$\tilde{S}_{exp}(Q,\omega) = \tilde{S}(Q,\omega) + \left(\tilde{S}_s(Q,\omega)\right)$$

$$\downarrow$$

$$\tilde{S}(Q,\omega) = \frac{4\pi}{\sigma^{sc}} \sum c_i c_j b_i b_j \tilde{S}_{ij}(Q,\omega) = \tilde{S}_{1phon}(Q,\omega) + \tilde{S}_{mult.ph.}(Q,\omega)$$

$$\downarrow$$

$$\tilde{S}_{1phon}(Q,\omega) = \frac{1}{\hbar\omega(e^{\beta/2}-e^{-\beta/2})} \frac{\hbar^2 Q^2}{\overline{M}} e^{-2\overline{W}} \underline{F(Q,\omega)}$$

$$F(Q,\omega) = \sum_{ij} e^{-(W_i+W_j)} b_i b_j \frac{c_i c_j}{\sqrt{M_i M_j}} f_{ij}(Q,\omega) \Big/ \sum_{ij} e^{-(W_i+W_j)} b_i b_j \frac{c_i c_j}{\sqrt{M_i M_j}}$$

Fig. 2. Rough sketch of the path from measured, normalized and partly corrected data to the total spectral function. The summation indices correspond for $S(Q,\omega)$ to those of eq.6

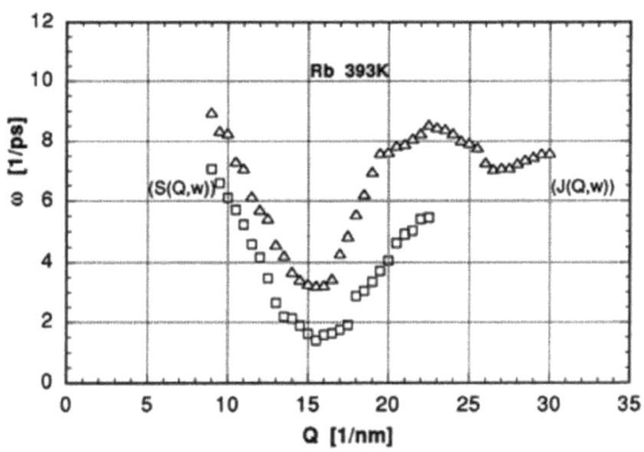

Fig. 3. Dispersions obtained for liquid Rb near its triple point from a fit of a model to the measured dynamic structure factor (lower dispersion) and from the maxima of the longitudinal current correlation function $J_l(Q,\omega)$ (upper dispersion)

broad and therefore shift to higher energies when multiplied by ω^2! This is shown at the example of the dispersions obtained from fits of a model to the dynamic structure factor and from the maxima of $J_l(Q,\omega)$ for liquid Rb [8] in Fig. 3.

The width of the excitations in disordered matter has at least two different contributions: there is the distribution of force constants due to the statistical distribution of distances between the vibrating units and the lifetime of the excitation is considerably shorter in disordered matter, especially in fluids, than in crystals, because of the topological disorder of the units, from which propagating excitations are scattered.

4 Collective Excitation in Fluids

For fluids, because of the impossibility to use the harmonic approximation, one has to start from a slightly enlarged equation compared to the one for the undamped harmonic oscillator used at the beginning of section 3. Damping of the modes via a friction constant f and external driving forces with an initial amplitude A_0 and frequency ω (in the liquid produced by the statistical fluctuating forces (braking and pushing at the same time) exerted by the environment on a tagged particle), have to be included

$$\ddot{z} + \frac{f}{m}\,\dot{z} + \frac{K}{m}\,z = A_0 e^{i\omega t}\ . \tag{24}$$

The Ansatz now contains the phase ϕ between the external driving force and the frequency of the resonator

$$z = a\, e^{i(\omega t - \phi)} \tag{25}$$

which leads to

$$(-\omega^2 + \omega_o^2) + i\,\delta\,\omega = \frac{A_o}{a}\, e^{i\phi} \tag{26}$$

where $\delta = f/m$. A very similar equation to the equation of motion above has been used by Langevin to describe the Brownian motion of a tagged particle in a fluid [5]:

$$\dot{v}(t) + \frac{f}{m}\,v(t) = \frac{F(t)}{m} \qquad \text{with}\, v(t) = \dot{r}(t)\ . \tag{27}$$

In order to include besides the diffusive motion also the excitations, restoring forces have to be included, which leads to the *generalized Langevin equation* (GLE) for a viscous *and* elastic medium at the frequencies of the collective excitations (THz-region)

$$\dot{a}(t) - i\,\Omega\, a(t) + \int_0^t dt'\, M(f,t') a(t-t') = F(t)\ . \tag{28}$$

Here a(t) is a vector containing the dynamic variable chosen, Ω now is a frequency matrix, containing the vibrational frequencies, F(t) is the statistical force acting on the particle under consideration and the damping now is included in the *memory function* $M(f,t')$, again a matrix, which includes the *history* of the physical processes which have influenced the 'life' of the particle during the time covered within the limits of the integral. In a loosely description: M remembers what the F(t) have 'done' to the particles, therefore M is proportional to the correlation of the statistical forces themselves:

$$M(T) = \frac{\int_0^\infty < F(0)\, F(t)\, dt}{< |a|^2 >} \ . \tag{29}$$

Solving the generalized Langevin equation in an approximate manner requires a model for the memory function. This approximative solution is usually started with a Laplace transformation from t- to s-space (mirror-space), where s now is a *complex* energy variable. For a dynamical variable a(t) this is

$$a(s) = \int_0^\infty a(t) e^{-st}\, dt \ . \tag{30}$$

One could have set up the GLE equally well for a correlation function ψ in place of dynamical variables. Using this one gets

$$\psi(s) = \int_0^\infty \psi(t)\, e^{-st}\, dt \ . \tag{31}$$

A purely formal solution of this equation leads to

$$\psi(s) = \frac{1}{(s - i\,\Omega + \Psi(s))} \tag{32}$$

where $\Psi(s)$ now is a new correlation function, which contains derivatives of ψ. For Ψ one can repeat the circle and arrives this way at the continued fraction representation of the approximate solution of the generalized Langevin equation:

$$\psi(s) = \cfrac{1}{s - i\,\Omega_o + \cfrac{\Delta_1}{(s - i\,\Omega_1 + \cfrac{\Delta_2}{(s - i\,\Omega_2 + \cfrac{\Delta_3}{(s - i\,\Omega_3 +}}}} \tag{33}$$

Usually for the unknown Δ_i the ratio of neighbouring even frequency moments of the dynamic structure factor (ω_0, ω_l, see below) are inserted and the continued fraction is closed at the level of the fourth frequency moment, the last one, which does not require three-particle or higher correlation functions and higher derivatives of the interaction potential than the second.

Besides the two dispersions , which can be obtained from the Q-dependent maxima in the dynamic structure factor and the spectrum of the longitudinal current-current correlation function, which, - exactly as in the case of the two dispersions obtained from $S(Q,\omega)$ and $F(Q,\omega)$ for amorphous solids -, are the same concerning their *shape* but not concerning their frequencies at the same Q (see Fig. 3), two further dispersions can be obtained from the ratio of two subsequent (even) frequency moments of the dynamic structure factor.

$$\langle \omega^n \rangle = \int_{-\infty}^{\infty} \omega^n \, S(Q,\omega) d\omega \tag{34}$$

The first three even frequency moments for the dynamic structure factor of a classical system are:

$$\langle \omega^0 \rangle = \int_{-\infty}^{\infty} \omega^0 \, S(Q,\omega) d\omega = S(Q) \tag{35}$$

i.e.the static structure factor.

$$\langle \omega^2 \rangle = \int_{-\infty}^{\infty} \omega^2 \, S(Q,\omega) d\omega = \frac{k_B T \, Q^2}{M} = (Q \, \mathrm{v}_0)^2 \tag{36}$$

which is the mean energy transfer from particles of mass M moving with a thermal velocity $v_0 = \sqrt{k_B T/M}$ at temperature T.

$$\begin{aligned} \langle \omega^4 \rangle &= \int_{-\infty}^{\infty} \omega^4 \, S(Q,\omega) d\omega \\ &= \langle \omega^2 \rangle \left[3 \langle \omega^2 \rangle + \left(\frac{n}{M} \right) \int_0^{\infty} g(r) \left(\frac{d^2\phi}{dz^2} \right) (1 - \cos\{Qz\}) d^3 r \right] \end{aligned} \tag{37}$$

ϕ is the pair potential and $g(r)$ the radial distribution function, the Fourier transform of the static structure factor . The second part corresponds to the Fourier transform of the second derivative of the pair potential, which in case of a harmonic solid would correspond to the elements of the dynamical matrix in the phonon calculations (see eq. 20). For the disordered system this has to be weighted with the probability g(r) to find atoms in a distance r.

The *adiabatic* dispersion ω_0, which is equal to the *isothermal* dispersion as long as $\gamma = c_p/c_v \approx 1$, is the ratio of the second to the zeroth frequency moment

$$\omega_0^2(Q) = \frac{\langle \omega^2 \rangle}{\langle \omega^0 \rangle} = \frac{k_B \, T \, Q^2}{M \, S(Q)} = \frac{(Q \, v_0)^2}{S(Q)} \, . \tag{38}$$

It should start off at smallest Q with a slope equal to the isothermal sound velocity and it continues with the shape of a dispersion curve. The variation of this shape however is completely determined by the (reciprocal) shape of the structure factor, as the rest is a smooth increasing function proportional to Q^2. The 'dispersion like' shape of the isothermal dispersion therefore stems from the *static* properties of the sample. These are also determined by the interatomic interactions, which however enter the isothermal dispersion in a rather indirect manner.

The *high frequency* dispersion:

$$\omega_l^2(Q) = \frac{\langle\omega^4\rangle}{\langle\omega^2\rangle} = 3\left\langle\omega^2\right\rangle + \left(\frac{n}{M}\right)\int_0^\infty g(r)\left(\frac{d^2\phi}{dz^2}\right)(1-\cos\{Qz\})\,d^3r \qquad (39)$$

the second term is the only part, which contains any information on the interatomic interaction, the restoring forces, via the effect of the second derivative of the pair potential on the dynamics of the atoms in the disordered system. 'High frequency' means here that the time of one vibration is too short for a complete relaxation of the viscosity η, i.e. $\omega\tau_\eta >> 1$. In the case of the adiabatic dispersion the viscosity has time to relax, i.e. $\omega\tau_\eta << 1$, however the time is too short for the relaxation to thermal equilibrium, i.e. $\omega\tau_T >> 1$. Only in the case of the isothermal dispersion, both relaxation can take place between subsequent vibrations, i.e. also $\omega\tau_T << 1$.

The high frequency dispersion contains two contributions, of which the first one is called the *free particle dispersion*, because it corresponds to the $\omega(Q)$ dependence of free particles, i.e. it increases proportional to Q^2 and to T. Only via the second part one can study the influence of the interatomic potential on the atomic dynamics. As an example, the isothermal and the high-frequency dispersions as obtained from the ratio of the experimentally determined frequency moments of the dynamic structure factor of liquid Rb is given (besides others) in Fig. 4.

These dispersions can be determined either from the ratio of the measured frequency moments obtained by integration of the dynamic structure factor, which requires a very accurate determination of $S(Q,\omega)$, or from a fit of a model to $S(Q,\omega)$, which contains these two quantities as Q-dependent fit parameters. As $\omega_l(Q)$ has two distinctly different contributions, one can use it also for the interpretation of dispersions determined in a different way.

5 Collective Excitation in Fluids: Some Examples

In the following some result obtained by NIS will be discussed, leaving aside the numerous results from computer simulations, in which either model potentials for the interactions in the molecular dynamics calculations or Car-Parrinello type of *ab initio* simulations have been applied. Concerning experiments, one has to keep in mind that in contrast to experiments with single crystals, where on can measure the same dispersion in different Brillouin zones, for disordered samples the dispersion (if it can be determined at all) does not repeat, but is different at each Q. The experiments are often done at lowest Q, because there one expects the largest probability to

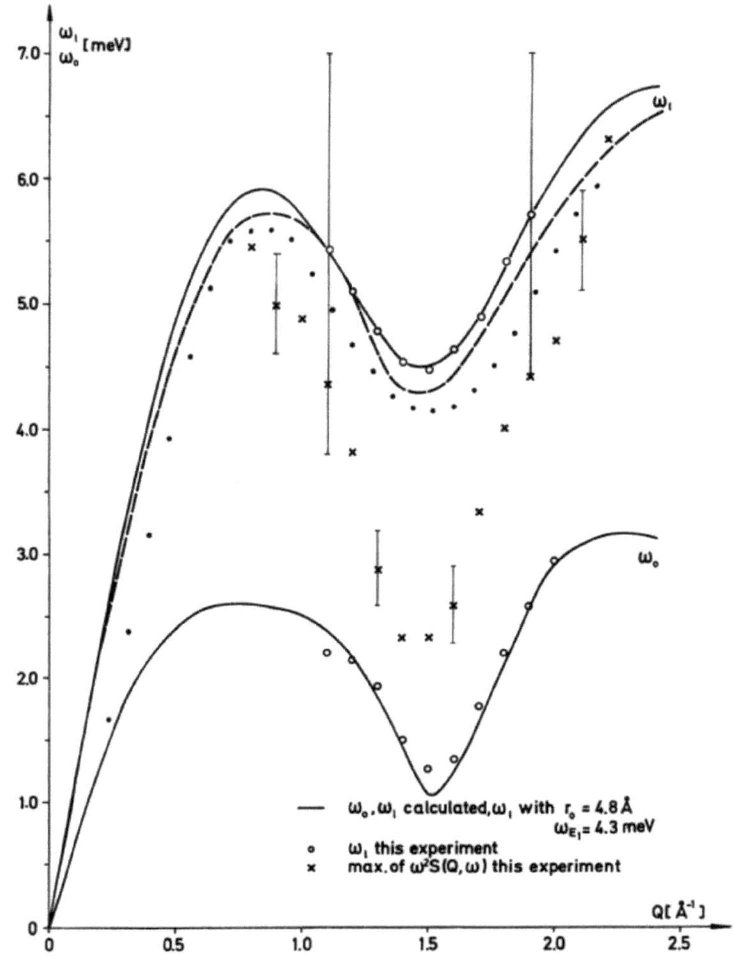

Fig. 4. Dispersions obtained for liquid Rb near its triple point from the ratio of the experimentally determined even frequency moments. Isothermal dispersion $(\langle\omega^2\rangle / \langle\omega^0\rangle)$(lower dispersion) and high frequency dispersion $\langle\omega^4\rangle / \langle\omega^2\rangle)$ (upper dispersion)

find collective excitations as peaks or shoulders in the inelastic part of the dynamic structure factor, - because in the hydrodynamic limit of small Q and ω one always finds such excitations there. For crystals the smallest Q are within the first Brillouin zone and one therefore speaks in such case of Brillouin-scattering. In disordered systems we do not have Brillouin zones, however the short range order (SRO) leads to what is called *pseudo-Brillouin zones* (of lowest order)(PBZ), in which the measured dispersions behave similar (see Fig.4) to the dispersion in the Brillouin zones of crystals. Here the maxima of the static structure factor at Q_P, Q_{2P} etc. act like 'smeared out'

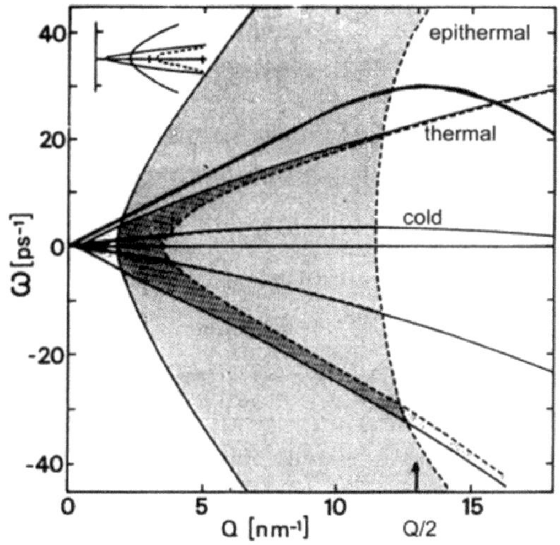

Fig. 5. Dynamical ranges of NBS experiments with cold, thermal and epithermal neutrons. Only in the latter case the dispersion of amorphous MgZn (*upper thick line*) is largely covered in the dynamical range of the NBS experiment

reciprocal lattice "points" and at half the distance between $Q = 0$ and Q_P one finds the zone boundary of the first PBZ. Experiments with Q-values below this zone boundary are therefore also called Brillouin scattering experiments, in the case of NIS: neutron Brillouin scattering or NBS experiments.

In order to *cover* a linear dispersion inside the dynamical region of such an experiment done at small Q, i.e. $\omega(Q) = cQ$, with c the velocity of the excitation, the velocity of the incident neutrons, v_0, corresponding to E_0, has to be slightly (optimal $\sqrt{2}$) larger than c. As the absolute resolution decreases rapidly with increasing E_0, such experiment are increasingly more difficult the higher c (see Fig. 5). This problem does not exist for X-ray inelastic scattering (XIS) as there the energy of the incident radiation is around 21 keV.

5.1 Observability of Collective Excitations in Fluids

To measure phonon dispersions in single crystals by NIS, one needs a sufficiently large crystal of good quality, i.e. sufficiently small mosaic width. Then one can measure phonon peaks in $S(Q, \omega)$ of similar quality in metallic crystals as well as in molecular, ionic or covalent crystals.

This is by no means the case for liquids, - the dispersion in amorphous solids has not been investigated well enough to draw any final conclusion there. For simple liquid metals one can obtain the dispersion relations from peaks and shoulders in the inelastic part of the dynamic structure factor up to Q-values of about 0.85 Q_P, Q_P being the Q-value corresponding to the main peak of the static structure factor S(Q). For molecular liquids like e.g.

the liquid heavier rare gases (see e.g. references in [9]) the visibility of such inelastic peaks cease at a Q-value of about 0.06 Q_P, thus about 15 times lower compared to Q_P. To our knowledge definite inelastic peaks have not been observed in an $S(Q,\omega)$ measured by neutron scattering from an ionic liquid, e.g. a molten salt. Likewise little information exists about inelastic peaks in the dynamic structure factor of a liquid with predominantly covalent bonds [10].

There have been several attempts to look for an explanation. From the many arguments put forward, here three of them will be briefly mentioned without any intention in this selection:

One of these explanations is based on a hand-waving but physically reasonable argument, put forward by Wolfgang Götze and collaborators. It states that the higher value of the static structure factor in the low Q-limit shows that argon near the triple point is more compressible than a metal,

$$\chi_T = (nk_BT)^{-1}S(0) \tag{40}$$

and thus collective modes should be damped more easily in Ar than in an Alkali metal. It seems difficult to apply this argument to molten salts, where even at the lowest Q-values of the investigations up to now no inelastic peak shows up in the inelastic part of $S(Q,\omega)$ without existence of an extraordinary compressibility.

Lewis and Lovesey [11] investigated this question by computer-simulations and established a necessary existence-criterion for collective modes in liquids, which says that one should be able to observe collective modes in the dynamic structure factor of a simple liquid up to Q-values for which $\omega_l(Q)$ does not exceed $3\omega_0(Q)$. That this criterion only reflects the *necessary* condition was shown by Söderström and collaborators shortly after the publication of this existence criterion, when they found in liquid lead collective excitations up to Q-values which exceeded far this limit of $3\omega_0(Q)$ [12].

Hahn and Mountain [13] investigated this question again by computer simulations, making molecular dynamics (MD) simulations for liquids using a Lennard-Jones potential, characteristic for the interaction in liquid rare gases, and then with an electronically screened Coulomb potential characteristic for the interaction in Rb, in which experimentally one finds in fact collective excitations (see Fig.3). In their liquid metal model, they find collective excitations to much higher Q-values in $S(Q,\omega)$ than in the Lennard-Jones liquid, and this is still the case, when they replace the hard slope r^{-12} in the Lennard-Jones potential by the softer repulsive part of the Rb model potential and use this for the "metal". They concluded from this that it is the repulsive part of the interaction potential, which decides up to which Q-values one is able to observe collective modes in the dynamic structure factor of simple liquids.

However most recently Livia Bove [14] and collaborators have show by their investigation of a mixture of K and Cs, which was chosen to be electronically equivalent to Rb, that one can *scale* the dispersion curves of liquid

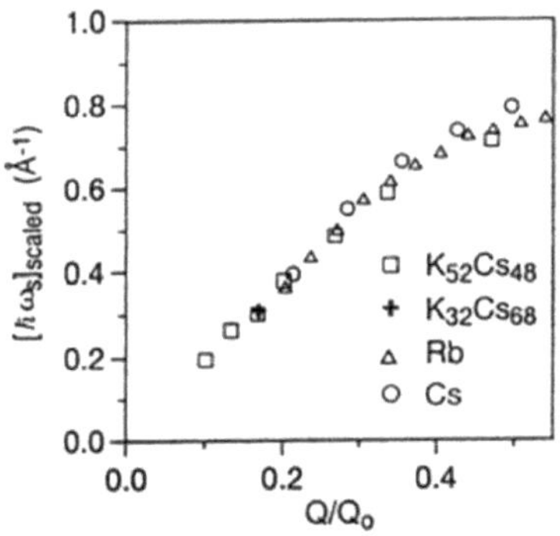

Fig. 6. Dispersions of some liquid alkali metals and a K-Cs mixture scaled by their effective masses and electronic screening

Rb, Cs and K-Cs (see Fig.6), after having taken their differences in effective masses and electronic screening into account. This puts the weight partly away from the repulsive core to the conduction electrons, which means that also the repulsive part of the potential has lost in weight in this argumentation.

The scaling is done using the model of Bohm and Staver , who treat the metallic liquid as an ionic gas screened by a homogeneous electron gas with dielectric function $\epsilon(Q)$. The collective excitations in the ion gas are given by the ion gas plasmons of frequency Ω_p

$$\omega^2(Q) = \frac{\Omega_p^2}{\epsilon(Q)} \,. \tag{41}$$

The plasmon frequency is given by

$$\Omega_p^2 = \frac{4\pi\, n\, e^2}{M} \,, \tag{42}$$

with n and e the electron density and the charge of the electron, respectively. For alloys, an effective mass has to be used instead of the atomic mass, M, of the pure metal

$$\Omega_p^2 = \frac{4\pi\, e^2}{M_{eff}} \tag{43}$$

$$M_{eff} = \frac{M_1\, M_2}{n_1\, M_1 + n_2\, M_2} < \overline{M} \tag{44}$$

and if $\epsilon(Q)$ is not the same for the constituents of the alloy also $\epsilon(Q)$ has to be scaled.

In view of the fact that collective excitations show up (or not) so differently in liquids with different types of interaction Jahn, Pratesi and Suck [15] have investigated the atomic dynamics in "liquids with competing interactions" , as we have named them, binary liquid alloys of an alkali element with a polyvalent element (Zintl systems), in which the dominant *type* of

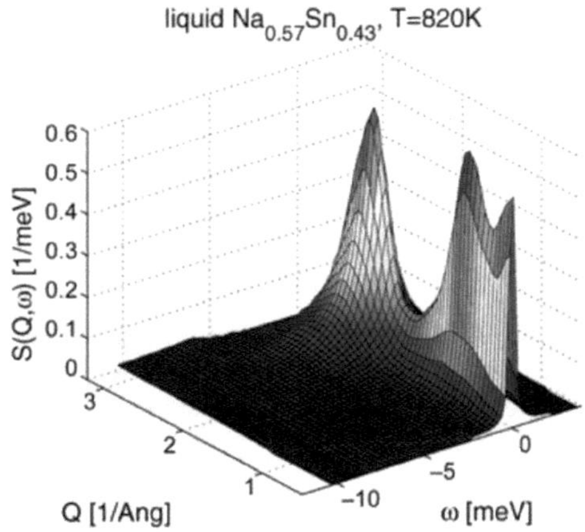

Fig. 7. Dynamic structure factor of liquid $Na_{57}Sn_{43}$ measured at 820 K

interaction changes with composition. The interaction is metallic at 95 at% alkali metal, starts to be ionic from approximately 85 to 80 at% alkali and becomes partly covalent from about 60 at% alkali. Thus one can approach the concentration regions, where ionic and covalent bonding is dominant, from the side, where collective excitations are still visible in $S(Q,\omega)$, i.e. in the Alkali rich region of the phase diagram. Systems investigated so far are RbSb, KSb and NaSn. As an example in Fig.7 the total dynamic structure factor of NaSn near the equiatomic composition, i.e. next to maximum of the covalent contribution to the binding forces, is shown. One clearly observes in front of the main peak of the - on energy scale - relatively smooth $S(Q,\omega)$ a strong prepeak due to some persistent medium range order and a small increase towards Q=0 due to the incoherent scattering from Na and possibly also from some concentration fluctuations. The - in energy - very sharp pre-peak demonstrates the long living medium range atomic configuration in this Zintl alloy. As a consequence of its sharpness this peak appears as a small hump in $S(Q)$ after integration at constant Q (zeroth energy moment).

The dispersions shown in Fig. 9 were obtained from $J_l(Q,\omega)$ [16], of which a cut (at different Na- concentrations) is shown in Fig. 8.

As long as the system is dominated by the metallic interaction, dispersions as for a pure liquid Alkali metal, here Na are observed. But as soon as ionic forces take over, it becomes most difficult, to determine the dispersions even from $J_l(Q,\omega)$. In the concentration range, where ionic and later partly covalent forces dominate, the dispersions are shifted to higher Q values as are the maxima of the static structure factors, indicating a stronger and with this shorter bonding, and they are shifted to higher energies, likewise demonstrating the stronger inter-atomic interactions in this concentration range.

For the first time a splitting of the dispersion obtained from the *total* $J_l(Q,\omega)$ could be observed. The ratio of the energies of the two corresponding dispersion branches are in reasonable agreement with the square root of the ratio of the masses of the two elements Na and Sn (see eq. (15)).

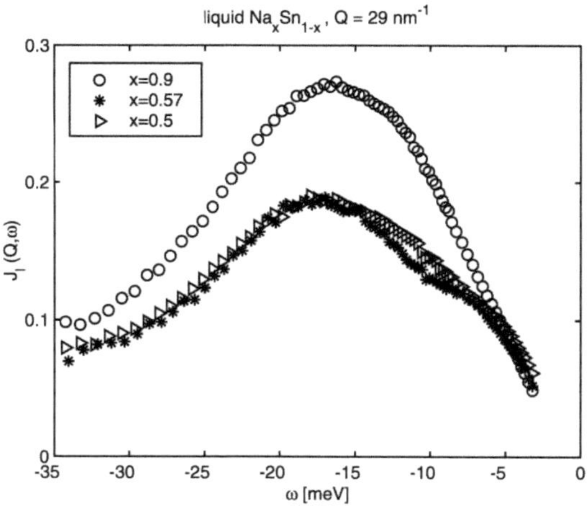

Fig. 8. Cut through $J_l(Q,\omega)$ at $Q = 29\,\mathrm{nm}^{-1}$ for liquid NaSn with different concentrations. The appearance of a second maximum in $J_l(Q,\omega)$ at lower energies in the concentration region of partially covalently bound Sn is obvious

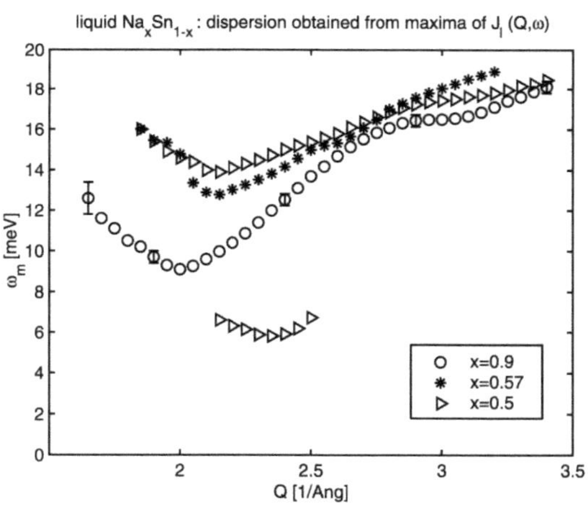

Fig. 9. Dispersions of NaSn at different concentrations. While the weighting factors of the partial dynamic structure factors change continuously with concentration, the abrupt changes of the dispersions with concentration show the change in the strength of the interaction when its character changes from dominant metallic to ionic to partially covalent

These observations for NaSn are strongly corroborated by the results obtained for the likewise polyvalent liquid RbSb [15], except that there, instead of a splitting of the dispersion curves at Sn-concentrations above 43 at%, for RbSb a second energy band is found at higher energy in the spectrum of the velocity auto- correlation function, when reaching the same concentration range of the polyvalent metal. Concerning the observability of collective modes as density fluctuations in $S(Q,\omega)$ these results do not yet answer the question, but just confirm what was known before: as soon as one loses the metallic interaction one loses the peaks and shoulders in the inelastic part of $S(Q,\omega)$. This questions obviously requires a very systematic investigations at the lowest Q-values, which can be reached within the necessary dynamical range of the experiment (see Fig. 5).

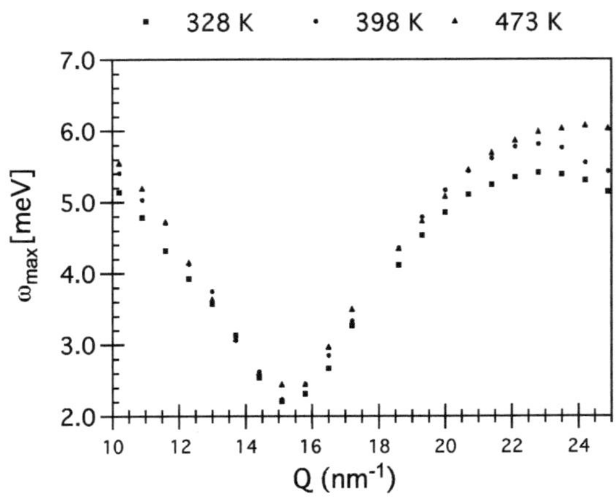

Fig. 10. Dispersions of liquid Rb measured at different temperatures. While the dispersions change little next to the their minimum (first maximum of S(Q)), the change of the free particle dispersions (at the highest Q-values of these experiments) is obvious

5.2 Temperature and Pressure Dependence of the Dispersion of Collective Excitations in Simple Metals

Pratesi, Suck and Egelstaff [17] have shown - in reanalysing previous data - that the dispersion of collective modes in different Q-regions is sensitive to different parameters: when one changes temperature or density, one part of the dispersion is rather little affected by this changes and an other part reacts strongly. This was demonstrated at the example of the dispersion of liquid Rb (as this has the most favourable ratio of the coherent to the incoherent scattering cross-section of all alkali metals), measured on a *grid* in the T-p phase diagram, i.e. for temperatures between 328K and 473K and densities between 1.394 and 1.486 g/cm^3.

The results obtained can be discussed using the high frequency dispersions (see eq. (39)): When one changes the temperature, the main reaction is in the region at highest Q-values (see Fig.10), because at highest Q the free particle dispersion , which is proportional to Q^2, is dominating the potential part and the free particle dispersion contains T as a parameter.

This strong influence of the temperature on the free particle dispersion continues, if one raises the temperature towards the critical temperature, as has been shown experimentally by Winter and collaborators [18] and via calculations by Kahl and collaborators [19].

In the opposite case, when one varies the pressure and with this the density and the inter-atomic distances and with this the inter-atomic potential, the high Q part of the dispersion is little affected. But the dispersion around and below the first peak of the static structure factor, i.e. at Q-values at and

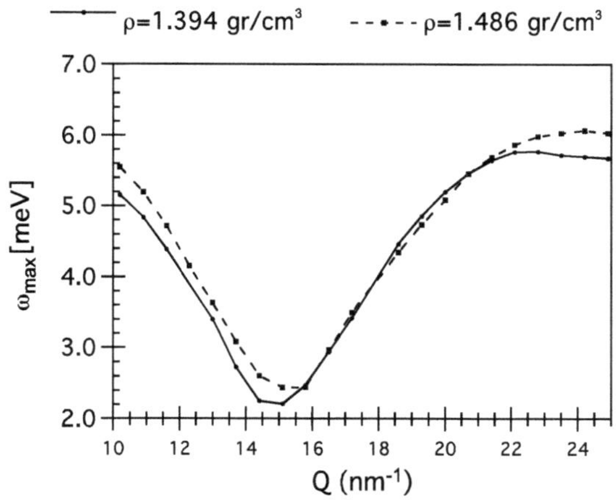

Fig. 11. Dispersions of liquid Rb measured at different densities. While the dispersions change little at highest Q they change clearly next to the their minimum (also because the first maximum of S(Q) shifts with density)

below Q_P, which was nearly unaffected by the temperature change, reacts considerably on the change of the potential, as is shown in Fig.11.

This result shows at the same time very clearly, that effects from the interatomic potential can be studied only at Q-values at and below Q_P, - and this explains the origin of the interest in such experiments. At $Q > Q_P$ already mean-field theories are quite successful in describing the atomic dynamics in simple liquids. Below Q_P, but still outside the hydrodynamic limit, even the most advanced theories like the kinetic theory and the mode coupling theory [5] are still unable to describe the collective modes in liquids in a *quantitative* manner.

Presently it is also not clear, how far down toward smaller Q this sensitivity of the dispersion to changes of the interatomic potential will go. Will it increase or decrease when one measures collective excitations down to much smaller Q-values, - still outside the hydrodynamic limit and below the critical point -, when the wavelength of the excitations becomes longer and longer and takes a mean value over many atoms? In principle there could even be a Q-range of maximal sensitivity to potential changes.

6 Collective Excitation in a Metallic Glass

Experimental investigations of the collective excitations in amorphous solids are still very rare. One reason for this is that such experiments are very difficult to perform, as the sound velocity is normally higher than in liquids, which forces to use higher incident neutron energies and with this reduced resolution. On the other hand one is not hampered by a broad quasielastic line

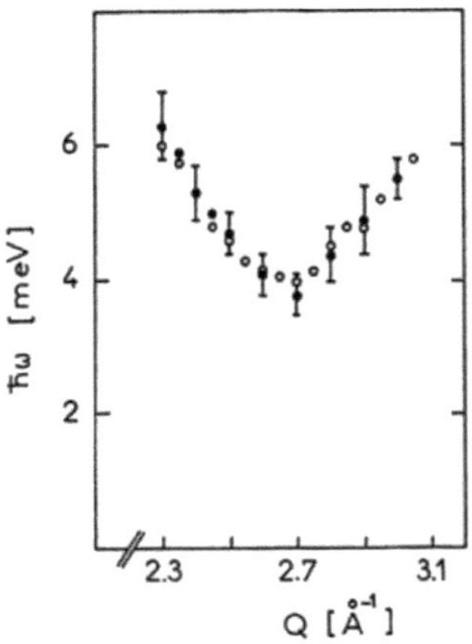

Fig. 12. Dispersions of collective excitations near Q_P in the metallic glass $Mg_{70}Zn_{30}$ showing a 'roton like' minimum

which tends to cover part of the inelastic part of $S(Q,\omega)$ in the investigation of liquids. To my knowledge, the first serious investigation of the dispersion of collective excitations in an amorphous solid, here the metallic glasses MgZn, was spread over many years. First results [20] were misinterpreted as being due to transverse modes, while they were longitudinal excitations. Computer experiments showed later [21, 22] that the transverse branch should hardly be visible in the Q-range of that experiment and it should show no sloping behaviour after the border of the first PBZ. Later the 'roton-like' dispersion around Q_P was found [23], which is shown in Fig.12 for the case of MgZn as an example. After that it took some time [24, 25] and improvement of experimental techniques, until the dispersion branch could be followed down into the first PBZ (see Fig. 13) [25]. Computer simulations of this glass [22] predict two dispersion branches, a branch for acoustic excitations and one for optic excitations. Up to now only one branch was found in the low Q region and further experiments at even lower wavevector transfers Q are necessary to find a complete answer. Towards higher Q-values, the measured dispersions [26] are in reasonable agreement with the afore mentioned computer simulations. The acoustic branch below and the optic branch above both agree reasonably well with data measured at 6K for the metallic glass $Mg_{70}Zn_{30}$ in the Q-range next to and above Q_P. Unfortunately however, these two branches do *not* correspond to the acoustic and optic excitations in the complete MgZn system, but to the excitations seen in the *partial* dynamic structure factors: up the lighter mass Mg-Mg correlations, down the heavier Zn-Zn correlations. Thus one does not see the excitations in the MgZn system but in the Mg- and in the Zn-*subsystem* separately. Only at the lowest Q-value accessible in the simulation, $3\,\mathrm{nm}^{-1}$, both elements vibrate together in phase or in counter-phase. The excitations can only there be described by plane waves.

Thus it would be very interesting to study in detail this transition from the excitations in the whole system to those in the subsystems, - but presently

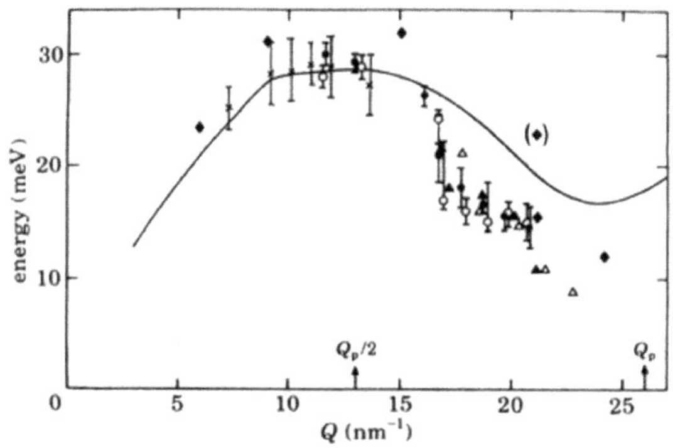

Fig. 13. Dispersions of collective excitations below Q_P down to Q-values within the first PBZ for the metallic glass $Mg_{70}Zn_{30}$

one is still a factor of 3 above this region! The calculations of Hafner show that the experiment should be possible as the excitations should become considerably narrower and correspondingly more intense, when one would intrude into this low-Q-region [22].

7 Conclusions

The results presented above show the difficulty in the investigations of collective excitations in disordered systems like fluids and amorphous solids and the incompleteness of our knowledge of these excitations up to now. While excitations in simple liquid metals have been investigated quite successfully already and first systematic results for metallic glasses have been obtained, corresponding results for liquids with ionic and covalent bonds are still sparse, and for most of the amorphous solids, even for most of the glasses, hardly any experimental results have been collected. Still, even if collective excitations do not give us the most complete information, like vibrational density-of-states do, they provide us with the *most detailed* information, and they are therefore extremely necessary for a detailed comparison with simulations and theoretical calculations in order to get an as complete understanding of the atomic dynamics of disordered systems as possible.

References

1. J.-B. Suck: Int. J. Mod. Phys. B **7**, 3003 (1993)
2. S.W. Lovesey: *Theory of Neutron Scattering from Condensed Matter*, (Oxford University Press 1984)
3. G.F. Syrykh, M.N. Khlopkin, M.G. Zemlyanov, A.S. Ivanov, H. Schober: Appl. Phys. A **74**, S951 (2002)

4. H. Uhlig, J.-B. Suck: priv. comm.
5. J.-P. Boon, S. Yip: *Molecular Hydrodynamics*, (McGraw-Hill New York 1980)
6. N.W. Ashcroft, N.D. Mermin: *Solid-State Physics*, 3rd edn (Saunders College Publishing New York 1976)
7. J. Hafner: *From Hamiltonian to Phase Diagrams*, (Springer, Berlin, Heidelberg, New York, London, Paris, Tokyo 1987)
8. P. Chieux, J. Dupuy-Philon, J.-F. Jal, J.-B. Suck: J. Non-Cryst. Solids **205-207**, 370 (1996)
9. J.-B. Suck: J.Phys.: Condens. Matter **3**, F73 (1991)
10. S. Jahn: The atomic dynamics of liquids with competing interactions. PhD Thesis, Technical University Chemnitz, Chemnitz (2003)
11. J.W.E. Lewis, S. Lovesey: J. Phys. C: Solid State Phys. **10**, 3221 (1978)
12. O. Söderström, J.R.D. Copley, J.-B. Suck, B. Dorner: J.Phys. F **10**, L151 (1980)
13. S. Hahn, R. Mountain, C.S. Hsu, A. Rahman: Phys. Rev. A **22**, 767 (1980)
14. L. Bove, F. Sacchetti, C. Petrillo, B. Dorner: Phys. Rev. Lett. **85**, 5352 (2000)
15. S. Jahn, G. Pratesi, J.-B. Suck: Appl. Phys. A **74**, S1664 (2002), Phys. Rev. Lett. **92** 185507 (2004)
16. S. Jahn, J.-B. Suck: J. Non-Cryst. Solids **312-314**, 134 (2002), Europhys. Lett. **67** 793 (2004)
17. G. Pratesi, J.-B. Suck, P.A. Egelstaff: J. Non-Cryst. Solids **250-252**, 91 (1999)
18. R. Winter, C. Pilgrim, F. Hensel, C. Morkel, W. Gläser: J. Non-Cryst. Solids **156-158**, 9 (1993)
19. G. Kahl, S. Kambayashi, G. Nowotny: J. Non-Cryst. Solids **156-158**, 15 (1993)
20. J.-B. Suck, H. Rudin, H.-J. Güntherodt, H. Beck: J. Phys. C **13**, L1045 (1980)
21. G. Grest, S. Nagel, A. Rahman: Phys. Rev. Lett. **49**, 1271 (1982)
22. J. Hafner: J. Phys. C: Solid State Phys. **16**, 5773 (1983)
23. J.-B. Suck, H. Rudin, H.-J. Güntherodt, H. Beck: Phys. Rev. Lett. **50**, 49 (1983)
24. J.-B. Suck P.A. Egelstaff, R.A. Robinson, D.S. Sivia, A.D. Taylor: Europhys. Lett. **19**, 207 (1992)
25. C.J. Benmore, S. Sweeney, P.A. Egelstaff, R.A. Robinson, J.-B. Suck: J. Phys.: Condensed Matter **11**, 7079 (1999)
26. J.-B. Suck, H. Rudin: Atomic dynamics of metallic glasses. In *Glassy Metals II*, ed by H. Beck, H.-J. Güntherodt (Springer, Berlin Heidelberg New York 1983) pp 217-260

Theories of the Structural Glass Transition

Rolf Schilling

Institut für Physik, Johannes Gutenberg-Universität Mainz,
Staudinger Weg 7, 55099 Mainz, Germany
Rolf.Schilling@uni-mainz.de

1 Introduction

Equilibrium phase transitions, e.g. the transition at 0°C from water to an ice crystal, are common phenomena in nature. Such phase transitions between a disordered high-temperature phase and an ordered low-temperature one are rather well understood and can be theoretically described within statistical mechanics. Given the interaction between the species (particles, spins, etc.) the partition function can be calculated, in principle. Its logarithm yields, e.g. for a canonical ensemble, the free energy which is singular at the equilibrium transition point. This allows us to fix this point from first principles.

Besides such order-disorder transitions there also exist transitions between disordered phases. Excluding liquid-liquid transitions, these will be called *glass transitions*. A prominent example is the transition from a supercooled melt of SiO_2 molecules to an amorphous phase which is the well-known "window glass". One distinguishes two types of glass transitions: spin glass transitions [1, 2] and structural glass transitions transition!structural glass [3, 4, 5, 6]. Their main difference is that the former occur mostly in systems with *quenched disorder* and for the latter the disorder is *self-generated*. We stress that this classification should not be taken to be strict, since there are also models *without* quenched disorder showing spin glass behavior [7]. Typical systems that undergo a structural glass transition are liquids, particularly molecular liquids, like the famous example of SiO_2. In recent years investigations of specific spin glass models, e.g. Potts glass and so-called p-spin models with $p \geq 3$ where p spins are coupled by randomly frozen-in (i.e. quenched) *infinite range interactions*, have revealed some similarities with structural glasses [8]. We will come back to this point in Sect. 4.

In contrast to conventional order-disorder transitions, glass transitions are less well understood. There is a broad consensus that a spin glass transition exists and that an appropriately disorder-averaged free energy is singular at the transition point in the case of mean field models, i.e. models with infinite-range interactions or infinite dimensions. But this is less obvious for short-range interactions. The situation for the structural glass transition is even less satisfactory, although substantial progress has been made in the last two decades.

In this article we will mainly focus on the structural glass transition. Spin glass behavior and spin glass transition will be discussed in this monograph by H. Horner. For further details and references on spin glasses, the reader may consult his contribution.

As mentioned above, structural glasses can be obtained by cooling a liquid. In order to bypass crystallisation one has to choose a *finite* cooling rate. For good glassformers, like SiO_2, this rate can be rather modest, whereas bad glassformers, like most metallic glasses, require extremely high cooling rates. Under such a cooling process the shear viscosity, $\eta(T)$, increases. Close to the so-called *calorimetric* glass transition temperature, T_g, the supercooled liquid falls out of equilibrium and becomes a glass. At T_g, thermodynamical quantities like density, $n(T)$, specific heat at constant pressure, $c_p(T)$, etc. show cross-over behavior, i.e. the slope of $n(T)$ and $c_p(T)$ make a more or less well pronounced jump, depending on the cooling rate. T_g itself depends on the cooling rate, too. Although T_g plays an important practical role, it is less interesting from a fundamental point of view, due to its cooling rate dependence. Besides T_g there are at least three more characteristic temperatures, T_0, T_K and T_c. For many glassformers, $\eta(T)$ can be fitted by the Vogel-Fulcher-Tammann law:

$$\eta(T) = C \exp \frac{A}{k_B(T - T_0)} \quad , T \geq T_0, \tag{1}$$

with $A > 0$ and $C > 0$. The shear viscosity diverges at T_0. Extrapolating the excess entropy $S_{\text{excess}}(T)$ of the supercooled liquid with respect to the crystalline phase to lower temperatures there is the so-called Kauzmann temperature, T_K, at which S_{excess} vanishes:

$$S_{\text{excess}}(T_K) = 0 \, . \tag{2}$$

Since it is argued that a disordered phase should not have a smaller entropy than the crystalline one, S_{excess} can not become negative. Therefore, the system has to undergo a *static* glass transition at T_K. This conclusion, however, is not compelling, since there exists inverse melting, i.e. liquids freeze when heated or crystals melt when cooled [9]. In that case the total entropy of the crystal is higher than that of the liquid. A recent discussion of the Kauzmann problem can be found in Ref. [10]. T_g, T_0 and T_K have played an essential role for many decades. In 1984 quite a new theoretical approach, the mode coupling theory [11], showed that there is a critical temperature, T_c, at which a *dynamical* glass transition takes place. One of the main features is that the nonergodicity parameters, $f(\boldsymbol{q}, T)$, which can be considered as glass order parameters, change discontinuously at T_c:

$$f(\boldsymbol{q}, T) = \begin{cases} 0 & , \quad T > T_c \\ > 0 & , \quad T \leq T_c \, . \end{cases} \tag{3}$$

Since then numerous experimental investigations and computer simulations were stimulated (see reviews [12, 13] and Ref. [14, 15]). They have shown new characteristic *dynamical features* close to the dynamical glass transition point, T_c, consistent with mode coupling theory (see also the contribution by R. Zorn and U. Buchenau in this monograph).

This short exposition of some of the characteristics of glassy behavior should have given a first impression on how diverse the phenomena in the glass-transition region can be. Therefore it is obvious that a successful theoretical description which covers all facets is extremely hard. There are mainly two possible theoretical approaches: phenomenological or microscopic ones. *Phenomenological theories* start from some of the phenomena of glasses, and are named thereafter. Based on these phenomena, a theoretical description is developed capable of describing the observed phenomena. In several cases appealing "physical pictures" are used. However, a couple of assumptions are made which are not proven. The predictive power of such phenomenological approaches is rather limited. This is quite different from a *microscopic theory*. By microscopic we mean that the physical quantities can be calculated from first principles if the interactions between the species are given. Since the glass transition region is located at rather high temperature, quantum effects can be neglected. Therefore, a microscopic theory starts from a classical N-body problem. The next section will discuss some of the phenomenological theories. The major part of this article is devoted to microscopic theories, the 3rd section describes mode coupling theory and the 4th section the replica theory for structural glasses.

Finally we want to stress that the present contribution presents a selection and does not aim to be complete. This holds mainly for the phenomenological models. We also do not discuss the "potential energy landscape" approach [16] which recently has led to new interesting results [17, 18]. Almost all of them were obtained from computer simulations. It would be desirable to complement these investigations with analytical theories.

2 Phenomenological Approaches

The presentation in this section will be rather short. More details and additional phenomenological approaches can be found in the monographs [3, 5, 6] and in the review [4].

2.1 Adam-Gibbs Theory

In 1965 Adam and Gibbs suggested a theory based on the assumption that dynamically cooperative regions occur when decreasing the temperature towards the glass transition point [19]. The particles in these regions perform cooperative motion, which leads to a reduction of the configurational degrees

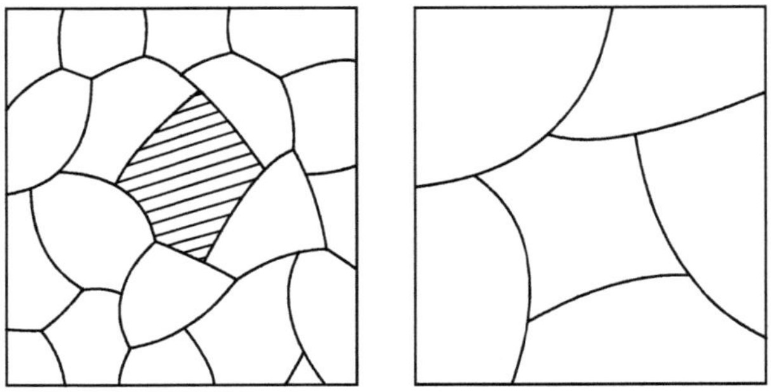

Fig. 1. Illustration of dynamically cooperative regions (one of them is shown as hatched). *Left part*: high temperature; *right part*: low temperature

of freedom. Here we follow the description of Adam-Gibbs theory as given in Ref. [4]. The basis of that theory is a number of assumptions:

1. For a given temperature, T, there are dynamically cooperative regions labeled by $1, \ldots, i, \ldots, n(T)$ with $N_1(T), \ldots, N_i(T), \ldots, N_{n(T)}(T)$ particles. The number of regions, $n(T)$, decreases with decreasing temperature (see Fig. 1). Due to the conservation of the total number of particles, N, it is:
$$\sum_{i=1}^{n(T)} N_i(T) = N \ . \tag{4}$$
2. These regions take two configurations, only. Accordingly the entropy, s, per region is:
$$s = k_B \ln 2 \quad . \tag{5}$$
The assumption that each region takes two configurations, only, is not crucial, but it should be a *finite* number of order one.
3. Fluctuations of $N_i(T)$ are small, that is:
$$N_i(T) \approx N_0(T) = N/n(T) \ , \tag{6}$$
because of Eq. (4).
With these three assumptions we can relate the average number of particles, $N_0(T)$, within a dynamically cooperative region to the configurational entropy:
$$S_c(T) = k_B \ln [\text{number of total configurations}] \ . \tag{7}$$
With $2^{n(T)}$ as the number of total configurations we get from Eq. (7) by use of Eq. (6):
$$N_0(T) = N k_B \ln 2 / S_c(T), \tag{8}$$

that is, $N_0(T)$, and therefore the size of those regions, is inversely proportional to the configurational entropy. This is plausible since the number of possible configurations decreases with increasing $N_0(T)$. The next important steps are to assume:

4. There is a temperature, T_K, at which $N_0(T)$ becomes infinite (for $N = \infty$); then Eq. (8) implies

$$S_c(T) = \begin{cases} > 0 & , \quad T > T_K \\ = 0 & , \quad T \leq T_K \end{cases} , \tag{9}$$

i.e T_K is the Kauzmann temperature [20].

5. The transition between both configurations of a region is an activated process with an activation energy:

$$E(T) = e_0 N_0(T), \tag{10}$$

where e_0 may be weakly T-dependent. Then the transition time $\tau(T)$ is given by:

$$\tau(T) \sim \exp(\beta E(T)) = \exp(\beta e_0 N_0(T)) \ . \tag{11}$$

Substituting N_0 from Eq. (8), we arrive at:

$$\tau(T) \sim \exp\left[\frac{e_0 N \ln 2}{T S_c(T)}\right] . \tag{12}$$

Assuming that $S(T)$ vanishes linearly at T_K, we obtain from Eq. (12) the Vogel-Fulcher-Tammann law close to T_K:

$$\tau(T) \sim \exp\left[\frac{e_0 N \ln 2}{T_K S_c'(T_K)(T - T_K)}\right] , \tag{13}$$

with $T_0 = T_K$.

This example demonstrates that the appealing picture of dynamically cooperative regions, in combination with several assumptions, allows us to derive the Vogel-Fulcher-Tammann law. But it also shows that no additional predictions are made. Furthermore, it is obvious that the growth of the dynamically cooperative regions would be accomplished by a divergent length scale, which, however, has never been found in experiments or computer simulations.

2.2 Free-Volume Theory

This phenomenological description has been made by Cohen and Turnbull in 1959 [21]. In order to illustrate their idea we choose a system of *hard* spheres with average density n. Let us fix the positions of all spheres except of the i-th sphere. Then the i-th sphere can move freely in the so-called *free volume*, $v_f(i)$ (see Fig. 2). Although $v_f(j)$ is correlated with $v_f(i)$ for $i \neq j$ one assumes:

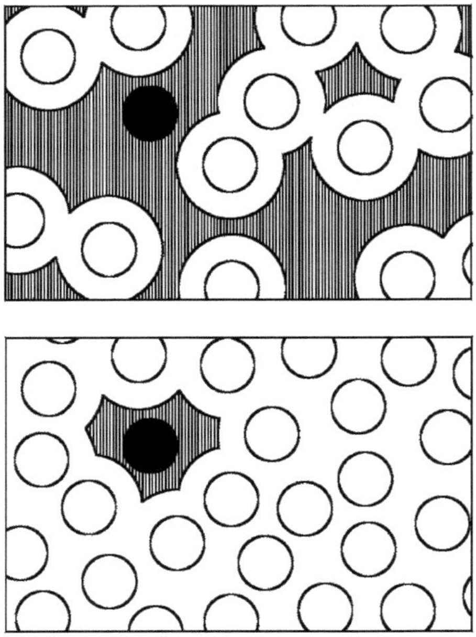

Fig. 2. Illustration of the free volume of one sphere (*full circle*) within a fixed configuration of other spheres (*open circles*). The free volume is shown as hatched regions. Upper panel: $T > T_0$, for which the free volume forms a percolation cluster. Lower panel: $T < T_0$, for which the free volume does not percolate but gets localized

1. Every sphere has a free volume, v_f:
$$i \to v_f(i). \tag{14}$$
2. $v_f(i)$ are independent random numbers with probability density $p(v_f)$ which is assumed to be exponential:
$$p(v_f) = \mathcal{N} \exp(-v_f/\bar{v}_f), \tag{15}$$
with the mean free volume per particle:
$$\bar{v}_f = \frac{1}{N} \sum_{i=1}^{N} v_f(i), \tag{16}$$
where $\mathcal{N}$ is a normalization constant.
3. It is obvious that $v_f(i)$ (and therefore $\bar{v}_f$) decreases with increasing density or decreasing temperature in the case of "soft" particles. An essential assumption is that there is a temperature $T_0 > 0$ such that the mean free volume vanishes at T_0:
$$\bar{v}_f(T) = \begin{cases} \alpha(T - T_0) & , \quad T \geq T_0 \\ 0 & , \quad T < T_0 \end{cases} \tag{17}$$
with α as the expansion coefficient.

4. The inverse shear viscosity is proportional to the probability that the free volume per particle is larger than a certain value v_f^*, i.e.:

$$\eta^{-1}(T) \sim \int_{v_f^*}^{\infty} dv_f p(v_f). \tag{18}$$

Substituting Eq. (15) and Eq. (17) into Eq. (18) we get the Vogel-Fulcher-Tammann law:

$$\eta(T) \sim \exp\left[\frac{v_f^*}{\alpha(T - T_0)}\right]. \tag{19}$$

2.3 Extended Free-Volume Theory

The free-volume theory has been extended by Cohen and Grest [22]. The main idea of this extension is to make a connection between the glass transition and percolation. If the density n (temperature T) is below n_g (above T_g), the individual free volumes $v_f(i)$ overlap or, in other words, the free volume *percolates* in such a way that a (macroscopic) *percolation cluster* exists (see upper panel of Fig. 2). In this case a particle can move macroscopic distances, i.e the diffusion constant is finite. Now, at increasing n (decreasing T) there may exist a critical value n_g (T_g) at which the percolation cluster disappears, which implies that the particles become localized in a glass phase with zero diffusion (see lower panel of Fig. 2). If this scenario is correct, the glass transition point would coincide with the percolation threshold of the free volumes $v_f(i)$. Since percolation is well understood [23] this relationship would imply several characteristic features, e.g. the fractal nature of the percolation cluster close to n_g (or T_g). However, there is no experimental evidence for such a fractal behavior. The main critique on the free volume theory is that the density used in the lower panel of Fig. 2 to demonstrate the localization of the free volume still corresponds to a liquid. Computer simulations show that the density of a liquid of, e.g. hard discs is already higher than that used in the lower panel of Fig. 2. Accordingly, the free volume, as defined above, is already localized in the liquid phase.

2.4 Gibbs–DiMarzio Theory

This approach is not really phenomenological. But the model which has been studied by Gibbs and DiMarzio is very special and has a limited range of applicability. Therefore it is included in this section. The model by Gibbs and DiMarzio is able to explain the vanishing of the configurational entropy [24]. A system with N polymers on a cubic lattice with N_s lattice sites is considered. Each polymer consists of l monomers and can form conformations labeled by $1, 2, \cdots, i, \cdots$. Let n_i be the number of polymers with conformation i

and $P(\{n_i\}, N, N_s)$ be the probability that N polymers on a lattice with N_s sites build a set of $\{n_i\}$ conformations. Associating an energy ε_i with each individual conformation, the total energy, E, is given by:

$$E = \sum_i n_i \varepsilon_i \, . \tag{20}$$

Then the number of configurations $\Omega(E, N, N_s)$ with energy E is:

$$\Omega(E, N, N_s) = \sum_{\{n_i\}} P(\{n_i\}, N, N_s) \; \delta(E - \sum_i \varepsilon_i n_i), \tag{21}$$

from which one obtains the configurational entropy:

$$S_c(E, N, N_s) = k_B \, \ln \Omega(E, N, N_s) \, . \tag{22}$$

Of course, the nontrivial problem is to determine P and to perform the sum over $\{n_i\}$ in Eq. (21). Making use of a mean field approximation due to Flory and Huggins [25] one finally gets a critical energy, E_K, which corresponds to the Kauzmann temperature, T_K, with:

$$S_c(E, N, N_s) = \begin{cases} > 0 & , \quad E > E_K \\ = 0 & , \quad E \leq E_K \end{cases} \quad . \tag{23}$$

Since mean field approximations tend to produce phase transitions even for systems that do not show such transitions, the validity of the results in Eq. (23) is not obvious.

3 Microscopic Theory: Mode Coupling Theory

We hope that the short presentation in Sect. 2 has demonstrated that the status of those phenomenological theories is not satisfactory. Although plausible "physical pictures" are involved, they are based on a couple of crucial assumptions. However, the validity of those assumptions remains completely unclear. This demands a *microscopic* approach based on *first principles*. Such an approach to structural glass transitions was made for the first time in 1984 by Bengtzelius, Götze and Sjölander [11]. Starting from the liquid side, these authors applied the mode coupling theory (MCT) developed by Kawasaki [26] in order to describe the critical slowing down close to a critical point to the relaxation of the density fluctuations in a supercooled liquid. The result is an equation of motion (see below) for the intermediate scattering function, $S(\boldsymbol{q}, t)$, for a *simple* liquid, which can be measured by neutron and light scattering (see the contributions by R. Zorn and U. Buchenau and J. B. Suck in this monograph). The properties of the solution, $S(\boldsymbol{q}, t)$, of the MCT equations were mainly investigated in great detail by Götze and his coworkers. These results can be found in the reviews [27, 28, 29, 12, 13] and in their references. Since most glassformers are molecular systems which involve translational *and* rotational degrees of freedom, MCT had been extended to a single linear molecule in a simple liquid by Franosch et al. [30] and to a liquid of linear molecules and arbitrary molecules by Scheidsteger and this author [31] and by Fabbian et al. [15], respectively. This was accomplished by the use of the tensorial formalism which allows the separation of translational and rotational degrees of freedom. Alternatively, one can also use a site-site representation for molecular systems. MCT based on such a description was worked out by Chong and Hirata [32] and Chong, Götze and Singh [33].

Before discussing MCT, let us anticipate that the derivation of the MCT equations requires some more or less strong approximations. Although these approximations can not be controlled, e.g. due to the lack of a smallness parameter, it is interesting to notice that the MCT equations for a simple liquid were also obtained by quite different approaches, which are the use of generalized fluctuating hydrodynamics by Kirkpatrick [34] and Das and Mazenko [35], density functional theory by Kirkpartick and Wolynes [36], and recently by the use of the equation of motion for the microscopic density, $\rho(\boldsymbol{q}, t)$, in conjunction with assuming the density fluctuations to be Gaussian by Zaccarelli et al. [37]. That these quite different approaches lead to the same mathematical structure of the MCT equations shows its "robustness". In addition, it has been argued that the MCT equations become exact even in infinite dimensions [36]. Further support for MCT comes from a spherical model (without quenched disorder) [38] and for spin glass models with infinite-range interactions (see Ref. [8, 39] and their references), for which the corresponding q-independent MCT equation has been proven to be exact.

Now we will turn to the discussion of MCT. This will be done in two sections. In the first one we present the derivation of the MCT equations and in the second one the properties of their solutions.

3.1 Derivation of the MCT Equations

We will restrict ourselves to a simple liquid of N identical particles with mass, m, in a finite volume, V. We assume two-body interactions, $v(\boldsymbol{x} - \boldsymbol{x}')$, such that the *classical* Hamiltonian is given by:

$$H(\{\boldsymbol{x}_i\}, \{\boldsymbol{p}_i\}) = \sum_n \frac{1}{2m}\boldsymbol{p}_n^2 + V(\{\boldsymbol{x}_i\}), \tag{24}$$

with the potential energy:

$$V(\{\boldsymbol{x}_i\}) = \frac{1}{2}\sum_{i \neq j} v(\boldsymbol{x}_i - \boldsymbol{x}_j). \tag{25}$$

$\{\boldsymbol{x}_i\}$ and $\{\boldsymbol{p}_i\}$ are the corresponding positions and momenta, respectively. Fixing an initial point $\{\boldsymbol{x}_i\}$, $\{\boldsymbol{p}_i\}$ in phase space, the phase space point

$\{\boldsymbol{x}_i(t)\}$, $\{\boldsymbol{p}_i(t)\}$ at time t is determined by Newton's equation of motion or equivalently by the Hermitean Liouville operator:

$$\boldsymbol{x}_i(t) = e^{i\mathcal{L}t}\boldsymbol{x}_i \quad , \quad \boldsymbol{p}_i(t) = e^{i\mathcal{L}t}\boldsymbol{p}_i \ . \tag{26}$$

Of course, one cannot solve these equations of motion for a macroscopic system. Therefore one has to use a theoretical framework which restricts itself to the relevant variables. For a liquid this is the *microscopic* density:

$$\rho(\boldsymbol{x},t) = \sum_n \delta(\boldsymbol{x} - \boldsymbol{x}_n(t)), \tag{27}$$

where the dependence of the positions $\boldsymbol{x}_n(t)$ on the initial point $\{\boldsymbol{x}_i\}$, $\{\boldsymbol{p}_i\}$ is suppressed. Now we recall that $\rho(\boldsymbol{x},t)$ will have a slowly varying part in the strongly supercooled regime, besides fast motions (vibrations) around the quasi-equilibrium positions. Therefore, we can consider $\rho(\boldsymbol{x},t)$ or its Fourier transform:

$$\rho(\boldsymbol{q},t) = \sum_n e^{i\boldsymbol{q}\boldsymbol{x}_n(t)} = e^{i\mathcal{L}t}\,\rho(\boldsymbol{q}), \tag{28}$$

as a *slow variable*. If $\rho(\boldsymbol{q},t)$ is slow, then the current density

$$\boldsymbol{j}(\boldsymbol{q},t) = \sum_n \frac{1}{m}\boldsymbol{p}_n(t)e^{i\boldsymbol{q}\boldsymbol{x}_n(t)}$$

is slow, too, since it is related to $\rho(\boldsymbol{q},t)$ through the continuity equation:

$$\dot{\rho}(\boldsymbol{q},t) \equiv i\mathcal{L}\rho(\boldsymbol{q},t) = i\boldsymbol{q}\cdot\boldsymbol{j}(\boldsymbol{q},t) \ . \tag{29}$$

We can continue taking time derivatives. Then $\ddot{\rho}(\boldsymbol{q},t)$ is given by:

$$\ddot{\rho}(\boldsymbol{q},t) = i\boldsymbol{q}\cdot\sum_n \frac{1}{m}\,\dot{\boldsymbol{p}}_n(t)\,e^{i\boldsymbol{q}\boldsymbol{x}_n(t)} + \text{kinetic part} \ . \tag{30}$$

Using $\dot{\boldsymbol{p}}_n/m = -\partial V(\{\boldsymbol{x}_i\})/\partial\boldsymbol{x}_n$ and that Eq. (25) can be rewritten as

$$\begin{aligned} V(\{\boldsymbol{x}_i\}) &= \frac{1}{2}\int d^3x \int d^3x'\ v(\boldsymbol{x}-\boldsymbol{x}')\,\rho(\boldsymbol{x})\rho(\boldsymbol{x}') \\ &= \frac{1}{2V}\sum_{\boldsymbol{q}} \tilde{v}(\boldsymbol{q})\rho^*(\boldsymbol{q})\,\rho(\boldsymbol{q}), \end{aligned} \tag{31}$$

it is easy to prove that:

$$\ddot{\rho}(\boldsymbol{q},t) = \frac{1}{2V}\sum_{\boldsymbol{q}_1,\boldsymbol{q}_2}{}' w_0(\boldsymbol{q},\boldsymbol{q}_1,\boldsymbol{q}_2)\,\rho(\boldsymbol{q}_1,t)\,\rho(\boldsymbol{q}_2,t) + \text{kinetic part}, \tag{32}$$

with the bare and time-independent vertex:

$$w_0(\boldsymbol{q},\boldsymbol{q}_1,\boldsymbol{q}_2) = \boldsymbol{q}\cdot[\boldsymbol{q}_1\tilde{v}(\boldsymbol{q}_1)+\boldsymbol{q}_2\tilde{v}(\boldsymbol{q}_2)]\,. \tag{33}$$

Here we have used that the Fourier transform $\tilde{v}(\boldsymbol{q})$ of the pair potential is real and $\sum'$ denotes summation, such that $\boldsymbol{q}_1+\boldsymbol{q}_2=\boldsymbol{q}$. The result in Eq. (32) reveals that the "force" $\ddot{\rho}(\boldsymbol{q},t)$ contains contributions from a pair of modes, which are coupled by the bare vertex w_0. It is rather obvious that the use of a r-body interaction would yield a contribution $\rho(\boldsymbol{q}_1,t)\rho(\boldsymbol{q}_2,t)\cdot\ldots\cdot\rho(\boldsymbol{q}_r,t)$. Consequently, the "force" $\ddot{\rho}(\boldsymbol{q},t)$ must be slow, as well. By continuing this procedure one obtains a set of slow variables. In the following, however, we will restrict ourselves to the two variables $\rho(\boldsymbol{q},t)$ and $j(\boldsymbol{q},t)=\boldsymbol{q}\cdot\boldsymbol{j}(\boldsymbol{q},t)/q(q=|\boldsymbol{q}|)$, the longitudinal current density. But we have to keep in mind that $\ddot{\rho}(\boldsymbol{q},t)$ also is slow. Having chosen $\rho(\boldsymbol{q})$ and $j(\boldsymbol{q})$ (at $t=0$) as slow variables, one can apply the Mori-Zwanzig projection formalism [40, 41] to derive an *exact* equation of motion for the normalized intermediate scattering function, $\Phi(\boldsymbol{q},t)=S(\boldsymbol{q},t)/S(\boldsymbol{q})$, with

$$S(\boldsymbol{q},t)=\frac{1}{N}\langle\rho(\boldsymbol{q},t)^*\rho(\boldsymbol{q})\rangle=\frac{1}{N}\langle\rho^*(\boldsymbol{q})e^{-i\mathcal{L}t}\,\rho(\boldsymbol{q})\rangle, \tag{34}$$

where we used the Hermiticity of $\mathcal{L}$. $S(\boldsymbol{q})$ is the static structure factor, which depends on the thermodynamical variables T, n, etc. The exact Mori-Zwanzig equation for $\Phi(\boldsymbol{q},t)$ reads:

$$\ddot{\Phi}(\boldsymbol{q},t)+\Omega_q^2\,\Phi(\boldsymbol{q},t)+\int\limits_0^t dt'\,M(\boldsymbol{q},t-t')\dot{\Phi}(\boldsymbol{q},t')=0, \tag{35}$$

with the memory kernel:

$$M(\boldsymbol{q},t)=\frac{1}{N}\,\frac{m}{k_BT}\,\frac{1}{q^2}\langle\ddot{\rho}(\boldsymbol{q})^*\,Q\,e^{-iQ\mathcal{L}Qt}\,Q\ddot{\rho}(\boldsymbol{q})\rangle, \tag{36}$$

and the microscopic frequencies:

$$\Omega_q=\left(\frac{k_BT}{m}\,\frac{q^2}{S(\boldsymbol{q})}\right)^{1/2}. \tag{37}$$

The initial condition is $\Phi(\boldsymbol{q},0)=1$, and $\dot{\Phi}(\boldsymbol{q},0)=0$ for all $\boldsymbol{q}$. The result in Eq. (35) makes it obvious that the problem to calculate $\Phi(\boldsymbol{q},t)$ has been shifted to the calculation of $M(\boldsymbol{q},t)$, which seems to be hopeless, as well. But this is not really true. In contrast to Φ, it is possible to approximate the memory kernel, M. $M(\boldsymbol{q},t)$ is the correlation function of the "forces" $Q\,\ddot{\rho}(\boldsymbol{q})$; however, with the *reduced Liouvillian*, $\mathcal{L}'=Q\,\mathcal{L}\,Q$. Q is the projector (see below) which projects perpendicular to both slow variables, $\rho(\boldsymbol{q})$ and $j(\boldsymbol{q})$. Since $\ddot{\rho}(\boldsymbol{q})$ contains a coupled pair of modes $\rho(\boldsymbol{q}_1)\,\rho(\boldsymbol{q}_2)$ (cf. Eq. (32)) and because $Q\rho(\boldsymbol{q}_1)\,\rho(\boldsymbol{q}_2)\neq 0$, the "force" $Q\ddot{\rho}(\boldsymbol{q})$ still contains a slow part. This suggests the use of the following approximation [27]:

$$Q\ddot{\rho}(\boldsymbol{q}) \to Q|\ddot{\rho}(\boldsymbol{q})\rangle \approx \mathcal{P}Q|\ddot{\rho}(\boldsymbol{q})\rangle, \tag{38}$$

where $\mathcal{P}$ is the projector onto pairs of modes:

$$\mathcal{P} = \sum_{\boldsymbol{q}_1\boldsymbol{q}_2,\boldsymbol{q}'_1\boldsymbol{q}'_2} g(\boldsymbol{q}_1\boldsymbol{q}_2;\boldsymbol{q}'_1\boldsymbol{q}'_2)|\rho(\boldsymbol{q}_1)\rho(\boldsymbol{q}_2)\rangle\langle\rho(\boldsymbol{q}'_1)^*\rho(\boldsymbol{q}'_2)^*|, \tag{39}$$

and g is determined such that $\mathcal{P}^2 = \mathcal{P}$. The reader should note that we have introduced a bra and ket notation, $\langle|$ and $|\rangle$, respectively, like in quantum mechanics. This can be done since the canonical average $\langle A^*B\rangle$ of two phase space functions A^* and B can be interpreted as the scalar product $\langle A^*|B\rangle$. It is the existence of this scalar product which allows us to introduce projectors. Substituting Eq. (38) and Eq. (39) into Eq. (36) relates $M(\boldsymbol{q},t)$ to the correlation function

$$\langle\rho(\boldsymbol{q}'_3)^*\rho(\boldsymbol{q}'_4)^*|e^{-i\mathcal{L}'t}|\rho(\boldsymbol{q}_1)\rho(\boldsymbol{q}_2)\rangle\ . \tag{40}$$

This relationship involves the *static* quantities $g(\boldsymbol{q}_1\boldsymbol{q}_2;\boldsymbol{q}'_1\boldsymbol{q}'_2)$, $g(\boldsymbol{q}_3\boldsymbol{q}_4;\boldsymbol{q}'_3\boldsymbol{q}'_4)$, and $\langle\ddot{\rho}(\boldsymbol{q})^*Q|\rho(\boldsymbol{q}_3)\rho(\boldsymbol{q}_4)\rangle\langle\rho(\boldsymbol{q}'_1)^*\rho(\boldsymbol{q}'_2)^*|Q\ddot{\rho}(\boldsymbol{q}_1)\rangle$. Now, the crucial approximation is the factorization of the correlator Eq. (40) and simultaneous replacement of $\mathcal{L}'$ by $\mathcal{L}$:

$$\begin{aligned}
&\langle\rho(\boldsymbol{q}'_3)^*\rho(\boldsymbol{q}'_4)^*|e^{-i\mathcal{L}'t}|\rho(\boldsymbol{q}_1)\rho(\boldsymbol{q}_2)\rangle \\
&\approx [\langle\rho(\boldsymbol{q}_1)^*|e^{-i\mathcal{L}t}|\rho(\boldsymbol{q}_1)\rangle\,\langle\rho(\boldsymbol{q}_2)^*|e^{-i\mathcal{L}t}|\rho(\boldsymbol{q}_2)\rangle\delta_{q'_3q_1}\,\delta_{q'_4q_2} \\
&\qquad +(1\longleftrightarrow 2)] \\
&= N^2S(\boldsymbol{q}_1)S(\boldsymbol{q}_2)\,\Phi(\boldsymbol{q}_1,t)\,\Phi(\boldsymbol{q}_2,t)[\delta_{q'_3q_1}\,\delta_{q'_4q_2}+\delta_{q'_3q_2}\,\delta_{q'_4q_1}]\ .
\end{aligned} \tag{41}$$

The condition $\mathcal{P}^2 = \mathcal{P}$ implies that $g(\boldsymbol{q}_1\boldsymbol{q}_2;\boldsymbol{q}'_1\boldsymbol{q}'_2)$ is the inverse of the "matrix" $(\langle\rho(\boldsymbol{q}'_1)^*\rho(\boldsymbol{q}'_2)^*\rho(\boldsymbol{q}_1)\rho(\boldsymbol{q}_2)\rangle)$. Since this "matrix" equals the correlator Eq. (40) at $t=0$, it is approximated by the right hand side of Eq. (41), at $t=0$. Using this approximation one immediately finds:

$$g(\boldsymbol{q}_1,\boldsymbol{q}_2;\,\boldsymbol{q}'_1,\boldsymbol{q}'_2) \approx (4N^2S(\boldsymbol{q}_1)S(\boldsymbol{q}_2))^{-1}[\delta_{q'_1q_1}\,\delta_{q'_2q_2}+\delta_{q'_1q_2}\,\delta_{q'_2q_1}]\ . \tag{42}$$

Using $Q = 1-\mathcal{P} = 1-(N\,S(\boldsymbol{q}))^{-1}|\rho(\boldsymbol{q})\rangle\langle\rho(\boldsymbol{q})^*|-(Nk_BT/m)^{-1}|j(\boldsymbol{q})\rangle\langle j(\boldsymbol{q})^*|$, the static correlation $\langle\ddot{\rho}(\boldsymbol{q})^*\,Q|\rho(\boldsymbol{q}_1)\rho(\boldsymbol{q}_2)\rangle$ can be expressed as follows [27]:

$$\begin{aligned}
\langle\ddot{\rho}(\boldsymbol{q})^*Q|\rho(\boldsymbol{q}_1)\rho(\boldsymbol{q}_2)\rangle &= N\frac{k_BT}{m}\,S(\boldsymbol{q}_1)\,S(\boldsymbol{q}_2)\{-n\boldsymbol{q}\cdot[\boldsymbol{q}_1c(\boldsymbol{q}_1)+\boldsymbol{q}_2c(\boldsymbol{q}_2)]+ \\
&+q^2[1-\langle\rho(\boldsymbol{q})^*\rho(\boldsymbol{q}_1)\rho(\boldsymbol{q}_2)\rangle/(N\,S(\boldsymbol{q})\,S(\boldsymbol{q}_1)\,S(\boldsymbol{q}_2))]\}\,\delta_{q_1+q_2,q}
\end{aligned} \tag{43}$$

where we introduced the direct correlation function $c(\boldsymbol{q})$ defined by $S(\boldsymbol{q}) = [1-nc(\boldsymbol{q})]^{-1}$. Substituting Eqs. (38)-(43) into Eq. (36) yields finally the MCT approximation for the memory kernel:

$$M(\boldsymbol{q},t) \approx \nu_q\,\delta(t) + \Omega_q^2 m(\boldsymbol{q},t), \tag{44}$$

$$m(\boldsymbol{q},t) = \frac{1}{2V} \sum_{\boldsymbol{q}_1,\boldsymbol{q}_2}{}' V(\boldsymbol{q},\boldsymbol{q}_1,\boldsymbol{q}_2)\, \Phi(\boldsymbol{q}_1,t)\, \Phi(\boldsymbol{q}_2,t), \tag{45}$$

with the *positive* vertices:

$$V(\boldsymbol{q},\boldsymbol{q}_1,\boldsymbol{q}_2) = \frac{1}{n}\, S(\boldsymbol{q})\, S(\boldsymbol{q}_1)\, S(\boldsymbol{q}_2)\, q^{-4}\{n\boldsymbol{q}\cdot[\boldsymbol{q}_1 c(\boldsymbol{q}_1) + \boldsymbol{q}_2 c(\boldsymbol{q}_2)] - \\ -q^2[1 - \langle\rho(\boldsymbol{q})^*\rho(\boldsymbol{q}_1)\rho(\boldsymbol{q}_2)\rangle/(N\, S(\boldsymbol{q})\, S(\boldsymbol{q}_1)\, S(\boldsymbol{q}_2))]\}^2. \tag{46}$$

The first term on the right hand side of Eq. (44) accounts for the fast part of the "force", $Q\ddot{\rho}(\boldsymbol{q})$, leading to a frictional contribution.

The Eqs. (35), (37) and (44)-(46) are the *mode coupling equations.* Due to the MCT approximation they are a *closed* set of integro-differential equations for the normalized correlator $\Phi(\boldsymbol{q},t)$, with initial conditions $\Phi(\boldsymbol{q},0) \equiv 1$ and $\dot{\Phi}(\boldsymbol{q},0) \equiv 0$, because of time reversal symmetry. As an input they only need the *static* two-point correlator, $S(\boldsymbol{q})$ (or equivalently the direct correlation function $c(\boldsymbol{q})$), the *static* three point correlator, $\langle\rho(\boldsymbol{q})^*\rho(\boldsymbol{q}_1)\,\rho(\boldsymbol{q}_2)\rangle$, and the frictional constants, ν_q. ν_q, which can only be determined from kinetic theory, does not influence the glassy behavior. Therefore it can be put to zero. The remaining *static* two- and three-point correlators, $S(\boldsymbol{q}) = \langle\rho(\boldsymbol{q})^*\rho(\boldsymbol{q})\rangle/N$ and $\langle\rho(\boldsymbol{q})^*\,\rho(\boldsymbol{q}_1)\,\rho(\boldsymbol{q}_2)\rangle$, can be calculated for a given potential energy $V(\{\boldsymbol{x}_i\})$. This is what makes MCT a microscopic first-principle theory. Here a comment is in order: MCT will be applied to the supercooled liquid, i.e. to a temperature regime where the stable thermodynamical phase is a crystal. In order to study the glass transition, one has to use the static correlators for the supercooled liquid and *not* for the crystal. MCT allows us to predict the time dependence of the density fluctuations, provided both static correlators are known. They can be obtained either from analytical approximation schemes [40] or from experiments and simulations. Application of the convolution approximation [40]:

$$\langle\rho(\boldsymbol{q})^*\rho(\boldsymbol{q}_1)\rho(\boldsymbol{q}_2)\rangle \approx N\; S(\boldsymbol{q})S(\boldsymbol{q}_1)S(\boldsymbol{q}_2)\; \delta_{q_1+q_2,q}, \tag{47}$$

leads to a further simplification of the vertices Eq. (46):

$$V(\boldsymbol{q},\boldsymbol{q}_1,\boldsymbol{q}_2) = nS(\boldsymbol{q})S(\boldsymbol{q}_1)S(\boldsymbol{q}_2)\,[\boldsymbol{q}\cdot(\boldsymbol{q}_1 c(\boldsymbol{q}_1) + \boldsymbol{q}_2 c(\boldsymbol{q}_2))]^2/q^4, \tag{48}$$

which involves $S(\boldsymbol{q})$ (or $c(\boldsymbol{q})$), only. For SiO_2 liquids, it has been demonstrated by Sciortino and Kob [42] that a satisfactory agreement of, e.g. the critical nonergodicity parameters determined from a MD simulation with those from MCT is only obtained with the vertices from Eq. (43) where the three-point correlator is *not* factorized. This is rather plausible, because SiO_2 is a covalent glassformer with bond-orientational correlations that are completely neglected by the convolution approximation in Eq. (47).

3.2 Solutions and Predictions of MCT

The main question which arises is: How can one detect a glass transition within MCT? The answer is rather simple. Let us use the *nonergodicity pa-*

rameters defined by:

$$f(\boldsymbol{q}) = \lim_{t\to\infty} \Phi(\boldsymbol{q},t) = -\lim_{z\to 0} z\hat{\Phi}(\boldsymbol{q},z) \ . \tag{49}$$

These parameters, which are just the infinite time limit of $\Phi(\boldsymbol{q},t)$ or the zero-frequency limit of its Laplace transform:

$$\hat{\Phi}(\boldsymbol{q},z) = i\int_0^\infty dt\Phi(\boldsymbol{q},t)e^{izt} \quad , \ \mathrm{Im}\, z > 0, \tag{50}$$

can be used as glass order parameters, because they vanish in an ergodic phase (= liquid phase) and are nonzero in a nonergodic one, which is interpreted as a glass phase. To be more precise, this is true for the correlation function of the density fluctuations $\delta\rho(\boldsymbol{q},t) = \rho(\boldsymbol{q},t) - \langle\rho(\boldsymbol{q},t)\rangle$, only. Since $\delta\rho(0,t) = 0$ and $\delta\rho(\boldsymbol{q},t) = \rho(\boldsymbol{q},t)$ for all $\boldsymbol{q} \neq 0$, it is sufficient to investigate $\Phi(\boldsymbol{q},t)$. Thus it is:

$$f(\boldsymbol{q}) = \begin{cases} 0 & , \ \text{liquid} \\ > 0 & , \ \text{glass} \end{cases} \ . \tag{51}$$

Since $\ddot{\Phi}(\boldsymbol{q},t) \to 0$ for $t \to \infty$, it is easy to show that Eq. (35) yields a nonlinear set of algebraic equations for $f(\boldsymbol{q})$:

$$\frac{f(\boldsymbol{q})}{1-f(\boldsymbol{q})} = \mathcal{F}[f(\boldsymbol{q})], \tag{52}$$

with:

$$\mathcal{F}[f(\boldsymbol{q})] = \frac{1}{2V}\sum_{\boldsymbol{q}_1,\boldsymbol{q}_2}{}' V(\boldsymbol{q},\boldsymbol{q}_1,\boldsymbol{q}_2) f(\boldsymbol{q}_1) f(\boldsymbol{q}_2) \ . \tag{53}$$

One can prove that the long-time limit, $f(\boldsymbol{q})$, is distinguished from other possible solutions of Eqs. (52), (53) by the following properties: [27]

(i) $f(\boldsymbol{q})$ is real (since $\Phi(\boldsymbol{q},t)$ is real)
(ii) $0 \leq f(\boldsymbol{q}) \leq 1$
(iii) If several solutions $f_1(\boldsymbol{q}) > f_2(\boldsymbol{q}) \geq f_3(\boldsymbol{q}) \geq \cdots$ exist, the long-time limit is the largest one: $f(\boldsymbol{q}) = f_1(\boldsymbol{q})$

The positiveness of the vertices is crucial for (iii).

It is obvious from Eqs. (52), (53) that $f(\boldsymbol{q}) \equiv 0$ is always a solution. In order to check whether a nontrivial solution exists, let us neglect the q-dependence for a moment. This leads to a so-called schematic model [27], the $\mathcal{F}_2$ model, where $\mathcal{F}[f] = vf^2, v \geq 0$, i.e. we have to solve:

$$\frac{1}{1-f} = vf^2 \ . \tag{54}$$

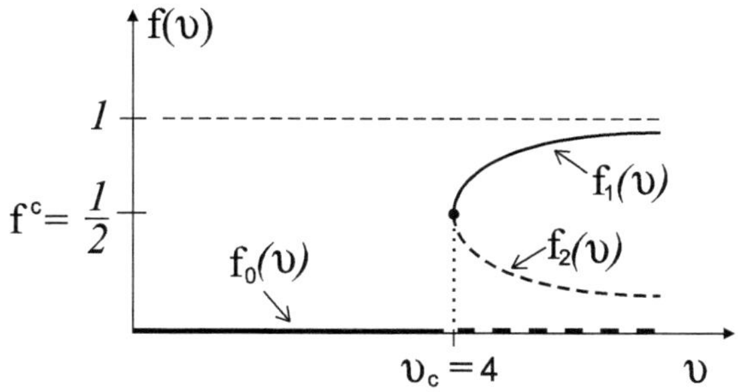

Fig. 3. Qualitative control parameter dependence of the nonergodicity parameter of the $\mathcal{F}_2$ model. $f_0(v) \equiv 0$ is the trivial solution. At $v_c = 4$ two new solutions, $f_1(v)$ and $f_2(v)$, bifurcate

For $f \neq 0$ this leads to a quadratic equation with solution:

$$f_{1/2} = \frac{1}{2}[1 \pm \sqrt{1 - 4/v}] \ . \tag{55}$$

Since $f_{1/2}$ is not real for $v < v_c = 4$, the only physical solution in this range is $f_0 = 0$. For $v \geq v_c$ two real solutions, $f_1 > f_2$, bifurcate from the trivial one. This scenario is illustrated in Fig. 3. Hence, we get for the F_2 model:

$$f(v) = \begin{cases} 0 & , \quad v < v_c \\ \frac{1}{2}[1 + \sqrt{1 - v_c/v}] & , \quad v \geq v_c \end{cases} \tag{56}$$

That is, at the critical control parameter $v_c = 4$, the nonergodicity parameter changes *discontinuously* from zero to the critical nonergodicity parameter $f^c = f(v_c) = 1/2$. This type of transition is called a *type-B transition*, in contrast to a type-A transition, which is *continuous* [27]. Such a *type-A transition* occurs, e.g. for the $\mathcal{F}_1$ model: $\mathcal{F}[f] = vf$, $v \geq 0$.

This simple example has shown that an ergodic-to-nonergodic transition can occur at a critical coupling constant v_c. Since the vertices $V(\boldsymbol{q}, \boldsymbol{q}_1, \boldsymbol{q}_2)$ are positive and $0 \leq f(\boldsymbol{q}) \leq 1$, $\mathcal{F}[f(\boldsymbol{q})]$ will diverge in the strong coupling limit $V(\boldsymbol{q}, \boldsymbol{q}_1, \boldsymbol{q}_2) \to \infty$. Therefore, Eq. (52) can only be fulfilled if the nontrivial solution converges to one. Therefore there must be a critical hypersurface in the control parameter space at which transitions from $f(\boldsymbol{q}) \equiv 0$ to $f(\boldsymbol{q}) \neq 0$ happen. Because the vertices depend, through $S(\boldsymbol{q})$, on the thermodynamic variables T, n etc. and increase with decreasing T (increasing n), there will be a critical temperature, T_c (critical density, n_c), at which the system undergoes an ergodic-to-nonergodic transition, i.e. a glass transition. From this we can conclude that MCT yields a *dynamical* glass transition, whereas static quantities, e.g. $S(\boldsymbol{q})$, are *not* singular at T_c (or n_c). Having established the existence of a glass transition singularity, one can study the dynamics close to it. For large times (small frequencies) compared to the microscopic time scale $\Omega_q^{-1}(\Omega_q)$, the Laplace transform of Eq. (35) yields with Eq. (44):

$$\frac{z\hat{\Phi}(\boldsymbol{q},z)}{1+z\hat{\Phi}(\boldsymbol{q},z)} = z\hat{m}(\boldsymbol{q},z) \,. \tag{57}$$

Here we used $\nu_q = 0$. The reader should notice that Eq. (57) does not involve anymore the microscopic frequencies Ω_q. The schematic representation of a solution of the MCT equations, Eqs. (35), (44)-(46), including the microscopic time regime, is shown in Figure 4. This figure clearly demonstrates the existence of a critical temperature, T_c, at which a jump of $f(\boldsymbol{q})$ occurs. The generic behavior close to T_c is given by (cf. also Eq. (56)):

$$f(\boldsymbol{q}) = \begin{cases} 0 \;\; , \quad T > T_c \\ f^c(\boldsymbol{q}) + \text{const}(T_c - T)^{1/2} \;, \; T \leq T_c \end{cases} \tag{58}$$

where the constant is q-dependent. We can observe from Figure 4 that *close* to T_c there is a time scale on which the correlator $\Phi(\boldsymbol{q},t)$ is close to the critical nonergodicity parameter $f^c(\boldsymbol{q}) = f(\boldsymbol{q})|_{T=T_c}$. This suggests the Ansatz:

$$\Phi(\boldsymbol{q},t) = f^c(\boldsymbol{q}) + h(\boldsymbol{q})G(t), \tag{59}$$

with $|G(t)| \ll 1$. It is important to realize that the q and t dependences are factorized. This is related to the type of bifurcation scenario for which the *non-degenerated* largest eigenvalue of the stability matrix of the linearized Eq. (52) becomes one. Therefore at T_c only *one* unstable eigenvector occurs. The critical amplitude, $h(\boldsymbol{q})$, is the amplitude of that eigenvector. Substituting Eq. (59) into Eq. (57) and expanding up to quadratic order in G, one obtains:

$$\sigma + \lambda\{-zLT[G^2(t)](z)\} - \{-z\widehat{G}(z)\}^2 = 0, \tag{60}$$

with the separation parameter

$$\sigma(T) \cong \sigma_0(T_c - T) = \begin{cases} < 0 \;\; , \quad \text{liquid} \\ > 0 \;\; , \quad \text{glass} \end{cases} \tag{61}$$

σ_0 is positive and the exponent parameter

$$\lambda = \lambda[S(\boldsymbol{q})|_{T=T_c}] \,. \tag{62}$$

The separation parameter is a measure of the distance from the transition point. The importance of λ will become clear below. Explicit expressions for σ and λ are given in Ref. [27].

At the transition point, i.e. for $\sigma = 0$, the solution of Eq. (60) is the *critical law*:

$$G(t) \sim t^{-a} \,, \tag{63}$$

where the exponent a is a solution of:

$$\lambda = \frac{\Gamma^2(1-a)}{\Gamma(1-2a)}, \tag{64}$$

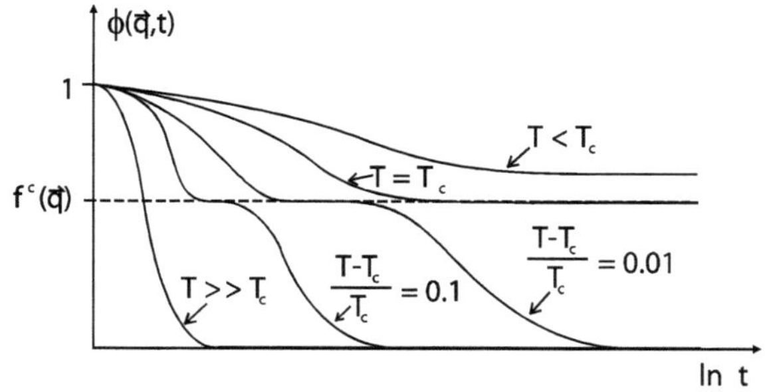

Fig. 4. Qualitative time and temperature dependences of the intermediate scattering function $\Phi(\boldsymbol{q}, t)$ for fixed $\boldsymbol{q}$. The values 0.1 and 0.01 for $(T - T_c)/T_c$ should not be taken literally. They should only indicate that T is chosen closer and closer to T_c

where $\Gamma(x)$ is the Gamma function. Since λ is uniquely determined by $S(\boldsymbol{q})$ at T_c it can be calculated microscopically, from which a follows. a is restricted to $0 < a \leq 1/2$. For $a = 0$ a higher order bifurcation scenario occurs [27]. To solve Eq. (60) for $T \neq T_c$ it is convenient to introduce a correlation scale $\sqrt{|\sigma|}$ by:

$$G(t) = \sqrt{|\sigma|}\; g_{\pm}(t/t_{\sigma}) \quad , \quad \sigma \gtrless 0, \tag{65}$$

where the time scale t_σ and the master functions $g_\pm$ can be obtained as follows. Introducing Eq. (65) into Eq. (60) and using $\hat{t} = t/t_\sigma$ and $\hat{z} = t_\sigma z$ leads to:

$$\pm 1/\hat{z} + \lambda LT[g_{\pm}^2(\hat{t})](\hat{z}) + \hat{z}(\hat{g}_{\pm}(\hat{z}))^2 = 0, \tag{66}$$

which has the solution:

$$g_{\pm}(\hat{t}) \sim \hat{t}^{-a}, \tag{67}$$

for $\hat{t} \ll 1$, i.e. $t \ll t_\sigma$, since $\pm 1/\hat{z}$ can be neglected due to $\hat{z} \gg 1$. Substituting Eq. (67) into Eq. (65) and taking into account that $G(t)$ must reduce for $\sigma \to 0$ to the critical correlator, Eq. (63) allows us to fix t_σ:

$$t_\sigma(T) \sim |\sigma|^{-\frac{1}{2a}} \sim |T - T_c|^{-\frac{1}{2a}} \quad , \quad T \lessgtr T_c\, . \tag{68}$$

For $\hat{t} \gg 1$ one obtains [27]:

$$g_{\pm}(\hat{t}) \cong \begin{cases} \sqrt{1-\lambda} & , \quad (+) \\ -B\hat{t}^{b} & , \quad (-) \end{cases} \tag{69}$$

with $B > 0$.

Again, the positive exponent b is determined by the exponent parameter λ:

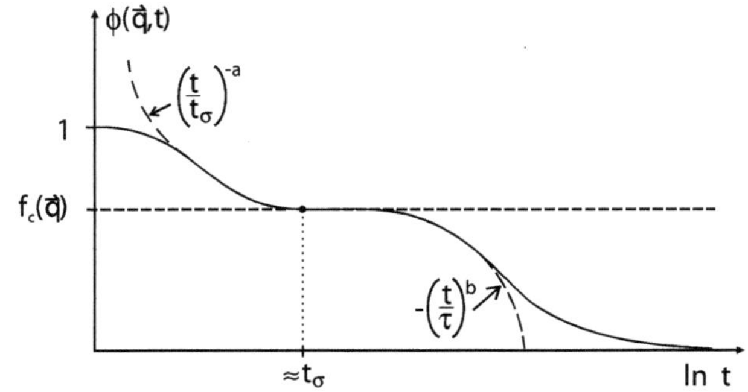

Fig. 5. $\Phi(\boldsymbol{q},t)$ for fixed $\boldsymbol{q}$ and a single temperature $T > T_c$ (*solid line*). Both power laws, the critical law and the von Schweidler law, are shown as *dashed lines.* The *horizontal dashed line* is the plateau height equal to the critical nonergodicity parameter $f_c(\boldsymbol{q})$

$$\lambda = \frac{\Gamma^2(1+b)}{\Gamma(1+2b)} \,. \tag{70}$$

In the liquid phase (minus sign in Eq. (69)) we obtain, besides the *critical law*, Eq. (63), a second power law, the so-called *von Schweidler law.* Substituting $g_-(\hat{t})$ from Eq. (69) into Eq. (65) leads to:

$$G(t) \sim (t/\tau(T))^b, \tag{71}$$

with the second time scale:

$$\tau(T) \sim |\sigma|^{-\frac{1}{2b}}\, t_\sigma \sim (T-T_c)^{-\gamma}\,, \quad T \geq T_c, \tag{72}$$

and the exponent

$$\gamma = \frac{1}{2a} + \frac{1}{2b} \,. \tag{73}$$

The von Schweidler law, Eq. (71), is valid for $\hat{t} \gg 1$, i.e. $t \gg t_\sigma(T)$ and $t \ll \tau(T)$ because of $|G(t)| \ll 1$. Both power laws are shown in Figure 5. The critical law describes the relaxation to the plateau values $f^c(\boldsymbol{q})$ and $f(\boldsymbol{q})$ above *and* below T_c, respectively, and the von Schweidler law the initial decay from the plateau for $T > T_c$. Both time scales, t_σ and τ, exhibit power-law divergence at T_c. τ diverges faster than t_σ, due to $\gamma > 1/(2a)$ (see Fig. 6).

These fascinating results proven by Götze in 1984 [43] and 1985 [44] were absolutely new. The time range $t_0 \ll t \ll \tau(T)$ (t_0 is a microscopic time scale $\sim \Omega_q^{-1}$) which includes both power laws has been called (fast) β-relaxation. It has a simple physical explanation. If the temperature is low enough, each particle feels a cage on a time scale t_σ. The first relaxation step (critical law) is a relaxation within the cage and the second one (von Schweidler law) is related to the "opening" of the cage for $t \gg t_\sigma$, which is the initial stage of the structural relaxation.

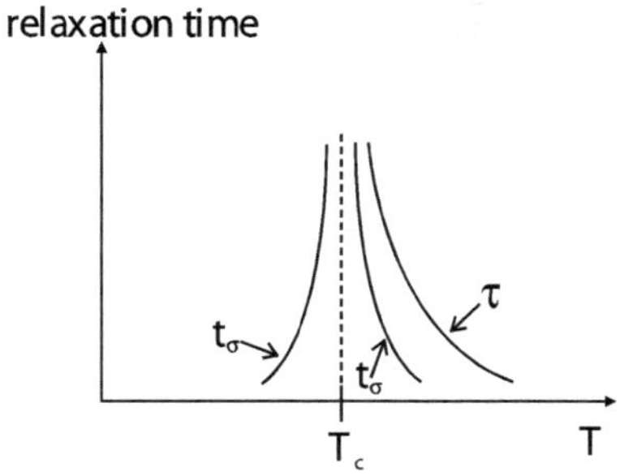

Fig. 6. Qualitative temperature dependence of both relaxation times $t_\sigma(T)$ and $\tau(T)$ which exhibit power law divergence at T_c

The fact that the q and t dependences of the correlators factorize in the β-regime is one of the highly nontrivial predictions of MCT which has not been found before in condensed matter physics. Numerous experiments [12] (see also the contribution by R. Zorn and U. Buchenau in this monograph) and computer simulations [13, 14, 15] have given consistent results with regard to these MCT predictions.

What remains is to study the dynamics for t on the order of or much larger than $\tau(T)$, which is called α-relaxation. There the factorization does not hold anymore. Therefore one has to solve Eq. (57). This can only be done numerically. But there is one important feature, which is the scale invariance of Eq. (57). This means that a scale transformation $t \to t \cdot y$, $\Phi(\boldsymbol{q}, t) \to \Phi(\boldsymbol{q}, t \cdot y)$ or $z \to z \cdot y^{-1}$, $\hat{\Phi}(\boldsymbol{q}, z) \to y\hat{\Phi}(\boldsymbol{q}, z \cdot y^{-1})$ leaves Eq. (57) invariant. This allows us to introduce a q-dependent master function $\Phi_q(\tilde{t})$ such that:

$$\Phi(\boldsymbol{q}, t; T) = \Phi_q(t/\tau(T)). \tag{74}$$

That is, the time and temperature dependences of Φ appear only in the combination $t/\tau(T)$. Equation (74) is well known in glass science as the time-temperature superposition principle. The master function $\Phi_q(\tilde{t})$ can be determined from a numerical solution of Eq. (57). The validity of Eq. (57) and Eq. (74) has also been tested experimentally [12] and by computer simulations [13, 15].

4 Microscopic Theory: Replica Theory

In 1996 Mézard and Parisi [45] presented a replica theory for the structural glass transition. This theory was inspired by spin glass theory [1, 2]. Kirkpatrick, Thirumalai and Wolynes first noticed the analogy between spin glass and structural glass transitions [8]. In 1987 they showed that mean-field spin glass models with one-step replica symmetry breaking [2] exhibit a *dynamical* (MCT-like) glass transition at T_c and a *static* one at T_s, which is below T_c. At T_s the configurational entropy vanishes. Hence T_s can be identified with T_K, the Kauzmann temperature. In the next section we will describe the physical picture behind this theory and its thermodynamic formulation. The second section contains the microscopic theory which can be considered as a first-principles approach of an earlier attempt by Singh, Stoessel and Wolynes to describe the glass transition by a density functional theory [46].

4.1 Thermodynamical Description

Let $\rho(\boldsymbol{x})$ be the local particle density of a liquid with N identical particles in a volume V. A theorem [47] guarantees that there exists a free energy functional $\mathcal{F}[\rho(\boldsymbol{x}), T]$ such that the equilibrium phases are obtained from:

$$\frac{\delta \mathcal{F}[\rho(\boldsymbol{x}), T]}{\delta \rho(\boldsymbol{x})} = 0. \tag{75}$$

$\mathcal{F}[\rho(\boldsymbol{x}), T]$ is not known exactly. In many cases one uses an approximation suggested by Ramakrishnan and Youssouff [48]. Let $\rho^{(\alpha)}(\boldsymbol{x})$ (T-dependence is suppressed), $\alpha = 1, 2, 3, \ldots$ be solutions of Eq. (75) (local minima) with free energy per particle $f_\alpha(T) = \mathcal{F}[\rho^{(\alpha)}, T]/N$ and let $\mathcal{N}(f, N, T)$ be the number of solutions with free energy $f = \mathcal{F}/N$ at T. The configurational entropy per particle $S_c(f, T)$ (cf. Sect. 1is defined by:

$$\mathcal{N}(f, N, T) = \exp[N S_c(f, T)] \, . \tag{76}$$

At high temperature the equilibrium phase is given by the uniform density solution $\rho(\boldsymbol{x}) \equiv n = N/V$ of Eq. (75). Inspired by mean-field spin glasses with a discontinuous transition, Mézard and Parisi [49, 50] assume that at the MCT temperature T_c an exponential number of solutions $\rho^{(\alpha)}(\boldsymbol{x})$ occur for f between $f_{\min}(T)$ and $f_{\max}(T)$, that is:

$$S_c(f, T) > 0 \, , \quad f_{\min}(T) < f < f_{\max}(T) \, . \tag{77}$$

Above $f_{\max}(T)$ and below $f_{\min}(T)$, $S_c(f, T) = 0$. This situation is illustrated in Figure 7. $S_c(f, T)$ varies smoothly with T and is concave in f, i.e. $\partial^2 S_c(f, T)/\partial f^2 < 0$. The *crucial* assumption is that $S_c(f, T)$ vanishes at $f_{\min}(T)$ with *finite* slope:

$$\frac{\partial S_c}{\partial f}(f_{\min}, T) < \infty \tag{78}$$

(see Figure 8). At low temperatures (below T_c) the partition function $Z(T, N)$ can be written as a sum over the individual local minima:

$$Z(T, N) = \exp[-\beta N \Phi(T)] \cong \sum_\alpha \exp[-\beta N f_\alpha(T, N)], \tag{79}$$

which becomes for N large:

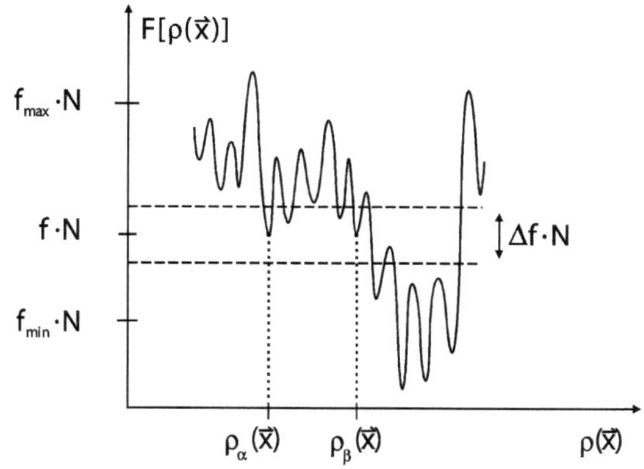

Fig. 7. Schematic illustration of the free energy landscape. For $f_{\min}(T) < f < f_{\max}(T)$ there are exponentially many local minima in an energy interval of width Δf (two of them, $\rho_\alpha(\boldsymbol{x})$ and $\rho_\beta(\boldsymbol{x})$, are shown explicitly). Above $f_{\max}(T)$ and below $f_{\min}(T)$ there are at most an algebraic number of minima

$$Z(T,N) \approx \int_{f_{min}(T)}^{f_{max}(T)} df \exp\{-N[\beta f - S_c(f,T)]\} . \tag{80}$$

The main contribution to this integral comes from the minimum solution $f^*(T)$ of the free energy:

$$\Phi(f,T) = f - TS_c(f,T) , \tag{81}$$

that is:

$$\Phi(T) = \min_f \Phi(f,T) = f^*(T) - TS_c(f^*(T),T) . \tag{82}$$

The Boltzmann constant is set to one. Now there are two possibilities. *First*, for temperatures below T_c but high enough, $f^*(T)$ will be within the interval $[f_{\min}(T), f_{\max}(T)]$. Then $f^*(T)$ follows from the solution of $\partial\Phi(f,T)/\partial f = 0$. Using Eq. (81) this yields:

$$\frac{1}{T} = \frac{\partial S_c}{\partial f}(f,T) \tag{83}$$

(see Fig. 8a). *Second*, $f^*(T)$ will decrease with decreasing T and will get stuck at $f_{\min}(T)$ (see Fig. 8b). Then $S_c(f^*(T),T) = 0$ and

$$\Phi(T) = f_{\min}(T). \tag{84}$$

If we denote by $s_0(T)$ the slope of $\partial S_c/\partial f$ at $f_{\min}(T)$:

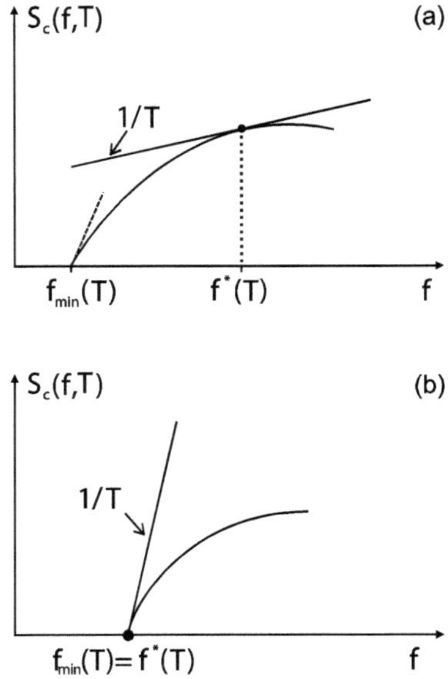

Fig. 8. Schematic f-dependence of the configurational entropy $S_c(f,T)$. (**a**) $T_K < T < T_c$ where $f^*(T)$ is between $f_{\min}(T)$ and $f_{\max}(T)$, (**b**) $T \leq T_K$ where $f^*(T)$ coincides with $f_{\min}(T)$. Note that the slope of the tangent (which must equal $1/T$) in (**a**) is smaller than in (**b**). The dashed line in (a) shows the slope of the tangent at $f_{\min}(T)$

$$s_0(T) = \frac{\partial S_c}{\partial f}(f_{\min}(T), T) < \infty, \tag{85}$$

this will happen at a temperature $T_s > 0$ with

$$\frac{1}{T_s} = s_0(T_s) \, . \tag{86}$$

Since S_c vanishes at T_s it is identical to the Kauzmann temperature T_K. The reader should note that the existence of a nonzero Kauzmann temperature and accordingly the existence of a static glass transition at T_K relies on the finite value of $s_0(T)$ for *all* temperatures, including zero.

Now the major goal is to calculate the free energy $\Phi(T)$ and the configurational entropy *below* T_K. This can be done by a trick. Instead of taking one system, one chooses m replicas which are weakly coupled to each other with coupling constant ε [49, 50, 51]. In the glass phase, i.e. below T_K, this small coupling will force all systems into the *same* local minimum α. Therefore the corresponding partition function is given by:

$$Z_m(T,N) \approx \int\limits_{f_{\min}(T)}^{f_{\max}(T)} df \exp\{-N[m\beta f - S_c(f,T)]\}, \tag{87}$$

in analogy to Eq. (80). The "saddle point" condition, Eq. (83), is:

$$\frac{m}{T} = \frac{\partial S_c}{\partial f}(f, T) . \tag{88}$$

If we allow m to take any real positive value, it is obvious that Eq.(88) has a solution $f_m^*(T) > f_{\min}(T)$ if $m < 1$ is small enough, even if $f_{m=1}^*(T) = f_{\min}(T)$. Then the free energy $\Phi(m, T)$ is given by:

$$\Phi(m, T) = \min_f [mf - TS_c(f, T)] . \tag{89}$$

Next, it is easy to prove that $\Phi(m, T)$ and the free energy per particle of the replicated system

$$\phi(m, T) = \frac{1}{m} \Phi(m, T) \tag{90}$$

allow us to calculate $f(T)$ and $S_c(T)$:

$$f(T) = \frac{\partial \Phi(m, T)}{\partial m}, \tag{91}$$

$$S_c(T) = \frac{m^2}{T} \frac{\partial \phi(m, T)}{\partial m} . \tag{92}$$

Hence, the knowledge of $\Phi(m, T)$ allows us to determine f and S_c. Increasing m towards one, there will be a critical value $0 \leq m^*(T) \leq 1$ at which $f_m^*(T) = f_{\min}(T)$. A schematic representation of the $m - T$ phase diagram is given in Fig. 9. There are two phases: first, a liquid phase above the solid line. For $T > T_K$ it is a liquid where the replicas become uncorrelated for $\varepsilon \to 0$, whereas for $T < T_K$ and $m < m^*(T)$ it is a "molecular liquid" where the particles of the m replicas form "molecules" even for $\varepsilon \to 0$. Second, below the solid line there is a glass phase. Since $S_c(T) = 0$ in the glass phase, Eq. (92) implies that $\phi(m, T)$ is *independent* of m, i.e. it is:

$$\phi(m, T) = \phi(1, T) \tag{93}$$

for all $m \geq m^*(T)$ and $T < T_K$. On the other hand, $\phi(m, T)$ is continuous at the liquid-glass phase boundary $m^*(T)$:

$$\phi_{\text{liquid}}(m^*(T), T) = \phi_{\text{glass}}(m^*(T), T). \tag{94}$$

Combining Eq. (93) and Eq. (94) one arrives at the important result:

$$\Phi(T) \equiv \phi(1, T) = \phi_{\text{liquid}}(m^*(T), T) . \tag{95}$$

Due to Eq. (95) one can calculate the free energy $\Phi(T)$ of the physical system from the free energy of the replica system in its *liquid* phase, despite $T < T_K$. Since there are powerful techniques for the calculation of the liquid free energy [40], the relationship in Eq. (95) allows us to calculate $\Phi(T)$ from $\phi(m, T)$ and the latter also allows us to determine the configurational entropy $S_c(f, T)$ from Eq. (91) and Eq. (92) by eliminating m.

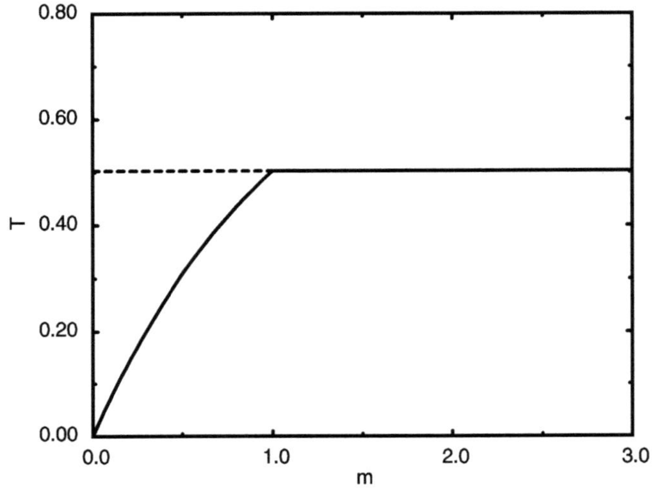

Fig. 9. Phase diagram of the replica system in the m-T space. The *solid line* separates the liquid from the glass phase (see also text). The *dashed line* within the liquid phase separates the "atomic" liquid for $T > T_K \simeq 0.5$ from the "molecular" liquid for $T < T_K$. The curved part of the *solid line* is $m^*(T)$

4.2 Microscopic Description

In the last section we have shown that the free energy $\Phi(T)$ and the configurational entropy $S_c(f,T)$ can be obtained from $\Phi(m,T)$, the free energy of a replica system in its *liquid* phase. Now we will describe how this can be performed from first principles.

The potential energy of a N-particle system in a volume V may be given by pair potentials $v(\boldsymbol{x})$:

$$V(\boldsymbol{x}_1, \ldots, \boldsymbol{x}_N) = \sum_{i<j} v(\boldsymbol{x}_i - \boldsymbol{x}_j) \; . \tag{96}$$

Let us consider m *identical* systems (replicas) which are weakly coupled by an attractive pair potential $w(\boldsymbol{x}_i^a - \boldsymbol{x}_j^b)$, which is considered to be short ranged. Then the total potential energy is given by:

$$V_m^\varepsilon(\{\boldsymbol{x}_i^a\}) = \sum_{a=1}^{m} V(\boldsymbol{x}_1^a, \ldots, \boldsymbol{x}_N^a) + \varepsilon \sum_{a<b} \sum_{i<j} w(\boldsymbol{x}_i^a - \boldsymbol{x}_j^b), \tag{97}$$

where $\boldsymbol{x}_i^a$ is the position of particle i in replica a and $\varepsilon \geq 0$ is an infinitesimal coupling constant. Note, that w breaks the replica permutational symmetry. It acts like a symmetry-breaking magnetic field in the case of a ferromagnet. For $\varepsilon > 0$ the thermodynamic limit $N \to \infty$ forces the replicas into the same local minima. Taking the limit $\varepsilon \to 0$ afterwards leaves the replicas in the same state for $T < T_K$ and makes them uncorrelated for $T > T_K$

(cf. discussion in section 4.1). Therefore for $T < T_K$ and $\varepsilon > 0$ it is (using an appropriate labeling of the particles):

$$\boldsymbol{x}_i^1 \approx \boldsymbol{x}_i^2 \approx \cdots \approx \boldsymbol{x}_i^m \,. \tag{98}$$

Therefore we can introduce center-of-mass coordinates $\boldsymbol{X}_i$ of an "m-atomic molecule" and relative coordinates $\boldsymbol{u}_i^a$ such that:

$$\boldsymbol{x}_i^a = \boldsymbol{X}_i + \boldsymbol{u}_i^a, \tag{99}$$

where

$$\boldsymbol{X}_i = \frac{1}{m}\sum_{a=1}^{m} \boldsymbol{x}_i^a, \tag{100}$$

and $\boldsymbol{u}_i^a$ has to fulfill the N constraints:

$$\sum_{a=1}^{m} \boldsymbol{u}_i^a = 0 \tag{101}$$

for all i.

The classical partition function (configurational part) is given by:

$$Z_m^\varepsilon(T,N) = \frac{1}{N!}\int \prod_{i,a} d^d x_i^a \exp[-\beta V_m^\varepsilon(\{\boldsymbol{x}_i^a\})], \tag{102}$$

where d is the spatial dimension. Making use of Eq. (99), this leads to:

$$Z_m^\varepsilon(T,N) = \frac{1}{N!}\int \prod_i d^d X_i \int \prod_{i,a} d^d u_i^a \tag{103}$$
$$\cdot \prod_{i=1}^{d} \Big(m^d \delta\Big(\sum_a \boldsymbol{u}_i^a\Big)\Big) \exp[-\beta V_m^\varepsilon(\{\boldsymbol{X}_i\},\{\boldsymbol{u}_i^a\})],$$

where the δ-function accounts for the constraints in Eq. (101). Due to Eq. (98), $|\boldsymbol{u}_i^a| \ll 1$. Therefore a harmonic approximation can be applied which yields:

$$Z_m^\varepsilon(T,N) = Z_m(T,N) \cong \frac{1}{N!}\int \prod_i d^d X_i \int \prod_{i,a} d^d u_i^a$$
$$\cdot \prod_i \Big(m^d \delta\Big(\sum_a \boldsymbol{u}_i^a\Big)\Big) \cdot \exp\Big\{-\beta\Big[mV(\{\boldsymbol{X}_i\})$$
$$+\frac{1}{2}\sum_a \sum_{i\mu,j\nu} M_{i\mu,j\nu}(\{\boldsymbol{X}_i\})(u_{i\mu}^a - u_{j\nu}^a)^2\Big]\Big\} \tag{104}$$

where $M_{i\mu,j\nu}$ is the Hessian matrix of V and $\boldsymbol{u}_{i\mu}^a$ is the μ-component of $\boldsymbol{u}_i^a$. Note that we were allowed to put $\varepsilon = 0$, because we already assumed that

Eq. (98) holds. Using an integral representation of the δ-function, the integrals with respect to $\boldsymbol{u}_{i\mu}^a$ are Gaussian and can be performed, which involves:

$$\left[\prod_{k=1}^{Nd} \lambda_k(\{\boldsymbol{X}_i\})\right]^{1/2} \left[\prod_{k=1}^{Nd} \lambda_k(\{\boldsymbol{X}_i\})\right]^{-m/2}$$

$$= \exp\left[-\frac{m-1}{2}\operatorname{Tr}\ln(M_{i\mu,j\nu}(\{\boldsymbol{X}_i\}))\right], \tag{105}$$

where λ_k are the eigenvalues of $(M_{i\mu,j\nu})$. The result can be brought into the following form:

$$Z_m(T,N) \cong (c_m(T))^N Z_1(T/m,N)$$
$$\cdot\Big\langle \exp\Big[-\frac{m-1}{2}\operatorname{Tr}\ln(M_{i\mu,j\nu}(\{\boldsymbol{X}_i\}))\Big]\Big\rangle(T/m), \tag{106}$$

where

$$c_m(T) = m^{\frac{d}{2}}(2\pi T)^{\frac{d}{2}(m-1)}. \tag{107}$$

$Z_1(T/m,N)$ is the partition function of the original system, but at a temperature T/m, and $\langle(\cdot)\rangle$ denotes averaging with respect to $\exp[-\beta m V(\{\boldsymbol{X}_i\})]/Z_1$ $(T/m,N)$. In the next step one uses the approximation:

$$\Big\langle \exp\Big[\cdots\Big]\Big\rangle \approx \exp\Big[-\frac{m-1}{2}\langle\operatorname{Tr}\ln(M_{i\mu,j\nu}(\{\boldsymbol{X}_i\}))\rangle\Big]. \tag{108}$$

The right hand side of Eq. (108) could be calculated if the density distribution of the eigenvalues were known, which is not the case. Therefore, more approximations are necessary. We will not present these technical manipulations, which can be found in Ref. [50], but we will present the idea. Let us skip for a moment the logarithm in Eq. (108). Then we have to calculate $\langle\operatorname{Tr}(M_{i\mu,j\nu}(\{\boldsymbol{X}_i\}))\rangle = \sum\limits_{i,\mu}\langle M_{i\mu,i\mu}(\{\boldsymbol{X}_i\})\rangle$. Since $M_{i\mu,i\mu}(\{\boldsymbol{X}_i\}) = \sum\limits_{j}(\partial^2 v/\partial r_\mu^2)(\boldsymbol{r} = \boldsymbol{X}_i - \boldsymbol{X}_j)$, this involves the average of the second derivatives of the pair potential $v(\boldsymbol{r})$. Assuming that $M_{i\mu,i\mu}$ do not fluctuate much one obtains finally:

$$\langle\operatorname{Tr}(M_{i\mu,j\nu}(\{\boldsymbol{X}_i\}))\rangle\left(\frac{T}{m}\right) \approx \frac{N}{d}\int d^d r g\left(\boldsymbol{r},\frac{T}{m}\right)\boldsymbol{\nabla}^2 v(\boldsymbol{r}). \tag{109}$$

That is, $\langle\operatorname{Tr}(M_{i\mu,j\nu}(\{\boldsymbol{X}_i\}))]\rangle(T/m)$ can be related to the pair distribution function at T/m. In a similar way one can express $\langle\operatorname{Tr}\ln(M_{i\mu,j\nu}(\{\boldsymbol{X}_i\}))]\rangle$ (T/m) by $g(\boldsymbol{r},T/m)$:

$$\langle\operatorname{Tr}\ln(M_{i\mu,j\nu}(\{\boldsymbol{X}_i\}))]\rangle\left(\frac{T}{m}\right) \approx N\mathcal{G}\left[g(\boldsymbol{r};\frac{T}{m})\right]. \tag{110}$$

Using these approximations we end up with the free energy $\Phi(m,T) = -T\lim\limits_{N\to\infty} N^{-1}\ln Z_m(T,N)$:

$$\beta\Phi(m,T) \approx -\ln c_m(T) + \Phi(T/m) + \frac{1}{2}(m-1)\mathcal{G}[g(\boldsymbol{r},T/m)] \,. \tag{111}$$

This result demonstrates what has been discussed in Sect. 4.1. The free energy, $\Phi(m,T)$, of the replica system is given by the free energy of the original system, $(m = 1)$, at T/m and by a functional, $\mathcal{G}$, of the pair distribution function of the original system, also at T/m. In the liquid phase $\Phi(T/m)$ can also be expressed by $g(\boldsymbol{r},T/m)$ [40]. Therefore, up to an irrelevant term $\ln c_m(T)$, one has succeeded in expressing $\Phi(m,T)$ by the pair distribution function at a temperature T/m. If T is below T_K, one has to choose m small enough in order to be in the liquid phase. As already mentioned above, there are powerful methods [40] to calculate $g(\boldsymbol{r},T/m)$ for the liquid phase. Finally, the obtained $\Phi(m,T)$ allows us to determine $\Phi(T)$ and $S_c(f,T)$ as described in Sect. 4.1. In Figure 10 we show $S_c(f,T)$ obtained by Mézard and Parisi [49] for a three-dimensional system $(d = 3)$ with density one and a soft-sphere pair potential $v(\boldsymbol{r}) = |\boldsymbol{r}|^{-12}$. The f-dependence has the qualitatively correct behavior, on which the discussion in Sect. 4.1 has been founded. Using the condition in Eq. (86) for the static glass transition point with $T_s \equiv T_K$, these authors obtained $T_K \cong 0.194$ or $\Gamma_K \cong 1.51$ for the dimensionless coupling parameter $\Gamma = nT^{-1/4}$. This value is in a reasonable range and is close to the values $\Gamma_K^{sim} \cong 1.60$ [52] and 1.46 [53] from a MD simulation of *binary* soft-sphere liquids. Since the authors of Ref. [53] claim that their results are compatible with MCT, it is not quite obvious whether the numerical values yield $\Gamma_c = nT_c^{-1/4}$ or not.

5 Summary

In this article we have reviewed phenomenological and microscopic theories for the structural glass transition. The phenomenological approaches rely on several assumptions which are not proven to be correct. Although they are connected with appealing "physical pictures", their predictive power is limited. However, one class of them has obtained a microscopic justification by the replica theory for structural glasses [45, 49, 50]. This theory, based on first principles, predicts a *static* glass transition at the Kauzmann temperature, T_K, where the configurational entropy, $S_c(T)$, vanishes, as stated in the Adams-Gibbs theory (Sect. 1). Although one can avoid the use of replicas [54], the replica theory has a certain beauty because the several equivalent glass phases with the same free energy can be described by copying the system m times and introducing a weak coupling between the copies (replicas). This coupling acts as a "symmetry" breaking field, similar to a magnetic field for a ferromagnet. Analytically continuing m to positive real numbers allows us to relate the thermodynamical properties of the glass phase, i.e. for $T < T_K$, to those of the liquid phase, provided m is taken small enough with respect to one.

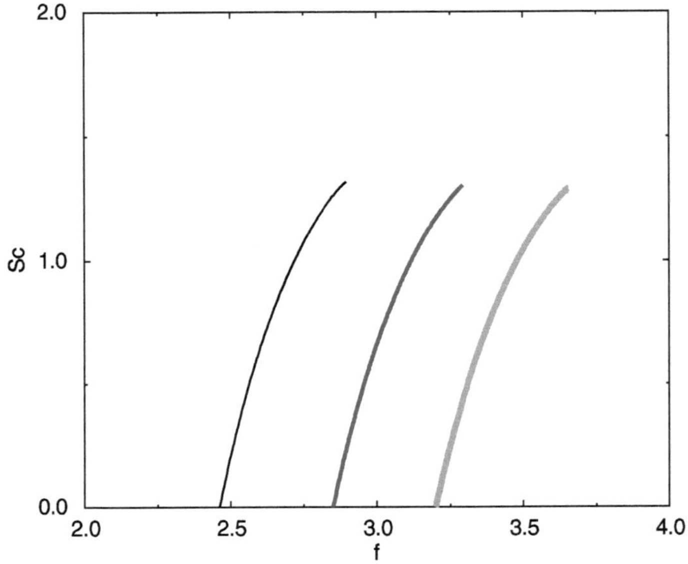

Fig. 10. Microscopic result for $S_c(f, T)$ for $T = 0.05, 0.1, 0.15$ (from left to right) for a system with soft-sphere potential (see also text)

Quite a different type of transition is obtained from mode coupling theory [27, 28, 29]. MCT is a dynamical theory, in contrast to replica theory. It provides an equation of motion for, e.g. the spatial Fourier transform $\Phi(\boldsymbol{q}, t)$ of the normalized density correlator of a simple liquid. Extension to molecular liquids is straightforward [15, 30, 31, 32, 33]. MCT predicts the existence of a *dynamical* glass transition at a temperature T_c where the dynamics change qualitatively. Above T_c the correlator Φ decays to zero, and it converges to the nonergodicity parameter, $f(\boldsymbol{q}) > 0$, below T_c. The nonergodicity parameters vary *discontinuously* at T_c and can be interpreted as glass order parameters. Besides this, MCT makes several new predictions. Close to T_c there exist two scaling laws, the *critical law* and the *von Schweidler law*. The corresponding time scales $t_\sigma(T)$ and $\tau(T)$ exhibit power law divergence at T_c. For times much larger than a typical microscopic time t_0 and much smaller than the α-relaxation time scale, $\tau(T)$, the $\boldsymbol{q}$- and t-dependences of $\Phi(\boldsymbol{q}, t)$ factorize, which is a very strong statement. Furthermore, all the exponents of those power laws can be obtained from only *one* parameter, λ, the exponent parameter. λ is determined by the static correlator $S(\boldsymbol{q})$ at T_c. This proves the very microscopic nature of MCT. Although λ and therefore the exponents are dependent on the physical system, i.e. on the interaction, they are *universal* for *all* correlators of one system which couple to the density fluctuations [27].

These two microscopic theories in some sense are complementary to each other and yield two different glass transition points. It is interesting that the existence of a static and a dynamical glass transition had also been found for mean-field spin glasses with discontinuous order parameter. There it has even been speculated that spin glasses are quite similar to structural glasses [8]. For the mean-field spin glasses, both transitions are related to *singularities* and occur at T_K and T_c with $T_K < T_c$. Therefore both transitions are *sharp*. Concerning MCT (as described in Sect. 3), sometimes called idealized mode coupling theory, it has been shown [35, 55, 56] that the singularity is removed, due to ergodicity-restoring processes. Nevertheless, if the time scale of these processes is much larger than $\tau(T)$, one can observe in a certain time and temperature window the dynamical behavior as predicted from the idealized theory [27, 28, 29]. Deviations start to emerge very close to T_c. This fact has some importance for the static glass transition. For mean-field spin glasses it was proven [57, 58] that the system does not relax to equilibrium below T_c. Therefore the static transition at T_K is masked. Due to the ergodicity-restoring processes this is not anymore true for systems with finite-range interactions. Despite that, it is by far not clear that there is a minimum free energy $f_{\min}(T)$ at which the configurational entropy $S_c(f,T)$ vanishes linearly with a slope $s_0(T) = (\partial S_c/\partial f)(f_{\min}(T),T) < \infty$. This is a very strong assumption which may not be fulfilled in general. For real systems, i.e. in finite dimensions, with short-range interactions, the physical picture described in Sect. 4.1 probably does not hold, i.e. there are not an exponential number of states with density $\rho^{(\alpha)}(\mathbf{x})$ and infinite lifetime. However, there might exist systems with large enough frustration to be close to the idealized situation, at least on a finite time scale. The same holds for MCT. For instance, the ergodicity-restoring processes seem to be extremely weak for colloidal systems. Indeed it has been shown that the dynamics of colloids can be described over many decades in time by MCT [59].

Independent of whether the singularities of both microscopic theories exist or not, the progress which has been made is significant; this is particularly true for MCT. The number of experiments and simulations [12, 13, 14, 15] (see also the contribution by R. Zorn and U. Buchenau in this monograph) that have found consistency with the MCT predictions in an appropriately chosen time and temperature interval is enormous, despite the deviations very close to T_c. One may hope that the few tests [60] of replica theory may be continued in order to check its validity in more detail. An extension including time dependence would be desirable as well.

Both theories probably describe idealized situations only. As mentioned above, MCT has been extended [55, 56] to include ergodicity-restoring processes. But this extended version has not really been tested, because it is rather involved. Describing the ergodicity-restoring processes by a single parameter δ, there was a comparison with experimental data which is satisfactory [61]. Whether replica theory can be extended in the case that the singularity is spurious is not clear. Such an extension might also lead to a much more complicated mathematical structure. In that case it might be much better to restrict to the idealized version, which holds for both theories.

Nevertheless there are still many challenging problems left. Let us mention some of them:

1. further investigations of similarities and dissimilarities between systems with and without quenched disorder
2. glass transition in lattice gas models [62]
3. connection between the "potential energy landscape" properties [16, 17, 18] with MCT and replica theory; e.g. it has been shown that the saddle index vanishes at the MCT temperature T_c [63]
4. investigation of models with trivial statics which may not exhibit a static glass transition but a MCT-like one [64]
5. investigations of models without or with weak static correlations [65], where replica theory and MCT (in its present form) do not yield a glass transition, although one is found in simulations [66]
6. nonequilibrium behavior (aging)

Acknowledgement

I gratefully acknowledge valuable comments on this manuscript by Pablo G. Debenedetti, Wolfgang Götze and Marc Mézard. Figures 9 and 10 were provided by Marc Mézard and Figs. 1 and 2 were produced by Michael Ricker. I also would like to thank both for their support.

References

1. K. Binder, A.P. Young, Rev. Mod. Phys. **58**, 801,(1986)
2. M. Mézard, G. Parisi, M.A. Virasoro "Spin Glass Theory and Beyond", World Scientific, Singapore 1987
3. J. Wong, C.A. Angell "Glass: Structure by Spectroscopy", Dekker, New York, 1976; C.A. Angell, Science **267**, 1924 (1995)
4. J. Jäckle, Rep. Prog. Phys. **49**, 171 (1986)
5. P.G. Debenedetti "Metastable Liquids: Concepts and Principles", Princeton University Press, Princeton, 1996
6. E. Donth "The Glass Transition: Relaxation Dynamics in Liquids and Disordered Materials", Material Science, Springer-Verlag, Berlin 2001
7. J.P. Bouchaud, M. Mézard, J. Phys. I (France) **4**, 1109 (1984); E. Marinari, G. Parisi, F. Ritort, J. Phys. A**27**, 7615 (1994); A**27**, 7647 (1994); L.F. Cugliandolo, J. Kurchan, G. Parisi, F. Ritort, Phys. Rev. Lett. **74**, 1012 (1995); P. Chandra, L.B. Ioffe, D. Sherrington, Phys. Rev. Lett. **75**, 713 (1995); L.B. Ioffe, A.V. Lopatin, J. Phys.: Condens. Matter **13**, L371 (2001)
8. T. R. Kirkpatrick, D. Thirumalai, Phys. Rev. Lett. **58**, 2091 (1987); Phys. Rev. B**36**, 5388 (1987); T.R. Kirkpatrick, P.G. Wolynes, Phys. Rev. B**36**, 8552 (1987)
9. A.L. Greer, Nature **404**, 134 (2000); F.H. Stillinger, P.G. Debenedetti, Biophysical Chemistry, in press; M.R. Feeney, P.G. Debenedetti, F.H. Stillinger, submitted to J. Chem. Phys.

10. F.H. Stillinger, P.G. Debenedetti, T.M. Truskett, J. Phys. Chem. B**105**, 11809 (2001)
11. U. Bengtzelius, W. Götze, A. Sjölander, J. Phys. C**17**, 5915 (1984)
12. W. Götze, L. Sjögren, Rep. Prog. Phys. **55**, 241 (1992); W. Götze, J. Phys.: Condens. Matter **11**, A1 (1999)
13. W. Kob, H.C. Anderson, Transp. Theory Stat. Phys. **24**, 1179 (1995); W. Kob in "Experimental and Theoretical Approaches to Supercooled Liquids: Advances and Novel Applications", eds. J. Fourkas, D. Kivelson, U. Mohanty, K. Nelson, ACS Books, Washington, 1997: p. 28; W. Kob, J. Phys.: Condens. Matter **11**, R85 (1999) W. Kob, Les Houches lecture notes 2002, cond-mat/0212344.
14. F. Sciortino, P. Gallo, P. Tartaglia, S.-H. Chen, Phys. Rev. E**54**, 6331 (1996); F. Sciortino, L. Fabbian, S.-H. Chen, P. Tartaglia, Phys. Rev. E**56**, 5397 (1997); C. Theis, F. Sciortino, A. Latz, R. Schilling, P. Tartaglia, Phys. Rev. E**62**, 1856 (2000); S. Kämmerer, W. Kob, R. Schilling, Phys. Rev. E**56**, 5450 (1997); E**58**, 2131, 2141 (1998); A. Winkler, A. Latz, R. Schilling, C. Theis, Phys. Rev. E**62**, 8004 (2000)
15. L. Fabbian, A. Latz, R. Schilling, F. Sciortino, P. Tartaglia, C. Theis, Phys. Rev. E**60**, 5768 (1999)
16. M. Goldstein, J. Chem. Phys. **51**, 3728 (1969); F.H. Stillinger, T.A. Weber, Phys. Rev. A**25**, 978 (1982); F.H. Stillinger, Science **267**, 1935 (1995); D. Sherrington, Physica D**107**, 117 (1997).
17. S. Sastry, P.G. Debenedetti, F.H. Stillinger, Nature **393**, 554 (1998); S. Büchner, A. Heuer, Phys. Rev. Lett. **84**, 2168 (2000); T.B. Schrøder, S. Sastry, J.C. Dyre, S.C. Glotzer, J. Chem. Phys. **112**, 9834 (2000); L. Angelani, R. Di Leonardo, G. Ruocco, A. Scala, F. Sciortino, Phys. Rev. Lett. **85**, 5356 (2000); K. Broderix, K.K. Bhattacharya, A. Cavagna, A. Zippelius, I. Giardina, Phys. Rev. Lett. **85**, 5360 (2000); A. Cavagna, Europhys. Lett. **53**, 490 (2001); T.S. Grigera, A. Cavagna, I. Giardina, G. Parisi, Phys. Rev. Lett. **88**, 055502 (2002); J.P.K. Doyle, D.J. Wales, J. Chem. Phys. **116**, 3777 (2002)
18. F. Sciortino, W. Kob, P. Tartaglia, Phys. Rev. Lett. **83**, 3214 (1999)
19. G. Adam, J.H. Gibbs, J. Chem. Phys. **43**, 139 (1965)
20. W. Kauzmann, Chem. Rev. **43**, 219 (1948)
21. M.H. Cohen, D. Turnbull, J. Chem. Phys. **31**, 1164 (1959)
22. M.H. Cohen, G.S. Grest, Phys. Rev. B**20**, 1077 (1979); ibid B**24** (1981)
23. D. Stauffer, A. Aharony "Introduction to Percolation Theory", Taylor and Francis Ltd., London, 1994
24. J.H. Gibbs, J. Chem. Phys. **25**, 185 (1956); J.H. Gibbs, E.A. Di Marzio, J. Chem. Phys. **28**, 373 (1958)
25. P.J. Flory, J. Chem. Phys. **9**, 660 (1941); M.L. Huggins, J. Chem. Phys. **9**, 440 (1941)
26. K. Kawasaki in "Phase Transitions and Critical Phenomena", eds. C. Domb, M.S. Green, Academic Press, London, 1976
27. W. Götze in "Liquids, Freezing and the Glass Transition", eds. J.P. Hansen, D. Levesque, J. Zinn-Justin, North Holland, Amsterdam 1991
28. R. Schilling in "Disorder Effects on Relaxational Processes", eds. R. Rickert, A. Blumen, Springer-Verlag, Berlin, 1994
29. H.Z. Cummins, J. Phys.: Condens. Matter **11**, A95 (1999)
30. T. Franosch, M. Fuchs, W. Götze, M.R. Mayr, A.P. Singh, Phys. Rev. E**56**, 5659 (1997)

31. R. Schilling, T. Scheidsteger, Phys. Rev. E**56**, 2932 (1997); R. Schilling, Phys. Rev. E**65**, 051206 (2002)
32. S.-H. Chong, F. Hirata, Phys. Rev. E**58**, 6188 (1998)
33. S.-H. Chong, W. Götze, A.P. Singh, Phys. Rev. E**63**, 011206 (2001)
34. T.R. Kirkpatrick, Phys. Rev. A**31**, 939 (1985)
35. S.P. Das, G.F. Mazenko, Phys. Rev. A**34**, 2265 (1986)
36. T.R. Kirkpatrick, P.G. Wolynes, Phys. Rev. A**35**, 3072 (1987)
37. E. Zaccarelli, G. Foffi, F. Sciortino, P. Tartaglia, K.A. Dawson, Europhys. Lett. **55**, 157 (2001)
38. S. Franz, J. Hertz, Phys. Rev. Lett. **74**, 2114 (1995)
39. J.-P. Bouchaud, L.F. Cugliandolo, J. Kurchan, M. Mézard, Physica A**226**, 243 (1996)
40. J.P. Hansen, I.R. McDonald, "Theory of Simple Liquids", 2nd edition, Academic Press, London, 1986
41. D. Forster "Hydrodynamical Fluctuations, Broken Symmetry and Correlation Functions", Benjamin, Reading, 1975
42. F. Sciortino, W. Kob, Phys. Rev. Lett. **86**, 648 (2001)
43. W. Götze, Z. Phys. B**56**, 139 (1984)
44. W. Götze, Z. Phys. B**60**, 195 (1985)
45. M. Mézard, G. Parisi, J. Phys. A**29**, 6515 (1996)
46. Y. Singh, J.P. Stoessel, P.G. Wolynes, Phys. Rev. Lett. **54**, 1059 (1985)
47. N.D. Mermin, Phys. Rev. **137**, A 1441 (1965)
48. T.V. Ramakrishnan, M. Youssouff, Phys. Rev. B**19**, 2775 (1979)
49. M. Mézard, G. Parisi, Phys. Rev. Lett. **82**, 747 (1999)
50. M. Mézard, G. Parisi, J. Chem. Phys. **111**, 1076 (1999)
51. R. Monasson, Phys. Rev. Lett. **75**, 2847 (1995); S. Franz, G. Parisi, J. Phys. I, (France) **5**, 1401 (1995)
52. B. Bernu, Y. Hiwatari, J.P. Hansen, J. Phys. C**18**, L371 (1985)
53. J.N. Roux, J.L. Barrat, J.P. Hansen, J. Phys.: Condens. Matter **1**, 7171 (1989)
54. M. Mézard, G. Parisi, J. Phys.: Condens. Matter **12**, 6655 (2000)
55. W. Götze, L. Sjögren, Z. Phys. B.**65**, 415 (1987)
56. R. Schmitz, J.W. Dufty, P. De, Phys. Rev. Lett. **71**, 2066 (1993)
57. H. Horner, Z. Phys. B**66**, 175 81987); A. Crisanti, H. Horner, H.J. Sommers, Z. Phys. B**92**, 257 (1993)
58. L.F. Cugliandolo, J. Kurchan, Phys. Rev. Lett. **71**, 173 (1993)
59. W. van Megen, S.M. Underwood, Phys. Rev. E**47**, 248 (1993)
60. B. Coluzzi, M. Mézard, G. Parisi, P. Verrocchio, J. Chem. Phys. **111**, 9039 (1999); B. Coluzzi, G. Parisi, P. Verrocchio, J. Chem. Phys. **112**, 2933 (2000); B. Coluzzi, G. Parisi, P. Verrocchio, Phys. Rev. Lett. **84**, 306 (2000)
61. H.Z. Cumnmins, W.M. Du, M. Fuchs, W. Götze, S. Hildebrand, A. Latz, G. Li, N.J. Tao, Phys. Rev. E**47**, 4223 (1993)
62. W. Kob, H.C. Andersen, Phys. Rev. E**48**, 4364 (1993); G. Biroli, M. Mézard, Phys. Rev. Lett. **88**, 025501 (2002); A. Lawlor, D. Reagan, G.D. McCullagh, P. De Gregorio, P. Tartaglia, K.A. Dawson, Phys. Rev. Lett. **89**, 245503 (2002)
63. L. Angelani et al., K. Broderix et al., T.S. Grigera et al. from Ref. [17]
64. K. Kawasaki, B. Kim, Phys. Rev. Lett. **86**, 3582 (2001); K. Kawasaki, B. Kim, J. Phys.: Condens. Matter **14**, 2265 (2002)
65. R. Schilling, G. Szamel, Europhys. Lett. **61**, 207 (2003); J. Phys.: Condens. Matter **15**, S967 (2003)
66. C. Renner, H. Löwen, J.L. Barrat, Phys. Rev. E**52**, 5091 (1995); S.P. Obukhov, D. Kobsev, D. Perchak, M. Rubinstein, J. Phys. I (France) **7**, 563 (1997)

Glassy Dynamics and Aging in Disordered Systems

Heinz Horner

Institut für Theoretische Physik, Ruprecht-Karls-Universität Heidelberg, Philosophenweg 19, 69120 Heidelberg, Germany
horner@tphys.uni-heidelberg.de

1 Introduction

A great number of disordered systems exhibit very long relaxation times as some critical temperature is approached. Below this temperature equilibrium is no longer reached within finite time and the behaviour of such systems becomes non ergodic. Among those are diluted alloys of magnetic ions in a non magnetic matrix, so called spin-glasses [1], supercooled liquids [2] entering some glassy state at low temperature or particles moving in a random potential exhibiting a transition from diffusion to creep or pinning [3, 4]. The temporal behaviour of such systems is often referred to as glassy dynamics. Below the characteristic temperature aging is observed [5]. If, for instance, a spin-glass is cooled in a magnetic field and kept for some waiting time, the complete decay of the magnetisation after the field has been switched off, is hindered for times of the order of the waiting time. Similar phenomena are observed for the deformation of glasses under the influence of applied forces.

Glassy dynamics can also be of relevance for systems and phenomena outside physics. For some combinatorial optimisation problems simulated annealing is used in order to find good solutions. This means that some stochastic dynamics, characterized by some "temperature", is used in order to find minima of a properly defined cost function. If such a system exhibits a freezing transition, it is not effective to spend much time in simulated annealing below this temperature and investigating the dynamics slightly above the freezing transition is much more efficient [6]. Other examples of systems of interest outside physics are models of neural networks or certain aspects of markets and other economical systems [7].

The systems mentioned above are classical. There is also interest in quantum mechanical disordered systems, for instance in certain spin-glasses or in interacting tunnelling systems being of relevance for glasses at very low temperatures. This lecture deals, however, with classical dynamical systems only.

A breakthrough of our understanding of the low temperature properties of glassy systems came from replica theory and the replica symmetry breaking scheme proposed by Parisi [8]. This theory focuses on the evaluation of the free energy averaged over some frozen disorder. The picture emerging from this theory is a decomposition of phase space into so called pure states or ergodic components, separated by barriers of infinite hight in the thermodynamic limit. Investigating the overlap among those states, an ultrametric organization reveals.

An alternative treatment of systems with quenched disorder is via dynamics [9, 10, 11, 12]. This will be the subject of this lecture. As it turns out there are many features and results common to both treatments, but there are also differences which will be discussed.

In Sect. 2 I present some of the systems of interest, in particular spin-glasses, supercooled liquids and glasses, drift, creep and pinning of a particle in a random potential, neural networks, graph partitioning as an example of combinatorial optimisation, the K-sat problem, a prototype problem from computer science, and finally the minority game as a model for the behaviour of agents on markets.

Section 3 deals with the dynamics of the spherical p-spin-glass with long ranged interactions. This model is a prototype for glassy dynamics. The equations of motion for correlation- and response-functions are derived and equilibrium solutions above the freezing temperature are investigated. Below the freezing temperature the system behaves for short time as if it were in equilibrium, for longer time its off equilibrium nature becomes apparent and correlation- and response-functions depend on waiting time. This is the manifestation of aging.

At the end of Sect. 3 I discuss the so called crossover region and aging in glasses. Within mode coupling theory, which is discussed in the contribution of R. Schilling [13] in this volume, and which shows great similarity to the p-spin model, a dynamical transition is found at some critical temperature T_c. This is sometimes referred to as "ideal glass transition". Comparison with experiments shows that this transition is actually smeared out and the fitted T_c is well above the actual glass transition region. Some smearing out also results from an extended version of the mode coupling theory, which applies, however, to equilibrium only. I am going to discus a modified version of the p-spin model, which is able to take account for this rounding as well as aging.

2 Examples of Disordered Systems

2.1 Spin-glasses

Typical spin-glasses are diluted alloys of magnetic ions in a non magnetic matrix, e.g. $Ag_{1-x}Mn_x$ with $x \sim 3\%$. The magnetic ions interact via the RKKY-interaction which is mediated through the conduction electrons. It is of the form

$$J(r) \sim r^{-3} \cos(k_f r) \tag{1}$$

shown in Fig. 1. For random distances r_{ij} between pairs of magnetic ions, their exchange interactions $J_{ij} = J(r_{ij})$ are also random variables including positive and negative values. This leads to frustration, as shown in the insert of Fig. 1 for a triple $i\,j\,k$ of ions with $J_{ij} > 0$, $J_{i,k} > 0$ and $J_{j,k} < 0$.

At high temperatures spin-glasses typically show a Curie-law for the magnetic susceptibility $\chi \sim 1/T$, as shown in Fig. 2 for $\underline{\mathrm{Cu}}\mathrm{Mn_x}$. Below some freezing temperature T_g the system is no longer in equilibrium. If the system is cooled in a small field, the magnetisation $M = \chi_{FC} B$ stays more or less constant (a and c in Fig. 2). The ratio $\chi_{FC} = M/B$ is called field cooled susceptibility. If, however, the system is cooled without applied field, and the field is applied only after some waiting time t_w, the resulting magnetisation is reduced, i.e. the zero field cooled susceptibility $\chi_{ZFC} = M/B < \chi_{FC}$ (b and d in Fig. 2). Approaching the critical temperature T_g critical slowing down can be observed, see Fig. 3.

The off equilibrium character of the low temperature phase is most clearly demonstrated by the decay of the remnant magnetisation. In this experiment the sample is rapidly cooled in a magnetic field to a temperature $T < T_g$. After some waiting time t_w the field is switched off, and the decay of the magnetisation is observed as function of time t elapsed after removal of the field.

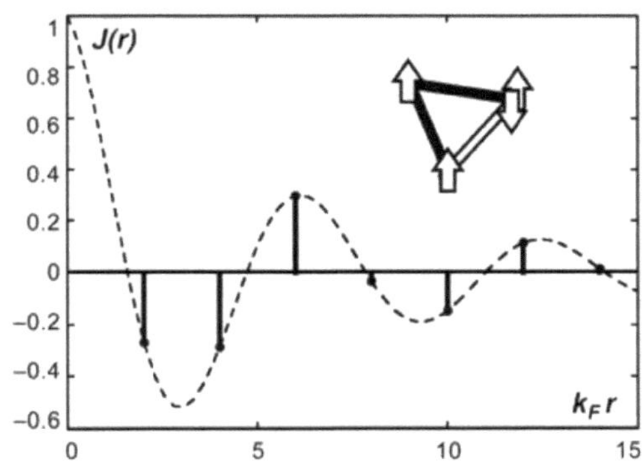

Fig. 1. RKKY-interaction

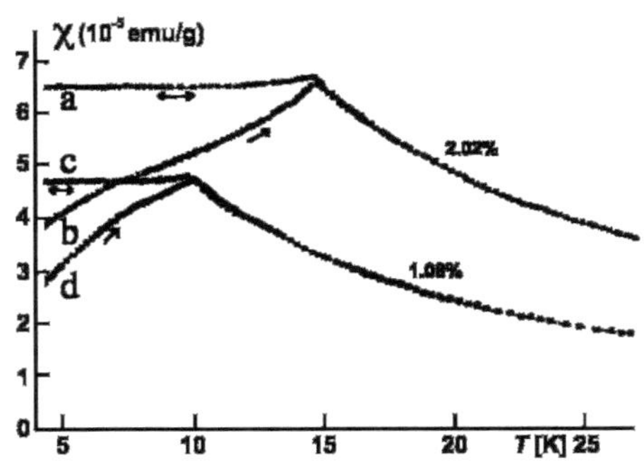

Fig. 2. Susceptibility of $\underline{\mathrm{Cu}}\mathrm{Mn_x}$ [14] for different values of the concentration x of the magnetic ions

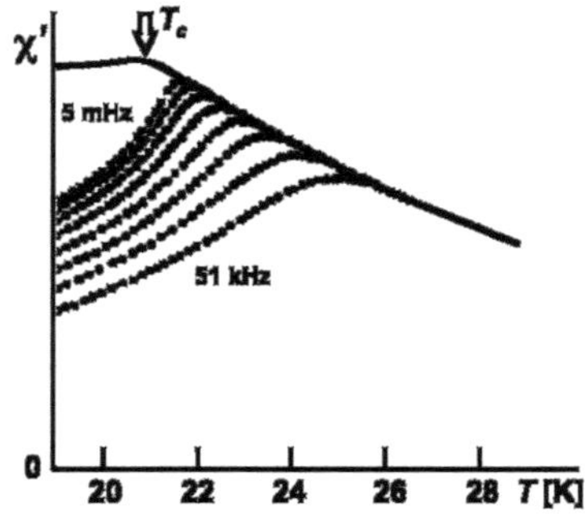

Fig. 3. Real part of the susceptibility $\chi'(\omega)$ for $Fe_{0.5}MN_{0.5}TiO_3$ [15]. The critical temperature is $T_g = 20.7\,K$

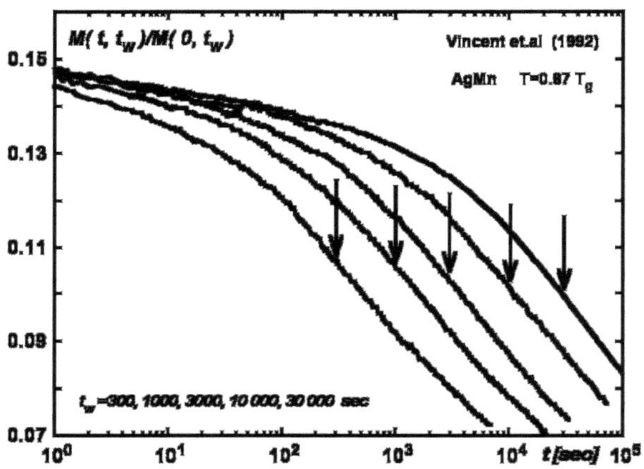

Fig. 4. Decay of the remnant magnetisation for various waiting times t_w indicated by *arrows* [16]

Following some rapid initial decay, not shown in Fig.4, the magnetisation stays almost constant up to time $t \sim t_w$.

Investigating more elaborate temperature programs $T(t)$ a variety of other striking memory and aging effects are observed [17]. Similar aging phenomena are also found in glasses [5].

Theoretical investigations are typically based on Ising models with random interactions. Their energy is

$$H = -\tfrac{1}{2} \sum_{i,j} J_{i,j} \sigma_i \sigma_j \qquad \sigma_i = \pm 1. \tag{2}$$

The interactions J_{ij} are assumed to be Gaussian distributed random variables with

$$\overline{J_{i,j}} = 0 \qquad \overline{J_{i,j} J_{k,l}} = \tfrac{1}{2} \{ \delta_{i,k} \delta_{j,l} + \delta_{i,l} \delta_{j,k} \} W_{i,j}. \tag{3}$$

Calculating e.g. the free energy

$$F = -k_B T \overline{\ln \Big(\sum_{\{\sigma=\pm 1\}} e^{-\beta H} \Big)}^J \tag{4}$$

the problem arises, that the logarithm of the partition function has to be averaged over the Gaussian disorder. Similar problems exist in evaluating expectation values, e.g.

$$\langle \sigma_i \sigma_j \rangle = \overline{\Big(\sum_{\{\sigma\}} \sigma_i \sigma_j e^{-\beta H} \Big) \Big/ \Big(\sum_{\{\sigma\}} e^{-\beta H} \Big)}^J . \tag{5}$$

This can be done using the replica trick [8].

As an alternative, one can examine dynamics in the form of stochastic processes having the Boltzmann distribution $\sim e^{-\beta H}$ as stationary solution. For an Ising model, Glauber dynamics can be used. It is is given by a single spin-flip master-equation such that

$$\frac{d}{dt} \langle \sigma_i(t) \rangle = \Big\langle \tanh \Big(\beta \sum_i J_{i,j} \sigma_j(t) \Big) \Big\rangle . \tag{6}$$

Using a path integral representation of this process, the average over the stochastic interactions can easily be performed. This will be subject of the second part of this lecture.

2.2 Supercooled Liquids and Glasses

If a liquid is cooled sufficiently fast, it may avoid crystallisation and enter the state of a supercooled liquid. With decreasing temperature the shear viscosity $\eta(T)$ increases and reaches a value of $\eta(T_g) \approx 10^{12}$ [Pa sec] defining the glass temperature T_g. At this value, plastic flow can hardly be observed in laboratory experiments. In some glasses, so called strong glasses, the viscosity follows more or less an Arrhenius law $\eta(T) \sim e^{\beta E_a}$ [18]. A typical strong glass former is SiO_2 with an activation energy $E_a \approx 4\,\text{eV}$ corresponding to the binding energy of a covalent bond. Fragile glass formers, on the other hand, show pronounced deviations from this law. Typical examples are organic molecular glasses, ionic glasses, polymers or proteins. Some examples are shown in Fig. 5.

There are other attempts to define a glass transition temperature, e.g. identifying a rapid drop in the specific heat or thermal expansion [20]. Actually the glass temperature defined in one way or another usually depends on the cooling rate, and it is not clear at all, if some finite T_g exists in real glasses. There is no need to discuss this further in this lecture, since theories of the structural glass transitions are covered in the contribution of R. Schilling [13] in this volume.

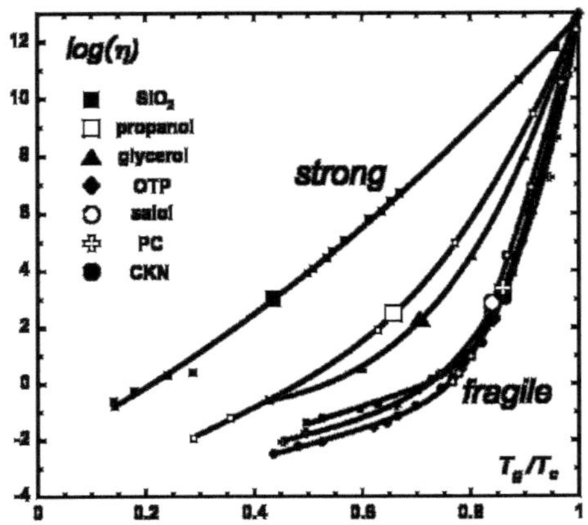

Fig. 5. Angell plot: Logarithm of the viscosity versus inverse temperature [18]. A straight line corresponds to Arrhenius behaviour. The fat symbols indicate the MCT-crossover temperature T_c [19]

There is, however, one aspect of interest in the present context. This is the "Ideal Glass Transition" found in mode coupling theory (MCT) [19, 13]. The resulting equations resemble those found for the dynamics of spherical p-spin interaction spin-glass [21, 22] to be discussed later in this lecture. This theory yields diverging time scales as some critical temperature T_c is approached, and non ergodic behaviour for $T < T_c$.

An onset of critical slowing down, usually referred to as α-relaxation, can be observed in fragile glasses, e.g. in the dielectric susceptibility $\chi''(\omega)$ of the ionic glass CKN [23] shown in Fig. 6.a. Fitting the critical temperature T_c and other parameters of mode coupling theory, as indicated in Fig. 6.b, yields the critical temperatures shown in Fig. 5.

Actually an extended version of the mode coupling theory [19, 24] yields a rounding of the singularity near T_c and equilibrium behaviour for temperatures below T_c, in contrast to the findings of the simplified mode coupling theory or the p-spin-glass. On the other hand, aging can be observed in glasses as well, indicating off equilibrium properties at low temperatures. An example is shown in Fig. 7. We shall come back to this point later.

Assuming pair interactions, the (potential) energy may be written as

$$H = \tfrac{1}{2} \int d\boldsymbol{r} d\boldsymbol{r}' \, V(\boldsymbol{r} - \boldsymbol{r}') \, n(\boldsymbol{r}) \, n(\boldsymbol{r}') \tag{7}$$

where $n(\boldsymbol{r})$ is the density at point $\boldsymbol{r}$.

If one is interested in slow dynamics only, it is sufficient to investigate the limit of overdamped motion, i.e. the Langevin equation

$$\frac{\partial}{\partial t} n(\boldsymbol{r}, t) = \boldsymbol{\nabla} n(\boldsymbol{r}, t) \cdot \boldsymbol{\nabla} \int d\boldsymbol{r}' \, V(\boldsymbol{r} - \boldsymbol{r}') \, n(\boldsymbol{r}', t) + \boldsymbol{\eta}(\boldsymbol{r}, t) \cdot \boldsymbol{\nabla} n(\boldsymbol{r}, t) \tag{8}$$

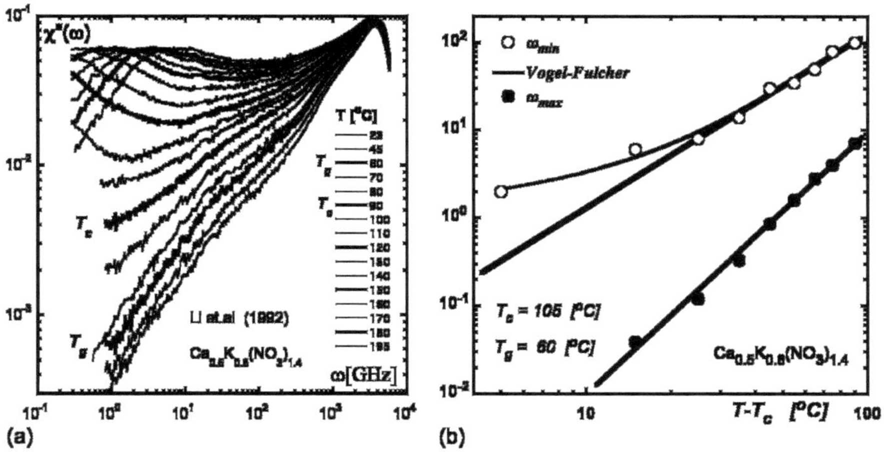

(a) (b)

Fig. 6. (**a**) Imaginary part of the dielectric susceptibility $\chi''(\omega)$ of the ionic glass CKN [23]. The α-relaxation shows up as the broad peak at the lower frequency. (**b**) The frequencies of the maximum $\omega_{\max}$ and minimum $\omega_{\min}$ of $\chi''(\omega)$ fitted to $\omega_{\max} \sim (T - T_c)^{\eta}$ and $\omega_{\min} \sim (T - T_c)^{\eta'}$

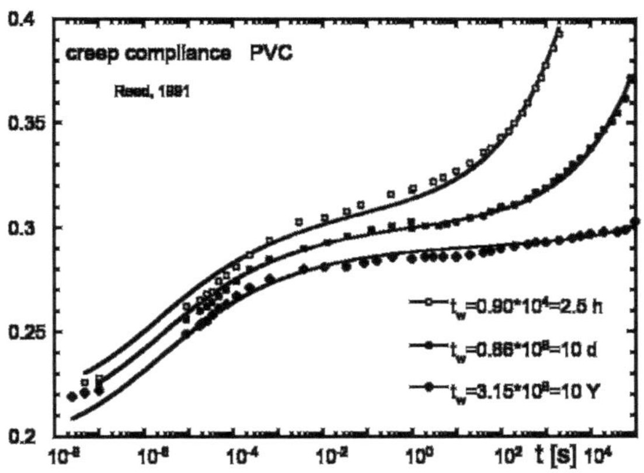

Fig. 7. Aging observed in PVC [25]. A shear stress is applied after some waiting time t_w following a quench from the liquid state. Then the resulting deformation is observed as function of time t

with fluctuating forces such that

$$\langle \boldsymbol{\eta}(\boldsymbol{r},t)\rangle = 0 \qquad \langle \boldsymbol{\eta}(\boldsymbol{r},t)\cdot\boldsymbol{\eta}(\boldsymbol{r}',t')\rangle = 2\,T\,\delta(\boldsymbol{r}-\boldsymbol{r}')\,\delta(t-t'). \tag{9}$$

Investigating time dependent correlation-functions, the nonlinear term in the above equation is treated via mode coupling theory. As mentioned, the resulting equations of motions are identical to those obtained for the p-spin-glass (in equilibrium). This is remarkable, because the Hamiltonian of the spin-

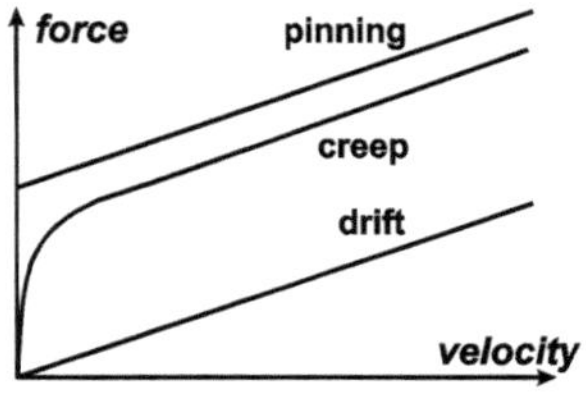

Fig. 8. Drift, creep and pinning for a particle moving in a random potential

glass has built in quenched disorder, whereas the Hamiltonian of the glass does not, and the disorder appears to be self generated.

The relevance of the findings of mode coupling for a glass transition at T_g is not completely obvious. Other concepts have been proposed to understand the glass transition, e.g. activated transitions among "inherent states" [26] or models with kinetically constraint dynamics [27, 28].

2.3 Motion of a Particle (Manifold) in a Random Potential

Another example for glassy dynamics is the motion of a particle in a correlated random potential [3, 4]. Using again the limit of strong damping, a Langevin equation may be used

$$\boldsymbol{v}(t) = \frac{d}{dt}\boldsymbol{r}(t) = -\boldsymbol{\nabla} V(\boldsymbol{r}(t)) + \boldsymbol{K} + \boldsymbol{\eta}(t) \tag{10}$$

where $V(\boldsymbol{r})$ is assumed to be a Gaussian random potential with moments

$$\overline{V(\boldsymbol{r})} = 0 \qquad \overline{V(\boldsymbol{r})\,V(\boldsymbol{r}')} = \frac{V_o}{(|\boldsymbol{r}-\boldsymbol{r}'|+a)^\gamma}\,, \tag{11}$$

and $\boldsymbol{K}$ is some external force pulling the particle. The thermal noise is again given by Eq. (9).

Depending on temperature T and the exponent γ, characterising the range of the random potential, various types of motion are found [4], see Fig. 8. For $T > T_c(\gamma)$ the average velocity $v(K) \sim K$, i.e. there is a finite friction constant. For $T < T_c(\gamma)$ and $\gamma < \gamma_c$ the particle moves only if the force exceeds some critical value $K > K_{\text{pinning}}(T,\gamma)$. For $T < T_c(\gamma)$ and $\gamma > \gamma_c$ creep is found, i.e. $v(K) \sim K^\eta$ with $\eta < 1$.

A similar freezing transition shows up for a chain or some other manifold moving in a random potential [29].

2.4 Neural Networks

Considerable effort has been put into the study of formal neural networks dealing with random patterns [30], e.g. the Hopfield model [31] as a prototype of an associative memory.

Representing the "on" and "off" state of a neuron by the two states of an Ising spin $\sigma_i = \pm 1$, the energy is assumed to be that of an Ising spin-glass Eq. (2). The couplings $J_{i,j}$ are, however, determined by the patterns to be memorised. Let $\xi_i^\mu = \pm 1$ be a set of random patterns, where $i = 1 \cdots N$ labels the neurons and $\mu = 1 \cdots A$ the patterns. Then the choice of the couplings $J_{i,j}$ such that

$$J_{i,j} = \tfrac{1}{N} \sum_\mu \xi_i^\mu \xi_j^\mu \tag{12}$$

ensures, that a fixed point of the dynamics, defined by Eq. (6), exists near each pattern μ at least as long as $A < \alpha_c N$ with $\alpha_c \approx 0.13$. Each fixed point is surrounded by some basin of attraction, i.e. if the initial state contains a sufficient part of a pattern, the complete pattern is reconstructed with a small number of errors. The size of the basin of attraction depends on the loading $\alpha = A/N$ [32]. For $\alpha > \alpha_c$ the memory is overloaded and looses its function completely. The dynamics becomes glassy, which is not too surprising, since the couplings $J_{i,j}$ can be positive as well as negative, resulting in frustration.

Methods developed for spin-glasses, replica theory and dynamics, have also been used to investigate this and other types of neural networks. In particular bounds of learning [30, 6] have been determined.

2.5 Combinatorial Optimisation Problems

Combinatorial optimisation problems show up in various technical applications. The problem is, to find the "best" out of a discrete set of states. Finding the ground state of an Ising spin-glass or solving the travelling sales man problem, i.e. finding the shortest path connecting a set of towns, are examples. In many cases the problem is NP-complete, i.e. the optimal solution can not be found with computational effort growing with the size N of the problem to some power. This means that for larger systems it is virtually impossible to find the optimal solution. In many applications it might, however, be sufficient to find good solutions in polynomial time. One of the standard procedures is simulated annealing [33]. It is implemented by constructing a cost function (energy), and using some stochastic dynamics, characterised by a temperature. This temperature is slowly decreased, and hopefully states with low values of the cost function are found.

Actually such a system might undergo a transition to glassy dynamics below some critical temperature T_c. If this is the case, extending simulated annealing to $T < T_c$ is not efficient, and searching for good solutions at $T \approx T_c$ is more effective [6].

A standard problem in this context is graph bipartitioning [34]. Assume electronic components $i = 1 \cdots N$ have to be placed on two chips A and B. The aim is to place them such that the number of connections between A and B is minimal. A cost function for this problem can be defined by mapping it to an Ising system with

$$\sigma_i = 1 \quad \text{if } i \text{ on } A \qquad \sigma_i = -1 \quad \text{if } i \text{ on } B. \tag{13}$$

and interactions

$$J_{ij} = 1 \quad \text{if } i \text{ and } j \text{ are connected} \qquad J_{i,j} = 0 \quad \text{else.} \tag{14}$$

The resulting cost function is

$$H = -\tfrac{1}{2} \sum_{i,j} J_{i,j}\, \sigma_i\, \sigma_j. \tag{15}$$

For the process of simulated annealing again Glauber dynamics, Eq. (6), may be used.

2.6 Random K-sat Problem

Another standard NP-problem is the K-sat problem [35]. This problem deals with N Boolean variables represented by $\sigma_i = \pm 1$. There are A clauses with K literals per clause among the Boolean variables or their negations, e.g. for $K = 3$: $R = \{\, \sigma_i \text{ OR } \sigma_j \text{ OR NOT } \sigma_k \,\}$. A clause $R_\mu(\{\sigma\})$ can be represented as

$$\xi_i^\mu = \begin{cases} 1 & \text{if} \quad \sigma_i \quad \text{in } R_\mu \\ -1 & \text{if} \quad \text{NOT}\, \sigma_i \quad \text{in } R_\mu \\ 0 & \quad \text{else}\,. \end{cases} \tag{16}$$

The clause $R_\mu(\{\sigma\})$ is not fulfilled if $\sum_{i=1}^{N} \xi_i^\mu \sigma_i = K$, and a cost function

$$H = \sum_{\mu=1}^{A} \delta\Big(\sum_{i=1}^{N} \xi_i^\mu \sigma_i - K \Big) \qquad \text{or} \qquad H = \sum_{\mu} \Big(\sum_{i=1}^{N} \xi_i^\mu \sigma_i - K \Big)^2 \tag{17}$$

may be used. Finding solutions of the K-sat problem, again simulated annealing can be used, and finding a ground state with $H = 0$ means that the corresponding choice of the Boolean variables satisfies all clauses. The formal similarity to the Hopfield model is obvious. For random clauses ξ_i^μ a critical $\alpha_c(K)$ exists, such that for $\alpha = A/K < \alpha_c(K)$ solutions can typically be found by the polynomial simulated annealing procedure.

2.7 Minority Game

Methods, developed for disordered systems, have also been applied to various problems in economy, especially concerning financial markets. Activities of this kind are usually subsummized under the notion econophysics. As an example the minority game [7] is discussed. This is a repeated game, where

N agents have to decide which of two actions, e.g. buy or sell, to take. The goal is to be in the minority. Each agent knows the outcome of the last L games and can select among two strategies, where the set of strategies is different for each agent. For each new game the agent selects the strategy which would have been most successful over the last L games. Again this problem can be mapped onto the dynamics of an Ising spin system with quenched disorder, the disorder being the random set of strategies for each agent. Depending on N and L freezing transitions are found.

3 Dynamics of the p-Spin Interaction Spin-Glass

3.1 Soft Spin Models

In the previous part of this lecture we have seen many different examples of disordered systems exhibiting transitions to glassy dynamics and off equilibrium behaviour. Some of these systems are based on Glauber dynamics with transition probabilities among a discrete set of states. Glauber dynamics is very appropriate for Monte-Carlo simulations, but difficult for analytic investigations [6]. In the following we therefore investigate a model based on continuous degrees of freedom. In equilibrium the resulting equations of motion are identical to those derived from mode coupling theory for glasses [13, 19]. The model is, however, more general in the sense that it allows to study off equilibrium properties and aging [5].

Actually we investigate a whole class of models. The degrees of freedom are N scalar real variables ϕ_i representing soft spins or other variables. The energy (Hamiltonian) is

$$H = \sum_i U(\phi_i) - \sum_i h_i\phi_i - \tfrac{1}{2}\sum_{i,j} J_{ij}\phi_i\phi_j - \tfrac{1}{3!}\sum_{i,j,k} J_{ijk}\phi_i\phi_j\phi_k - \cdots \qquad (18)$$

where $U(\phi)$ is some local potential constraining the distribution of the ϕ_i. Ising spins for instance can be approximated by choosing $U(\phi)$ as deep double well potential with minima at $\phi = \pm 1$, e.g. $U(\phi) = \overline{U}\,(1-\phi^2)^2$. For a spherical model $U(\phi) = \frac{1}{2}\mu\phi^2$ with μ such that $\langle\phi^2\rangle = 1$.

We allow for multi spin interactions $J_{ijk}\cdots$ involving two and more spins or particles. The interactions are Gaussian random variables with

$$\begin{aligned}
&\overline{J_{ij}} = 0 \qquad && \overline{J_{ij}J_{kl}} = \tfrac{1}{2}\{\delta_{i,k}\delta_{j,l} + \delta_{i,l}\delta_{j,k}\}\, W_{ij}^{\{2\}} \\
&\overline{J_{ijk}} = 0 \qquad && \overline{J_{ijk}J_{lmn}} = \tfrac{1}{3!}\{\delta_{i,l}\delta_{j,m}\delta_{k,n} + \cdots\}\, W_{ij}^{\{3\}} \\
&\cdots && \cdots
\end{aligned} \qquad (19)$$

For real spin-glasses the range of the interactions is finite, and one may choose $W_{ij}^{(2)} = W$ if site i and j are next neighbours and $W_{\cdots}^{(p)} = 0$ else (Edwards-Anderson model [36]). For the SK-model (Sherrington-Kirkpatrick [37]) on the other hand interactions with unlimited range are used. Models with long ranged interactions (or models in infinite dimensions) have the advantage that mean field theory yields the exact solution. For other problems like neural networks or combinatorial optimisation problems space has no meaning and interactions among all elements are appropriate. The strength of the interactions has to scale with the number N of elements in order to allow for a non trivial thermodynamic limit $N \to \infty$

$$W^{(p)}_{\cdots} = N^{-p/2}\, W_p. \tag{20}$$

Models of this kind are often referred to as mean field models.

3.2 Replica Theory

Equilibrium properties can be derived from the free energy averaged over the disorder

$$F(T,h) = -k_B\, T\, \overline{\ln\, Z(T,h,J)}^{J}. \tag{21}$$

The problem is to perform the average of the logarithm of the partition function instead of averaging the partition function itself and taking the logarithm afterwards. The replica trick [8, 37] uses the identity

$$\ln(x) = \lim_{n\to 0}\Big(x^n - 1\Big). \tag{22}$$

For integer n the partition function Z^n may be computed by replicating the system n times

$$Z^n = \prod_{\alpha=1}^{n} \int d\phi_1^\alpha \cdots d\phi_N^\alpha\, e^{-\beta \sum_\alpha H(\phi^\alpha)}. \tag{23}$$

This quantity can now be averaged easily over the Gaussian disorder. Since the disorder is the same for each of the replica, the resulting effective Hamiltonian

$$-\beta\mathcal{H} = -\beta \sum_{\alpha} \overline{H(\phi_\alpha)}^{J} + \tfrac{1}{2}\beta^2 \sum_{\alpha,\beta} \Big\{ \overline{H(\phi_\alpha) H(\phi_\beta)}^{J} - \overline{H(\phi_\alpha)}^{J}\, \overline{H(\phi_\beta)}^{J} \Big\} \tag{24}$$

couples different replica. The partition function Z^n can be evaluated for mean field models introducing the order parameters

$$q_{\alpha,\beta} = \overline{\Big\langle \phi_i^\alpha, \phi_i^\beta \Big\rangle}^{J}. \tag{25}$$

The problem is symmetric with respect to permutations of the replica. At high temperatures $T > T_c$ the order parameters are symmetric with respect to permutations as well, i.e.

$$q_{\alpha,\beta} = q_0 \delta_{\alpha,\beta} + q \left(1 - \delta_{\alpha,\beta}\right). \tag{26}$$

At low temperatures $T < T_c$ two different replica symmetry breaking schemes have been proposed [8], a one step symmetry breaking and alternatively a hierarchical symmetry breaking in p steps with $p \to \infty$, the so called full symmetry breaking. At the end the limit $n \to 0$ has to be taken extrapolating the results obtained originally for integer n to real n. The results for the low temperature state with full symmetry breaking are interpreted in terms of a state composed of distinct pure states with ultrametric organisation with respect to their overlap. Pure states are interpreted as regions of phase space, valleys, surrounded by barriers of infinite hight. Within each valley the system is in equilibrium. A corresponding interpretation is obtained from the following investigation of dynamics. For further details on the replica theory and results see [8, 13].

3.3 Langevin Dynamics and Path Integrals

An investigation of dynamics yields additional information about the temporal behaviour of correlations and especially about the critical slowing down at the freezing transition. Below the transition temperature $T < T_c$ signatures of non equilibrium are expected, in particular aging.

Specifying a Hamiltonian for a classical system does not specify its dynamics uniquely. The dynamics has, however, to be chosen such that the equilibrium state $\sim e^{-\beta H}$ is stationary. The simplest form is a Langevin equation corresponding to the overdamped motion of particles or spins

$$\frac{d\phi_i(t)}{dt} = -\frac{\delta H(\{\phi(t)\})}{\delta \phi_i(t)} + \eta_i(t) = F_i(\{\phi(t)\}) + \eta_i(t) \tag{27}$$

where the friction constant has been put to 1. The fluctuating forces, representing a heat bath, are assumed to be Gaussian distributed random variables with

$$\langle \eta_i(t) \rangle = 0 \qquad \langle \eta_i(t)\eta_j(t') \rangle = 2\,T\,\delta_{i,j}\,\delta(t-t'). \tag{28}$$

In contrast to the quenched disorder of the interactions $J_{...}$, the fluctuating forces $\eta_i(t)$ depend on time. Averages over the $J_{...}$ and the thermal motion, are distinguished by different notations, $\overline{\cdots}^J$ and $\langle \cdots \rangle$, respectively.

It is convenient, especially in view of the average over the quenched disorder, to introduce a path integral representation [38]. In the following the derivation is outlined, dropping the site index i for a moment. Let

$$\phi(t) = \phi(t; \{\eta\}; \phi_0) \tag{29}$$

be solution of Eq. (27) with initial condition $\phi(t_0) = \phi_0$. The thermal average of a product of ϕ-variables at different times is then

$$\langle \phi(t)\phi(t')\cdots\rangle = \int d\phi_0 P_0(\phi_0) \iint_{t_0} \mathcal{D}\{\eta\}\, e^{\mathcal{W}(\{\eta\};t_0)} \phi(t;\{\eta\};\phi_0)\, \phi(t';\{\eta\};\phi_0)\cdots \tag{30}$$

with

$$\mathcal{W}(\{\eta\};t_0) = -\tfrac{1}{4T} \int_{t_0} dt\, \eta(t)^2. \tag{31}$$

$P_0(\phi_o)$ is some distribution of initial values.

Instead of integrating over the fluctuating forces $\eta(t)$, it would be more convenient to have a path integral over the dynamical variables $\phi(t)$. This can be done by introducing imaginary auxiliary variables $\hat{\phi}(t)$, and then

$$\langle \phi(t)\phi(t')\cdots\rangle = \int d\phi_0 P_0(\phi_0) \iint_{t_0} \mathcal{D}\{\hat{\phi},\phi,\eta\}\, e^{\mathcal{W}(\{\hat{\phi},\phi,\eta\};t_0)} \phi(t)\, \phi(t')\cdots \tag{32}$$

with

$$\mathcal{W}(\{\hat{\phi},\phi,\eta\};t_0) = -\int_{t_0} dt\, \Big\{ \tfrac{1}{4T}\eta(t)^2 + \hat{\phi}(t)\Big[\dot{\phi}(t) - F(\phi(t)) - \eta(t)\Big]\Big\}. \tag{33}$$

The path integral over the auxiliary imaginary variables $\hat{\phi}(t)$ can be viewed as a product of δ-functions insuring the fulfilment of the equation of motion, Eq. (27), at all time.

The final step is to integrate over the fluctuating forces $\eta(t)$, which can easily be done by completing the square in Eq. (33), and now

$$\langle \phi(t)\phi(t')\cdots\rangle = \int d\phi_0 P_0(\phi_0) \iint_{t_0} \mathcal{D}\{\hat{\phi},\phi\}\, e^{\mathcal{W}(\{\hat{\phi},\phi\};t_0)} \phi(t)\, \phi(t')\cdots \tag{34}$$

with

$$\mathcal{W}(\{\hat{\phi},\phi\};t_0) = \int_{t_0} dt\, \Big\{ T\hat{\phi}(t)^2 - \hat{\phi}(t)\Big[\dot{\phi}(t) - F(\phi(t))\Big]\Big\}. \tag{35}$$

The path integral extends over real functions $\phi(t)$ with initial condition $\phi(t_0) = \phi_0$, and imaginary functions $\hat{\phi}(t)$ with unrestricted initial conditions. The path integral goes over functions in the interval $t_0 < t < t_1$ with t_1 greater than the latest time argument in the expectation value Eq. (34).

The auxiliary fields have a physical interpretation. Allowing for a time dependent external field $h(t)$ in Eq. (18), response functions can be expressed as expectation values of the original and auxiliary fields, e.g.

$$\frac{\delta\, \langle \phi(t)\rangle}{\delta h(t')} = \Big\langle \phi(t)\hat{\phi}(t')\Big\rangle. \tag{36}$$

The expectation value on the right hand side is of the form Eq. (34). The auxiliary fields are therefore denoted as response fields.

Response functions have to be causal, i.e. Eq. (36) has to vanish for $t' \geq t$. This is in accordance with the Ito calculus for the stochastic equation of motion Eq. (27), assuming that the action of the forces is retarded [38, 22].

The formulation given above is completely general in the sense that it is not restricted to equilibrium or small deviations from equilibrium. In general correlation- and response-functions are not related. In equilibrium, however, they have to obey fluctuation-dissipation-theorems (FDT)

$$\beta \frac{d}{dt'} q(t,t') = r(t,t') \tag{37}$$

where correlation- and response-function are defined as

$$q(t,t') = \langle \phi(t)\phi(t') \rangle \qquad r(t,t') = \left\langle \phi(t)\hat{\phi}(t') \right\rangle . \tag{38}$$

A sketch of the proof is as follows: The first step is to show that for equilibrium initial condition

$$P_0(\phi_0) = Z^{-1} e^{-\beta H(\phi_0)} \tag{39}$$

the initial time is arbitrary as long as it is earlier than the earliest time in expectation values of the type Eq. (34) or Eq. (36). Assume that the initial distribution at some time $t_0' < t_0$ is given by Eq. (39). For $t_0' < t < t_0$ one replaces $\hat{\phi}(t) \to \hat{\phi}(t) + \beta\dot{\phi}(t)$ in the path integral Eq. (34) and in the action Eq. (35). This results in

$$\begin{aligned} \mathcal{W}(\{\hat{\phi},\phi\};t_0) - \mathcal{W}(\{\hat{\phi},\phi\};t_0') = \int_{t_0'}^{t_0} dt \Big\{ T\hat{\phi}(t)^2 + \hat{\phi}(t)\Big[\dot{\phi}(t) + F(\phi(t))\Big]\Big\} \\ - \beta H(\phi(t_0)) + \beta H(\phi(t_0')). \end{aligned} \tag{40}$$

The last two terms in Eq. (40) originate from integrating a contribution $\beta\dot{\phi}(t)F(\phi(t)) = -\beta dH(\phi(t))/dt$ which appears if the above substitution is performed. Evaluating the path integral over $\phi(t)$ or actually $\dot{\phi}(t)$ for $t_0' < t < t_0$ the only contributions come from $\hat{\phi}(t) = 0$. The last two terms in Eq. (40) replace the initial equilibrium condition Eq. (39) at t_0' by the corresponding one at t_0. If the system is ergodic, the limit $t_0' \to -\infty$ may be taken and, as a consequence, the initial condition becomes irrelevant.

Consider now a small change of the external field $\delta h(t) = \delta h_0$ for $t < t_0$ and $\delta h(t) = 0$ for $t > t_0$. Assuming equilibrium

$$\frac{\delta \langle \phi(t) \rangle}{\delta h_0} = \int_{-\infty}^{t_0} dt' \, r(t,t') = \beta q(t,t_0). \tag{41}$$

For the first expression the initial time $t_0' \to -\infty$ is used. The second is obtained by differentiating the equilibrium initial condition at t_0 with respect to h. The FDT is obtained by differentiation both expressions of Eq. (41) with respect to t_0.

Deriving this result no explicit time dependence of the Hamiltonian is allowed and the system has to be ergodic.

3.4 Average over Quenched Disorder

As pointed out already in the context of the replica theory, the calculation simplifies considerably for mean field models with long ranged interactions, specified in Eq. (20). In the following we investigate models of this kind [9, 10, 21, 22].

Provided the initial condition does not depend on the disorder, the average over the interactions $J^{(p)}_{\dots}$ can easily be performed using the general relation

$$\overline{e^{ax}} = e^{a\overline{x}+\frac{1}{2}\overline{(x-\overline{x})^2}} \tag{42}$$

valid for Gaussian distributions. This yields, for the Hamiltonian Eq. (18) and the disorder specified in Eq. (19) and Eq. (20), the effective action

$$\begin{aligned}\overline{\mathcal{W}(\{\hat{\phi},\phi\})}^{J} &= \int\! dt \sum_i \Big\{T\hat{\phi}_i(t)^2 - \hat{\phi}_i(t)\Big[\dot{\phi}_i(t) - h - U'(\phi_i(t))\Big]\Big\} \\ &+ \int\! dt \int^t\! dt' \sum_i \Big\{\hat{\phi}_i(t) W'\Big(\tfrac{1}{N}\sum_j \phi_j(t)\phi_j(t')\Big)\hat{\phi}_i(t') \\ &\quad +\hat{\phi}_i(t)\tfrac{1}{N}\sum_j \phi_j(t) W''\Big(\tfrac{1}{N}\sum_k \phi_k(t)\phi_k(t')\Big)\hat{\phi}_j(t')\phi_i(t')\Big\}\end{aligned} \tag{43}$$

with

$$W(x) = \sum_p \tfrac{1}{p!}\, W^{(p)}\, x^p. \tag{44}$$

Correlation- and response- functions calculated with this action are now disorder averaged quantities

$$q(t,t') = \tfrac{1}{N}\sum_i \overline{\langle\phi_i(t)\phi_i(t')\rangle}^{J} \qquad r(t,t') = \tfrac{1}{N}\sum_i \overline{\Big\langle\phi_i(t)\hat{\phi}_i(t')\Big\rangle}^{J}. \tag{45}$$

3.5 Dynamic Mean Field Theory for Spherical Models

Computing local correlation- and response-function, i.e. averages involving $\hat{\phi}_i$ and ϕ_i at a single site i only, a saddle point evaluation of the corresponding path integral is possible. The local time dependent functions are obtained from an effective single site action

$$\begin{aligned}\overline{\mathcal{W}_{\text{eff}}(\{\hat{\phi}_i, \phi_i\})}^J &= \int dt \Big\{ T\hat{\phi}_i(t)^2 - \hat{\phi}_i(t) \Big[\dot{\phi}_i(t) - h - U'\big(\phi_i(t)\big) \Big] \Big\} \\ &+ \int dt \int^t dt' \Big\{ \hat{\phi}_i(t)\, W'\Big(q(t,t')\Big)\, \hat{\phi}_i(t') \\ &\quad + \hat{\phi}_i(t)\, r(t,t') W''\Big(q(t,t')\Big)\, \phi_i(t') \Big\}\end{aligned} \tag{46}$$

which has to be determined selfconsistently. This is exact for mean field models in the thermodynamic limit $N \to \infty$.

Compared to conventional mean field theory, e.g. for magnets, the present theory is much richer because the order parameters are the correlation- and response-functions and therefore functions of two time arguments.

For spherical models with $U(x) = \frac{1}{2}\mu x^2$ the effective action is quadratic, which simplifies the calculation further. In particular it allows to write down the resulting dynamical mean field equations in the following closed form [21, 22]

$$\begin{aligned}\left(\frac{d}{dt} + \mu(t)\right) q(t,t') &= h\, m(t') + \int_{t'}^{t} ds\, K(t,s) q(s,t') \\ &+ \int_{t_0}^{t'} ds \Big\{ M(t,s) r(t',s) + K(t,s) q(t',s) \Big\}\end{aligned} \tag{47}$$

$$\left(\frac{d}{dt} + \mu(t)\right) r(t,t') = \int_{t'}^{t} ds\, K(t,s) r(s,t') \tag{48}$$

$$\left(\frac{d}{dt} + \mu(t)\right) m(t) = h + \int_{t_0}^{t} ds\, K(t,s)\, m(s) \tag{49}$$

with

$$\begin{aligned}K(t,t') &= W''\big(q(t,t')\big)\, r(t,t') \\ M(t,t') &= W'\big(q(t,t')\big).\end{aligned} \tag{50}$$

The spherical constraint $q(t,t) = 1$ yields

$$\mu(t) = hm(t) + T + \int_{t_0}^{t} ds \Big\{ K(t,s) q(t,s) + M(t,s) r(t,s) \Big\}. \tag{51}$$

The above dynamical mean field equations are a set of coupled non linear integro-differential-equations for the correlation- and response-functions. Initial conditions are $q(t,t) = 1$ and $r(t,t) = 1$. It should be stressed that the above equations do not require equilibrium. They are therefore suited to deal with off equilibrium properties and aging. In general the correlation- and response-functions depend on both time arguments t and t' and not only on the difference $t - t'$.

3.6 Equilibrium Dynamics in the Ergodic Phase

Above the freezing temperature T_c, in equilibrium, the correlation- and response-functions depend on the difference $t - t'$ only and they obey fluctuation-dissipation-theorems, Eq. (37). The above equations of motion simplify considerably. The memory terms obey fluctuation-dissipation-theorems as well

$$K(t) = -\beta\,\dot{M}(t) \tag{52}$$

and Eq. (47) reads

$$\left(\frac{d}{dt} + \overline{\mu}\right) q(t) = h\,\overline{m} + \beta\Big\{M(t) - M(\infty)q(\infty)\Big\} - \beta\int_0^t ds\,\dot{M}(t-s)q(s) \tag{53}$$

with $M(t) = W'\big(q(t)\big)$. Magnetisation $\overline{m} = m(t\to\infty)$ and $\overline{\mu} = \mu(t\to\infty)$ are given by

$$\overline{\mu}\,\overline{m} = h + \beta\,\big\{W'(1) - W'(\overline{q})\big\}\,\overline{m} \tag{54}$$

with $\overline{q} = q(t\to\infty)$, and

$$\overline{\mu} = h\overline{m} + T + \beta\,\big\{W'(1) - \overline{q}\,W'(\overline{q})\big\}. \tag{55}$$

For $h = 0$ one has $\overline{m} = \overline{q} = 0$ and $\overline{\mu} = T + \beta\,W'(1)$. The following discussion will be restricted to $h = 0$. The general case with $h \neq 0$ is discussed in [22] and results will be shown later.
For general t, Eq. (53) can be solved numerically. Since we are dealing here with a purely dissipative system, $\dot{q}(t) \leq 0$ is required for all t. Replacing $q(s)$ by $q(t)$ in the integrand of Eq. (53) the following inequality is obtained

$$0 \leq -\dot{q}(t) \leq q(t) - \beta\,W'\big(q(t)\big)\,\big(1 - q(t)\big) \tag{56}$$

or

$$T\,q(t) \geq W'\big(q(t)\big)\,\big(1 - q(t)\big). \tag{57}$$

Left and right hand side of Eq. (57) are shown in Fig.9. Assuming $W(q) \sim q^p$ for $q \to 0$ two cases have to be distinguished:

A) $p = 2$: For $h = 0$ a second solution q_c branches off for $T < T_c$. The stability criterion is, however, violated for $q(t) < q_c$. This means that a phase transition into an off equilibrium phase has to take place at $T = T_c$. Since this solution branches off continuously from $\overline{q} = 0$ at T_c this transition is referred to as continuous transition. For $h \neq 0$ the straight line representing the left hand side cuts the $q{=}0$−axis at some negative value and there is no transition in finite field [39].
B) $p > 2$: For $T > T_c$ again a single solution $\overline{q} = 0$ exists. At $T_c(h)$ a new solution q_c fulfilling

$$W'(q_c)\,\big(1 - q_c\big)/q_c = T_c \tag{58}$$

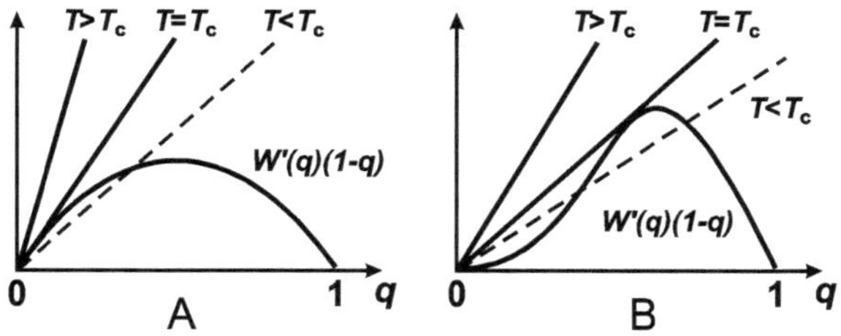

Fig. 9. Stability criterion Eq. (57) for $h = 0$. (A): $p = 2$ (B): $p > 2$. The left hand side is plotted for various temperatures. For $T < T_c$ the stability criterion is violated

shows up. This means that the resulting phase transition is discontinuous. For $T = T_c$ and $q = q_c$ the slope of the left and right hand side of Eq. (57) has to be the same resulting in

$$T_c = W''(q_c)\,(1 - q_c) - W'(q_c). \tag{59}$$

The two Eqs.(58) and (59) determine T_c and q_c.

This discussion can be extended to $h \neq 0$ [22]. The resulting phase diagram is shown in Fig. 10 for the $p =$ 3-spin-glass with $W(q) = \frac{1}{3!}q^3$.

For $h < 0.41$ the transition is discontinuous. The transition temperatures $T_c(h)$ obtained from dynamics is higher than the one found in replica theory. This surprising result might be explained by the proposal that the dominant states counted for in replica theory are not reached by the dynamics investigated [40].

For $h > 0.41$ the transition is continuous and both theories give the same transition temperature.

The results of a numerical integration of Eq. (53) for $h = 0$ is shown in Fig. 11 for temperatures slightly above and at the critical temperature T_c.

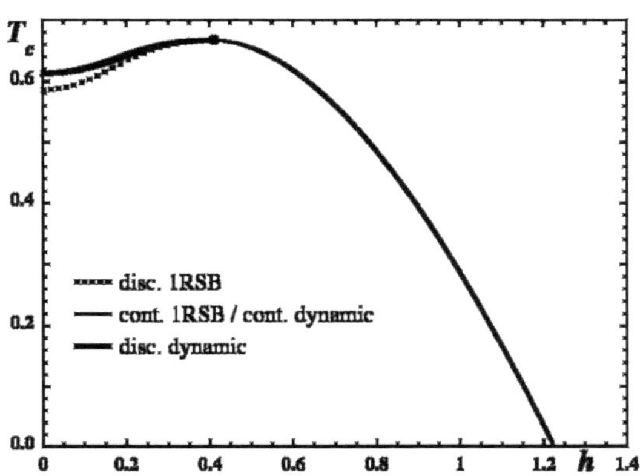

Fig. 10. Phase diagram of the spherical p-spin interaction spin-glass [22]

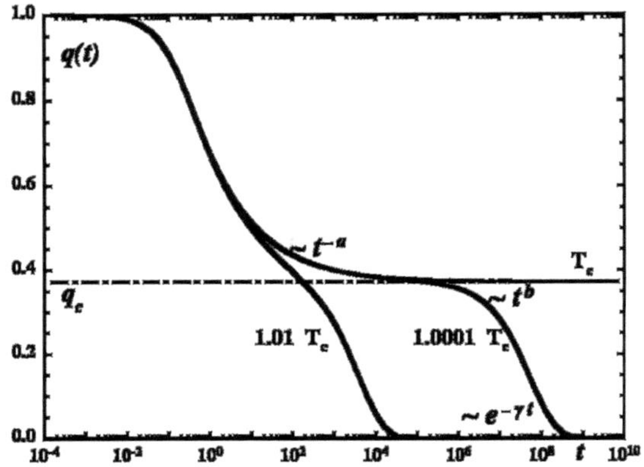

Fig. 11. Correlation function $q(t)$ for $T \geq T_c$ [22]

The asymptotic behaviour $q(t)$ close to T_c can be analysed by identifying appropriate scaling functions for different regions in time and by matching [19, 21, 22]. This resembles very much the crossover scaling analysis near multi critical points [41]. From Fig. 11 one can read off two characteristic temperature dependent time scales: A plateau time $\tau_p(T)$ can be defined as $q(\tau_p) = q_c$ and a second time scale $\tau_a(T)$ characterising the final decay. It may be defined as $q(\tau_a) = \frac{1}{2} q_c$.

For $t \ll \tau_p(T)$ the correlation-function approaches some universal function $q(t) \to \hat{q}_o(t)$ as $T \to T_c$. For $1 \ll t \ll \tau_p(T)$ Eq. (53) is solved by

$$\hat{q}_o(t) \to q_c + c_o\, t^{-a}. \tag{60}$$

The dynamical critical exponent a is solution of

$$\frac{\Gamma(1-a)^2}{\Gamma(1-2a)} = \frac{(1-q_c)W'''(q_c)}{2\,W''(q_c)}. \tag{61}$$

This result is obtained by inserting Eq. (60) into Eq. (53) with $T = T_c$. For $t \to \infty$ the leading contribution is $\mathcal{O}(1)$. The next to leading contributions $\mathcal{O}(t^{-a})$ yield Eq. (58) and Eq. (59) respectively. Collecting terms $\mathcal{O}(t^{-2a})$ Eq. (61) is obtained.

The scaling function ansatz for $t \sim \tau_p$ is

$$q(t) \approx q_c + \tau_p^{-a} \hat{q}_p(t/\tau_p). \tag{62}$$

The scale factor τ_p^{-a} follows from matching Eq. (60) and Eq. (62) at some $1 \ll t \ll \tau_p$ and $\hat{q}_p(\tau) \to c_o \tau^{-a}$ for $\tau \to 0$.

The contributions $\mathcal{O}(T - T_c)$ to Eq. (60) for $t \sim \tau_p$ give

$$\tau_p \sim (T - T_c)^{-1/2a}. \tag{63}$$

For $\tau_p \ll t \ll \tau_a$ Eq. (53) is solved by

$$\hat{q}_p(\tau) \to -c_p\,\tau^b \tag{64}$$

with b obeying

$$\frac{\Gamma(1+b)^2}{\Gamma(1+2b)} = \frac{(1-q_c)W'''(q_c)}{2\,W''(q_c)} = \frac{\Gamma(1-a)^2}{\Gamma(1-2a)}. \tag{65}$$

This means that the two dynamical critical exponents a and b are not independent [19, 22].

The last scaling regime applies for $t \sim \tau_a$. The scaling ansatz is

$$q(t) = \hat{q}_a(t/\tau_a). \tag{66}$$

Matching to Eq. (62) yields

$$\hat{q}_a(\tau) \to q_c - c_p\,\tau^b \quad \text{for} \quad \tau \to 0 \tag{67}$$

and

$$\tau_p = \tau_a^{b/(a+b)} \tag{68}$$

and with Eq. (63)

$$\tau_a \sim (T - T_c)^{-(a+b)/2ab}. \tag{69}$$

This means that only a single independent critical exponent exists. The scaling functions $\hat{q}_o(t)$, $\hat{q}_p(\tau)$ and $\hat{q}_a(\tau)$ have to be determined numerically.

The imaginary part of the frequency dependent susceptibility is

$$\chi''(\omega) = \int_0^\infty dt\, r(t)\, \sin(\omega t) = \beta\,\omega \int_0^\infty dt\, q(t)\, \cos(\omega t) \tag{70}$$

where the second expression is derived from the FDT, Eq. (37). Near T_c the main contributions are due to $t \sim 1$ and $t \sim \tau_a$, respectively. Inserting the scaling discussed above

$$\begin{aligned}\chi''(\omega) &= \beta\omega \int_0^\infty dt\, q_0(t)\, \cos(\omega t) + \beta\tfrac{\omega}{\omega_a} \int_0^\infty d\tau\, \hat{q}_a(\tau)\, \cos(\tfrac{\omega}{\omega_a}\tau) \\ &= \chi_o''(\omega) + \chi_a''(\tfrac{\omega}{\omega_a})\end{aligned} \tag{71}$$

with

$$\omega_a = 1/\tau_a \sim (T - T_c)^{(a+b)/2ab}. \tag{72}$$

With Eq. (60) for $\omega \to 0$

$$\chi_o''(\omega) \to \beta c_o \Gamma(1-a)\cos(\tfrac{1}{2}\pi(1-a))\,\omega^a \tag{73}$$

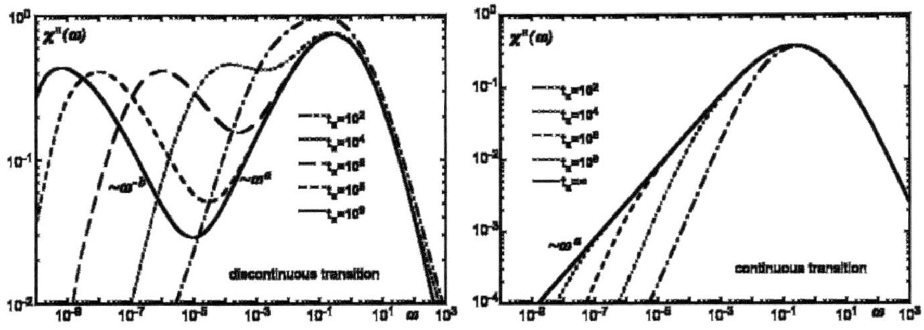

Fig. 12. Susceptibility $\chi''(\omega)$ computed from the numerical solution of Eq. (53) for $h = 0$ and various temperatures characterised by $\tau_a(T)$, Eq. (69). For comparison the susceptibility for finite h in the region of continuous transitions is also shown

is found. For $\omega \ll \omega_a$ the second contribution with Eq. (64) gives

$$\chi''_a(\omega) \to \beta c_a \Gamma(1+b)\cos(\tfrac{1}{2}\pi(1+b))\,\omega^{-b}. \tag{74}$$

Combining both expressions $\chi''(\omega)$ shows a minimum at a frequency

$$\omega_p \sim \tau_a^{b/(a+b)} \sim \tau_p^{-1} \sim (T - T_c)^{1/2a}. \tag{75}$$

The complete susceptibility is plotted in Fig.12. The similarity with the experiments on glasses, shown in Fig.6a, is obvious, and indeed the data on CKN and other glasses have been fitted to the equivalent results of mode coupling theory [19, 23, 24] reproducing the data quite well.

3.7 Off Equilibrium Dynamics and Aging in the QFDT-Phase

The investigation of the dynamics for $T < T_c$ requires to specify some long time scale τ_∞, and to consider the limit $\tau_\infty \to \infty$ eventually. There have been several proposals for such a time scale:

- For finite systems with N elements the equilibration time t_∞ is finite [9]. For large N, $\tau_\infty \sim e^{N^{1/3}}$ [42].
- Relaxing interactions $\overline{J(t)\,J(t')} \sim e^{(t-t')/\tau_\infty}$ [10].
- Slow cooling with $T(t) = (1 - t/\tau_\infty)\,T_c$ [22, 43].
- Fast cooling, aging, with $\tau_\infty = t_w$ [44, 45].

In the following we concentrate on aging. The system is rapidly cooled from high temperatures to some $T < T_c$ at $t = 0$. Alternatively a strong magnetic field could be switched off at $t = 0$. The requirement is, that the initial state is not correlated with the disorder.

Results from a numerical integration [43, 45] of the dynamical mean field equations (47–51) are shown in Fig.13.

Because the system is not in equilibrium at $T < T_c$, correlation- and response-functions depend on both time variables t and t', and not on

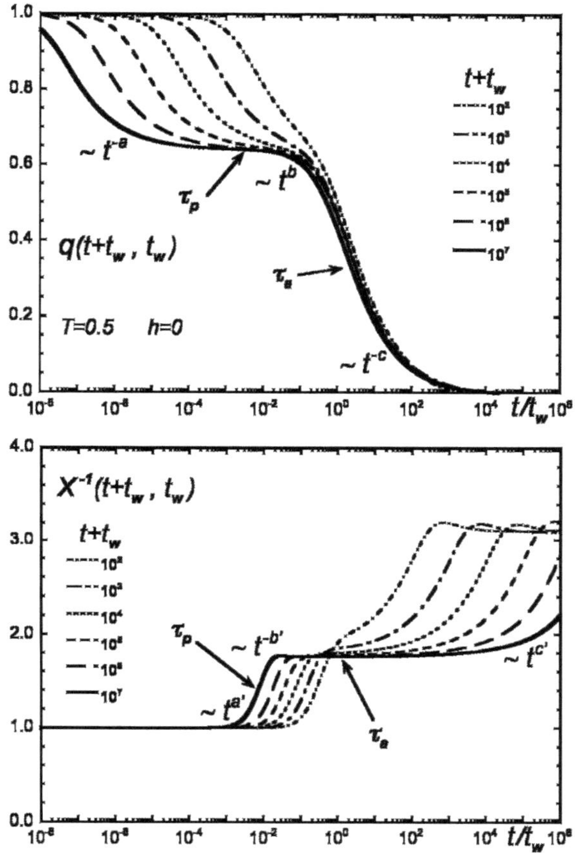

Fig. 13. Correlation function $q(t+t_w, t_w)$ and FDT-violation $X(t+t_w, t_w)$ for $T < T_c$ and $h = 0$. The curves are plotted for fixed $t + t_w$ as functions of t/t_w. The range $t/t_w \sim t + t_w$ corresponds to $t_w \sim 1$

$t - t'$ only as in equilibrium. Furthermore fluctuation dissipation theorems, Eq. (37), are not fulfilled. A measure for FDT-violation is

$$X(t,t') = T \frac{r(t,t')}{\frac{\mathrm{d}}{\mathrm{d}t'} q(t,t')} \tag{76}$$

which is shown in the lower part of Fig.13. This quantity can be relates to an effective temperature [46]

$$T_{\text{eff}}(t,t') = X^{-1}(t,t')\, T. \tag{77}$$

For short time $X(t+t_w, t_w) \approx 1$ is found. This means that the FDT is fulfilled, while $q(t+t_w, t_w)$ decays from 1 to q_c. A distance in phase space can be defined as

$$D(t,t_w) = \sqrt{\tfrac{1}{2}\left\langle \left(\phi_i(t+t_w) - \phi_i(t_w)\right)^2 \right\rangle} = \sqrt{1 - q(t+t_w,t_w)}\,. \tag{78}$$

This distance grows until $q(t+t_w,t_w) \approx q_c$ is reached, and the resulting $D_c = \sqrt{1-q_c}$ can be interpreted as the size of a typical valley in phase space. The system apparently equilibrates within such a valley, and escapes only after some characteristic time $\tau_p(t_w)$. This escape is irreversible, which is indicated by the deviation of $X(t+t_w,t_w)$ from 1 starting at $t \sim \tau_p(t_w)$.

For $t > \tau_p(t_w)$ the parameter $X(t+t_w,t_w)$ reaches a plateau at some X_c, which extends up to $t \gg t_w$. In this regime a modified fluctuation dissipation theorem (QFDT) holds. A solution of this kind was first observed in the context of learning in neural networks exhibiting a discontinuous transition as well [6]. It appears to be a common feature of discontinuous freezing transitions [4, 22].

Investigating the asymptotic behaviour of the solutions of the dynamical mean field equations, Eqs.(47-51), the parameters q_c, $\overline{q}$, $\overline{\mu}$, $\overline{m}$ and X_c can be evaluated analytically [22]. Especially q_c is given by $q_c = \beta W'(q_c)\,(1-q_c)$, which is of the form of Eq. (58), indicating that the solution is marginally stable at all temperatures $T < T_c$.

The crossover scaling analysis is similar to the one discussed for ergodic dynamics in the previous section, but the time scales $\tau_p(t_w)$ and $\tau_a(t_w)$ are now determined by the waiting time t_w.

Again dynamical critical exponents are introduced. They obey

$$\frac{\Gamma(1-a)^2}{\Gamma(1-2a)} = X_c\,\frac{\Gamma(1+b)^2}{\Gamma(1+2b)} = \frac{(1-q_c)W'''(q_c)}{2\,W''(q_c)} \tag{79}$$

which is identical to the result Eq. (65) for $T > T_c$, except for the factor X_c. The behaviour of $X(t+t_w,t_w)$ for $1 \ll t \ll \tau_p$ and $\tau_p \ll t \ll \tau_a$, respectively, is also ruled by power laws, as indicated in the lower part of Fig.13. The resulting exponents fulfil [43, 45]

$$a' = 3a + 1 \qquad b' = 3b - 1\,. \tag{80}$$

Obviously for the existence of the QFDT-phase $b' > 0$ has to be fulfilled. $b' = 0$ therefore marks a phase transition from the QFDT-phase ($\mathcal{A}$-phase) to a phase with a hierarchy of long time scales ($\mathcal{B}$-phase). Another criterion to be fulfilled is $X(t,t') \geq 1$. This gives rise to yet another phase ($\mathcal{C}$-phase) which covers only a tiny portion of the phase diagram.

The complete phase diagram is shown in Fig. 14. Qualitatively the same phase diagram is found for the mean field model of a particle moving in a correlated random potential [4].

In phase $\mathcal{A}$ the plateau-time τ_p and the characteristic time scale for aging τ_a are again related by Eq. (68) with the modified relationship, Eq. (79), among the exponents.

For $t \sim t_w$ it has been argued [44, 12] that

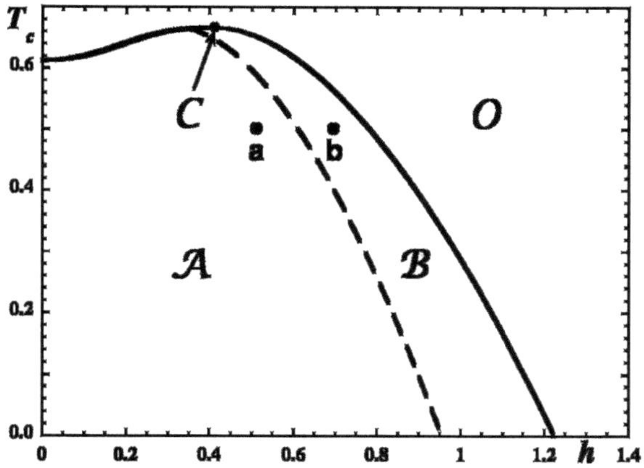

Fig. 14. Phase diagram of the spherical $p=3$–spin interaction spin-glass [45]. $\mathcal{O}$: Ergodic phase. $\mathcal{A}$: QFDT-phase. $\mathcal{B}$ and $\mathcal{C}$: Phases with a hierarchy of long time scales. The points **a** and **b** are discussed later

$$q(t+t_w, t_w) = Q\big(h(t+t_w)/h(t_w)\big) \tag{81}$$

as long as $X(t+t_w,t_w) \approx X_c$. The function $h(t)$ has to be monotonic, but is otherwise not determined. The arbitrariness in $h(t)$ resembles the invariance postulated for the long time behaviour of the SK-Model [9, 10].

$Q(x)$ can be computed numerically and $Q(x) - q_c \sim (x-1)^b$ for $x \to 1$. Especially for zero field the exponent $b=1$ and

$$Q(x) = \frac{q_c}{x} \tag{82}$$

is found. The postulate of scaling with $\tau_a(t_w) = t_w$ yields $h(t) = t^c$. The numerical results [45] shown in Fig. 13 agree reasonably well with this form of scaling and in particular with $c \approx 1/2$.

For the analogous case of the drift of a particle in a random potential [4], $\tau_a \sim \tau_\infty^\mu$, with $\mu = 1 - a(3b-1)/\big(2(a+b)+a(3b-1)\big)$, has been found in the $\mathcal{A}$-phase. In this problem the time scale $\tau_\infty = 1/v_{\text{drift}}$ plays the same role as t_w for aging. Typical values are $\mu \sim 0.9 \cdots 0.95$. Scaling of the form given in Eq. (66) with $\tau_a(t_w) \sim t_w^\mu$ is obtained from $h(t) = e^{(t^\eta - 1)\,c/\eta}$ with $\eta = 1-\mu$. The numerical solutions [45] indicate $\eta > 0$, but the longest time investigated is not long enough, to decide on the actual value of η.

3.8 Off Equilibrium Dynamics and Aging in the $\mathcal{B}$-Phase

We now turn to the $\mathcal{B}$-phase. Figure 15 shows correlation-function $c(t+t_w,t_w) = q(t+t_w,t_w) - m(t+t_w)\,m(t_w)$ and FDT-violation parameter $X(t+t_w,t_w)$ for field and temperature values indicated as points **a** and **b** in Fig. 14. The qualitative difference between $\mathcal{A}$- and $\mathcal{B}$-phase is clearly visible.

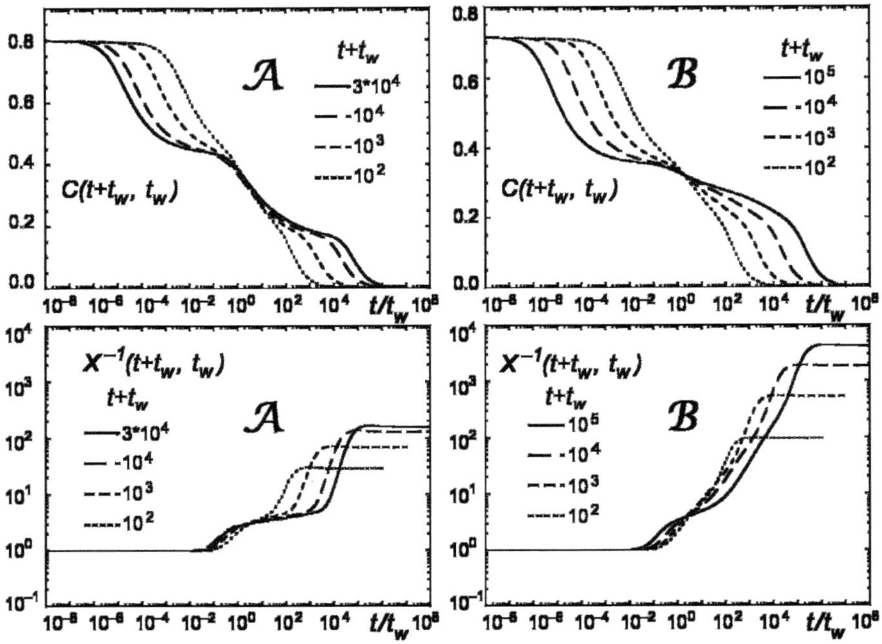

Fig. 15. Correlation function $c(t+t_w,t_w) = q(t+t_w,t_w) - m(t+t_w)\,m(t_w)$ and FDT-violation parameter $X(t+t_w,t_w)$ for $T=0.5$, $h=0.5$ ($\mathcal{A}$-phase) and $h=0.7$ ($\mathcal{B}$-phase) [45] corresponding to point a and b in Fig.14. The curves are plotted for fixed $t+t_w$. The range $t/t_w \sim t+t_w$ corresponds to $t_w \sim 1$

In the $\mathcal{A}$-phase, for $t \gg \tau_p$ and $t_w \gg 1$, the correlation function approaches a scaling form

$$c(t+t_w,t_w) \to \hat{c}_a(t/\tau_a) \qquad \text{with} \quad \tau_a \approx t_w\,, \tag{83}$$

whereas no such scaling is found in the $\mathcal{B}$-phase. This is in some analogy to one step versus full replica symmetry breaking [8].

Within replica theory the p=3-spin-glass does, however, not exhibit a full replica symmetry broken phase at all. This is another difference in the phase diagrams obtained from dynamics and replica theory. For the same range of time, the FDT-violation parameter $X(t+t_w,t_w)$ develops a plateau in the $\mathcal{A}$-phase, consistent with the QFDT-solution, whereas no such plateau shows up in the $\mathcal{B}$-phase.

3.9 Mean Field Dynamics of Systems with Short Ranged Interactions

The slowing down of the dynamics near T_c in a disordered system is not expected to be associated with a diverging length scale, at least not on the level considered here. A diverging length scale is expected only for certain four point correlation functions, which are factorized in the dynamic mean field theory or in mode coupling theory, e.g.

$$C(R_{ij},t-t') = \overline{\langle\phi_i(t)\phi_j(t)\phi_i(t')\phi_j(t')\rangle}^J - \overline{\langle\phi_i(t)\phi_i(t')\rangle\,\langle\phi_j(t)\phi_j(t')\rangle}^J . \quad (84)$$

Dynamic mean field equations for short ranged interactions can be derived adopting a factorization property

$$q_{i,j}(t,t') = f_{i,j}\cdot q(t,t') \qquad r_{i,j}(t,t') = f_{i,j}\cdot r(t,t'). \quad (85)$$

The corresponding factorization has been proposed and tested within mode coupling theory for supercooled liquids [19, 13]. The resulting equations for the time dependent parts, $q(t,t')$ and $r(t,t')$, are of the same form as those derived for mean field models, Eqs.(47-51). The definition of $W(q)$, Eq. (44), has to be modified, taking into account the short range nature of the interactions and the form factors $f_{i,j}$. The crucial assumption is, to neglect four point correlations of the form given above.

It should be mentioned, that a rather different picture, based on a droplet model description, has been developed [47]. Both pictures, mean field models and the droplet model, reproduce certain features of short ranged spin-glasses, and no clear cut evidence in favour or against one of the theories seems to exist at present.

3.10 *p*-Spin-Glass with Relaxing Bonds and Cage Relaxation in Supercooled Liquids

In a supercooled liquid one may distinguish two kinds of motion. Each particle in a sense is surrounded by a cage formed by other particles. Within this cage it may vibrate with amplitudes much smaller than the interparticle spacing. This picture is of course simplified in the sense, that the cage is also vibrating, and each particle is also part of the cage of other particles. It is probably more appropriate to view this motion as a collective localized anharmonic vibration. The second type of motion is a rearrangement of the particles forming the cage, or an escape of a particle from its cage. This type of motion involves jumps over distances of the order of the interparticle spacing, and this is supposed to be an activated process [20, 26].

The mode coupling theory for supercooled glasses [19, 13] deals with density fluctuations, and the resulting equations are identical to those derived for the $p=3$-spin-glass in equilibrium, Eq. (53). The initial decay of the correlation function towards the plateau (see Fig.11) is usually interpreted as being due to the motion within the cages, whereas the ultimate decay is supposed to result from the escape of a particle from its cage. On the other hand, activated processes are not contained in the standard mode coupling theory. The extended theory [48] captures those processes, and yields a rounding of the singularities found in the ideal mode coupling theory at T_c. This extended theory is still an equilibrium theory and can not describe aging [25].

In the following I sketch a slightly different picture [45] resulting in identical mode coupling type of equations above T_c. This formulation allows, however, to deal with off equilibrium properties. It is based on a distinction of the two types of motion described above.

We neglect for a moment the activated process of rearrangement of the cages and consider the anharmonic vibrations only. The position of a particle at time t may be written as

$$\boldsymbol{x}_i(t) = \boldsymbol{R}_i + \boldsymbol{u}_i(t) \tag{86}$$

where $\boldsymbol{u}_i(t)$ is the momentary displacement of particle i from its mean position $\boldsymbol{R}_i$. The configuration of the $\boldsymbol{R}_i$ can be viewed as inherent state [26]. The dynamics of the displacements $\boldsymbol{u}_i(t)$ is treated within the selfconsistent anharmonic phonon theory [49], developed originally for quantum crystals or strongly anharmonic crystals at high temperatures. The main idea of this approach is, to replace the harmonic and anharmonic coupling constants by their averages over thermal or quantum fluctuations, e.g.

$$\langle V\rangle_{ij} = \int dr\, V''(r)\, g_{ij}(r). \tag{87}$$

$V(r)$ is the pair potential and $g_{ij}(r)$ the static pair correlation function between particles i and j. Employing the factorization property Eq. (85) and retaining cubic anharmonicities, one ends up with equations identical to those of the dynamic mean field theory, Eqs.(47-51).

The coupling constants $\langle V\rangle_{i\dots}$ are calculated for fixed mean positions $\boldsymbol{R}_i$. These positions are, however, random and the coupling constants are therefore random variables as well. They play the role of the $J_{i\dots}$ in Eq. (18). Taking into account the activated motion of the $\boldsymbol{R}_i(t)$, the random couplings $J_{i\dots}(t)$ now depend on time as well. Under the assumption that they are Gaussian distributed random variables, Eq. (19) holds in the modified form

$$\overline{J_{i\dots}(t)J_{j\dots}(t')} = \tfrac{1}{p!N^{p/2}}\{\delta_{i,j}\cdots+\cdots\}W_pG_J(t-t'). \tag{88}$$

Since the couplings $J_{\dots}(t)$ are dynamical variables, corresponding correlation functions have to be introduced

$$\overline{J_{i\dots}(t)\hat{J}_{j\dots}(t')} = \tfrac{1}{p!N^{p/2}}\{\delta_{i,j}\cdots+\cdots\}W_pF_J(t-t'). \tag{89}$$

The resulting mean field equations are unchanged, except for the modified memory terms, Eq. (50),

$$\begin{aligned} K(t,t') &= W''\big(q(t,t')\big)\, G_J(t-t')\, r(t,t') + W'\left(q(t,t')\right) F_J(t-t') \\ M(t,t') &= W'\left(q(t,t')\right) G_J(t-t'). \end{aligned} \tag{90}$$

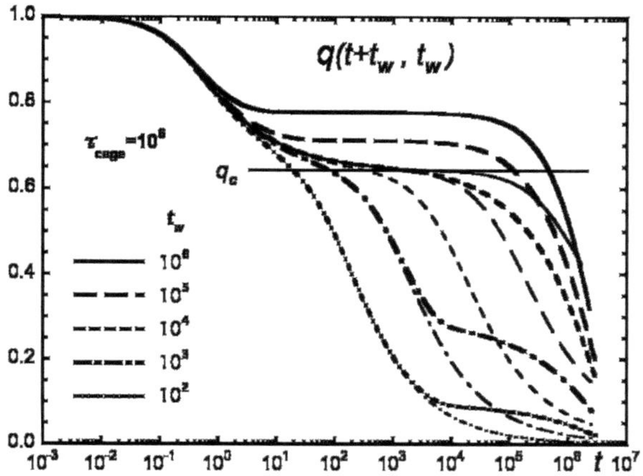

Fig. 16. Correlation-function $q(t+t_w, t_w)$ for $T < T_c$, cage relaxation time $\tau_{\text{cage}} = 10^6$ and various waiting times. For comparison the correlation function for quenched disorder ($\tau_{\text{cage}} = \infty$) is also shown (*thin lines*)

Assuming an activated process for the cage relaxation,

$$G_J(t) = e^{-t/\tau_{\text{cage}}} \qquad \text{and} \qquad F_J(t) = \frac{\beta_J}{\tau_{\text{cage}}} e^{-t/\tau_{\text{cage}}} \tag{91}$$

is chosen.

The choice $\beta_J = 0$ and $\tau_{\text{cage}} \to \infty$ is a realisation of quenched disorder [10]. The results for $t_w \gg \tau_{\text{cage}}$ are similar to those obtained for aging in Sect.3.7, with τ_{cage} replacing t_w. This holds in particular for the algebraic decay of the correlation function $q(t) - q_c \sim t^{-a}$.

The situation for $\beta_J = \beta$ is quite different. This means that the dynamics of the cages (bonds) can also equilibrate, and that fluctuation-dissipation-theorems hold for $t_w \gg \tau_{\text{cage}}$. Numerical integration of the dynamic mean field equations yields results shown in Fig.16.

There are now two long time scales, the waiting time t_w and the cage relaxation time τ_{cage}. There is yet another characteristic time scale $\hat{\tau}_{\text{cage}} = \tau_p(\tau_{\text{cage}})$ associated with the cage relaxation time, where τ_p is the plateau time introduced in Sect.3.7.

For $t + t_w < \hat{\tau}_{\text{cage}}$ the results found for quenched disorder are recovered. This means that the approach and departure of the correlation-function from the plateau q_c is ruled by power laws, fluctuation-dissipation-theorems are violated for $t > \tau_p(t_w)$ and aging is observed.

For $t_w \gg \tau_{\text{cage}}$ the system equilibrates, a new plateau value $> q_c$ is found. The ultimate decay of the correlation function is now ruled by τ_{cage}. No critical behaviour, i.e. no power laws in t, are found. The exponential decay

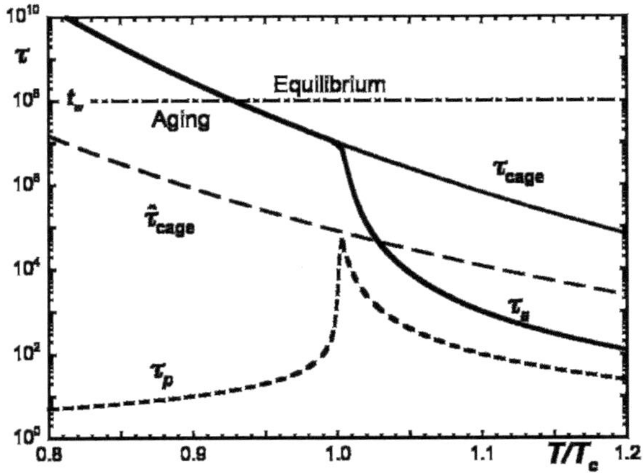

Fig. 17. Temperature dependent time scales for the p-spin-glass with relaxing bonds. For the waiting time t_w indicated in the figure the equilibrium and aging regimes are marked

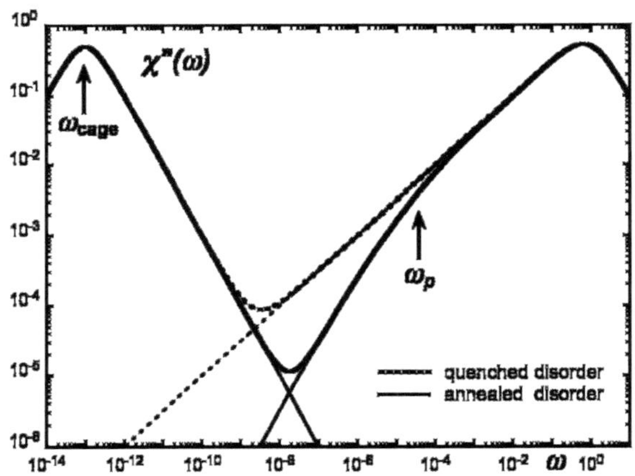

Fig. 18. Susceptibility $\chi''(\omega)$ for the p-spin-glass with relaxing bonds. "Quenched disorder" calculated with $\beta_J = 0$, "annealed disorder" calculated with $\beta_J = \beta$

of the correlation function towards the new plateau value is associated with a time scale $\tau_p(T) \sim (T_c - T)^{-1}$.

The susceptibility $\chi''(\omega)$, calculated for the p-spin-glass model with relaxing bonds, depends on the assumptions made for $F_J(t)$ or β_J. For equilibrated cage relaxation $\beta_J = \beta$ a "knee" shows up at $\omega_p \sim 1/\tau_p$, and $\chi''(\omega) \sim \omega$ for $\omega < \omega_p$ whereas $\chi''(\omega) \sim \omega^a$ for $\omega > \omega_p$. Such a knee has actually been observed in CKN [23] but this observation was later withdrawn. The knee is absent assuming non equilibrated cage dynamics with $F_J = 0$.

It has to be mentioned that the relaxation of the cage configurations has been put in by hand. In particular it is not influenced by the critical

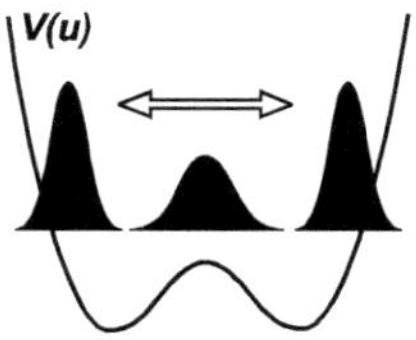

Fig. 19. Coupled 1-2-phonon oscillation in an anharmonic potential

slowing down of the anharmonic vibrations. This is a consequence of the mean field model and the N-dependent scaling of the interactions, Eq. (20). More realistic models with short range interactions would yield a coupling of the dynamics of the interactions J and the anharmonic vibrations, even within the mean field approximation discussed in Sect.3.9. This has, however, not been worked out.

The picture outlined in this section attributes the dynamics of a supercooled liquid, on a time scale shorter than the life-time of cages, to anharmonic vibrations. The instability at T_c is attributed to a softening of a coupled oscillation of the center of mass and the width of the thermal cloud, representing the motion of a particle in its cage. This motion is sketched in Fig. 19. The idea of a coupled 1-2-phonon process was originally developed for quantum crystals [49]. The critical slowing down near T_c within this picture is not directly related to the glass transition, and the actual glass transition at T_g has to be attributed to a slowing down of the activated cage relaxation.

The standard interpretation [21, 13, 12] is, to identifying the dynamic transition temperature T_c with the crossover region in glasses, and the onset temperature of replica symmetry breaking T_{RSB} with the Kauzmann temperature, which has to be lower than the glass transition temperature T_g. A comparison of T_c and T_g for various glass formers, as done in Fig.5, reveals that both temperatures can be rather far apart, whereas model calculations [50] result in differences between T_c and T_{RSB} much smaller than the values observed.

4 Outlook

In this chapter I have concentrated on the dynamics of disordered systems treated in mean field theory. I have discussed models within physics but also models for various questions outside physics. Among those are optimization problems, neural networks and problems related to economics. Especially for the applications outside physics, where spatial dimension is not relevant, mean field theory is in many cases exact. The dynamic mean field theory is quite rich because the order parameters are functions of two time variables, the correlation- and response-functions. At high temperature or strong noise level, the systems are ergodic. Lowering the temperature or noise level, a freezing transition shows up and the system becomes non ergodic.

Even on the level of mean field dynamics there are open questions, for instance the arbitrariness in the time parametrisation function $h(t)$ discussed in Sect. 3.7, or the scaling form of correlation- and response-functions in the aging regime of the $\mathcal{B}$-phase, discussed in Sect. 3.8. A more thorough understanding of the relationship between dynamics and replica theory is desired. In particular there is no obvious prescription of how to calculate entropy or free energy from dynamics. These quantities are, however, of prime interest in replica theory.

The coarse features of aging in spin-glasses are reasonably well described by dynamic mean field theory. There are, however, more elaborate experiments investigating special cooling and heating schedules [17]. It is not clear whether they can be understood on the basis of dynamic mean field theory.

Glasses and spin-glasses are systems with short ranged interactions. Some of their properties are captured by dynamic mean field theory or replica theory. Spatial correlations are not taken into account and questions concerning the existence of diverging length scales can not be answered. On the other hand real space formulations, e.g. the droplet model for spin-glasses [47] or models with kinetically constraint dynamics for glasses [27, 28], are concerned with spatial aspects and a better understanding of discrepancies and common features would be desired.

After all, it is surprising that the dynamic mean field theory for disordered systems can be applied to so many systems in physics and in other disciplines.

Acknowledgement

Partially supported by the ESF-programme SPHINX.

References

1. K. Binder, A.P. Young, Rev. Mod. Phys. **58**, 801 (1986);
   K. Fischer, J. Hertz, *Spin-glasses*, (Cambridge Univ. Press., Cambridge, 1991).
2. M.D. Ediger, C.A. Angell, S.R. Nagel, J. Phys. Chem. **100**,13200 (1996);
   P.G. Debenedetti, F.R. Stillinger, Nature **410**, 259 (2001).
3. Y.G. Sinai, Theor. Probab.Its Appl. **27**, 247 (1982);
   P. le Doussal, V.M. Vinokur, Physica C, **254**, 63 (1995);
   S. Scheidl, Z.Physik **B 97**, 345 (1995);
   L.F. Cugliandolo, P. Le Doussal, Phys. Rev. **E 53**, 152 (1996).
4. H. Horner, Z. Phys. **B 100**, 243 (1996).
5. L.C.E. Struick, *Physical Aging in Amorphous Polymers and Other Materials* (Elsevier, Houston, 1978);
   E. Vincent, J. Hammann, M. Ocio, J.P. Bouchaud, L. Cugliandolo, in *Complex Behaviour of Glassy Systems*, M. Rubi Editor, Lecture Notes in Physics (Springer Verlag, Berlin, 1997), Vol.**492**, 184.
6. H. Horner, Z. Phys. **B86**, 291 (1992) and Z. Phys. **B87** 371 (1992).

7. D. Challet, Y-C. Zhang, Physica **A 246**, 407, (1997);
D. Challet, M. Marsili, Phys. Rev. **E 60** R6271 (1999) and cond-mat/9904071
D. Challet, http://www.unifr.ch/econophysics/minority.
8. M. Mézard, G. Parisi, M.A.Virasoro, *Spin glass theory and beyond*, World Scientific (Singapore 1987).
9. H. Sompolinsky, A. Zippelius, Phys. Rev. **B 25**, 6860 (1982).
10. H. Horner: Z. Phys. **B57**, 29 (1984), Z. Phys. **B57**, 39 (1984) and Z. Phys. **B66**, 175 (1987).
11. J.P. Bouchaud, L. Cugliandolo, J. Kurchan, M. Mézard, Out of Equilibrium dynamics in spin-glasses and other glassy systems, in 'Spin-glasses and Random Fields', A.P. Young Editor, (World Scienti c, Singapore, 1998).
12. L. Cugliandolo, Les Houches Lecture Notes 2002, cond-mat/0210312.
13. R. Schilling, this Volume.
14. S. Nagata et.al., Phys.Rev. **B 19**, 1633 (1979).
15. K. Gunnarson et.al., Phys.Rev. **B43**, 8199 (1991).
16. P. Refregier, E. Vincent, J. Hammann, M. Ocio, J.Phys(France) **48**,1533 (1987).
17. E. Vincent, V. Dupuis, M. Alba, J. Hammann, J.P. Bouchaud, Europhys. Lett., **29**,1 (1999).
18. C.A. Angell, *Relaxation in Complex System*, K.L. Ngai, G.B. Wright (Eds.) (US Dept. Commerce, Spring eld, 1985).
19. U. Bengtzelius, W. Götze, A. Sjölander, J. Phys. **C 17**, 5915 (1984);
W. Götze, in *Liquids, Freezing and the Glass Transition*, eds. J.P. Hansen, D. Levesque, J. Zinn-Justin, North Holland, Amsterdam 1991;
W. Götze and L. Sjögren, Rep. Prog. Phys. **55**, 241 (1992).
20. E. Dont, *The Glass Transition*, Springer Series in Materials Science, Vol **48** (2001).
21. T.R. Kirkpatrick, D. Thirumalai, Phys. Rev. Lett. **58**, 2091 (1987); Phys. Rev. **B 36**, 5388 (1987). T.R. Kirkpatrick, P. Wolynes, Phys. Rev. **B 36**, 8552 (1987).
22. A. Crisanti, H. Horner, H.J. Sommers, Z.Phys. **B 92**, 257 (1993).
23. G. Li, W.M. Du, X.K. Chen, H.Z. Cummins, N.J. Tao, Phys. Rev. **A 45**, 3867 (1992).
H.C. Barshilia, G. Li, G.Q. Shen, H.Z. Cummins, Phys. Rev. **E 59**, 5625 (1999).
24. W. Götze, L. Sjögren, Z. Phys. **B 65**, 415 (1987) and J. Phys. **C 21**, 3407 (1988).
W. Götze, J. Phys. Condens. Matter, **11**, A1 (1999).
25. B.E. Reed, J.Non-Cryst.Sol. **131**, 408 (1991).
26. F.H. Stillinger, T.A. Weber, Phys. Rev. **A 25**, 978 (1982);
S. Sastry, P.G. Debenedetti, F. Stillinger, Nature, **393**, 554 (1998).
27. J. Jäckle, S. Eisinger, Z. Phys. B 84, 115 (1991).
28. L. Berthier, J.P. Garrahan, Phys. Rev. **E 68**, 041201 (2003) and cond-mat/0306469.
29. H. Kinzelbach, H. Horner, J. Phys. I (France) **3**, 1329 (1993) and J. Phys. I (France) **3**, 1901 (1993).
30. J. Hertz, A. Krogh, R.G. Palmer, *Introduction to the Theory of Neural Computation*, Addison Wesley (Redwood City (1991).
D.J. Amit, *Modeling Brain Function — The World of Attractor Neural Networks*, Cambridge University Press (Cambridge 1989).
31. J.J. Hopfield, Proc.Nat.Acad.Sci. USA, **81**, 2554 (1982).
32. H. Horner, D. Bormann, M. Frick, H. Kinzelbach, A. Schmidt, Z. Phys. **B 76**, 381 (1989).

33. S. Kirkpatrick, C.D. Gelatt, M.P. Vecchi, Science **220**, 671 (1983).
34. Y. Fu, P. W. Anderson, J. Phys. **A 19**, 1605 (1986).
35. R. Monasson, R. Zecchina, Phys. Rev. Lett. **76**, 3881 (1996) and Phys. Rev. **E 56** 1357 (1997).
    O.C. Martin, R. Monasson, R. Zecchina, Theoretical Computer Science **265** 3 (2001) or cond-mat/0104428.
36. S.F. Edwards, P.W. Anderson, J. Phys. **C 5**, 965 (1975).
37. D. Sherrington, S. Kirkpatrick, Phys. Rev. Lett. **35**, 1972 (1975).
38. R. Bausch, H.K. Janssen, H. Wagner, Z. Phys. **B 24**, 97 (1976).
39. W. Zippold, R. Kühn, H. Horner, Eur.Phys.J. **B 13**, 531 (2000).
40. A. Barrat, R. Burioni, M. Mezard, J. Phys. **A 29**, L81 (1996).
41. I.D. Lawrie, S. Sarbach, *Phase Transitions and Critical Phenomena, Vol. 9* C. Domb, J.L. Lebowitz, eds. (Academic, London, 1984).
    J.M. Yeomans, *Statistical mechanics of phase transitions* (Clarendon, Oxford, 1992).
42. H. Kinzelbach, H. Horner, Z.Phys. **B84** (1991).
43. H. Horner, Europhys. Lett. **2**, 487 (1986).
    M. Freixa-Pascual, H. Horner, Z. Phys. **B80**,95 (1990).
44. L.F. Cugliandolo, J. Kurchan, Phys. Rev. Lett. **71**, 173 (1993).
45. H. Horner, (to be published).
46. L.F. Cugliandolo, J. Kurchan, L. Peliti, Phys. Rev. **E55**, 3898 (1997).
47. D.S. Fisher, D.A. Huse, Phys. Rev. Lett **56**, 1601 (1986); Phys. Rev. **B 38**, 373 (1988).
48. M. Fuchs, W. Götze, S. Hildebrand, A. Latz: J. Phys. Cond. Matter **4**, 7709 (1992).
49. H. Horner, Z. Phys. **205**, 72 (1967); in *Lattice Dynamics*, Eds. G.K. Horton, A.A. Maradudin (North-Holland , Amsterdam 1972).
50. V. Krakoviack, C. Alba-Simionescoy, Europhys. Lett. 51, 420 (2000) and cond-mat/9912223.

Nonlinear Dynamics in Aperiodic Crystals

Ted Janssen

Institute for Theoretical Physics, University of Nijmegen
Toernooiveld, 6525 ED Nijmegen, The Netherlands
ted@sci.kun.nl

For aperiodic crystals non-linear dynamics plays a double role. In the first place non-linear interactions are essential in the phase transitions leading to aperiodic phases. And, second, they become relevant for the low frequency excitations which may appear as a consequence of the aperiodicity. An overview is given of the various types of aperiodic crystals, the vibrational excitations in such systems and the role of non-linear terms. The phenomena are illustrated on simple models such as the frustrated ϕ^4 model, the double chain model and the generalised Frenkel-Kontorova model.

1 Introduction

1.1 Quasi-periodicity

The most common crystals have lattice periodicity. There is a unit cell that is repeated in three independent directions, and they have diffraction patterns with sharp Bragg peaks lying on the nodes of a three-dimensional reciprocal lattice. Real crystals of this type also show some diffuse scattering due to the always present disorder, and the Bragg peaks are not exactly mathematical delta peaks because of finite size effects.

However, not all crystals are of this type. There are also solids for which the diffraction pattern also shows sharp Bragg peaks, but not all at the nodes of a three-dimensional lattice. However, the positions belong in that case still to a so called module. (Fig. 1) The positions of the peaks can be written as

$$\boldsymbol{k} = \sum_{i=1}^{n} h_i \boldsymbol{a}_i^*, \tag{1}$$

with integer indices h_i. Here the vectors $\boldsymbol{a}_i^*$ are the basis vectors. The set of vectors $\boldsymbol{k}$ is called a vector module or Fourier module. If $n = 3$, and the basis vectors are not co-planar, the basis vectors span a three-dimensional reciprocal lattice and the structure in direct space is lattice periodic. If $n > 3$ (for a three-dimensional space) the structure is aperiodic. Structures with such a diffraction pattern as in eq. (1) are called quasi-periodic, whatever the

Fig. 1. Diffraction pattern of a decagonal quasicrystal

value of n. A lattice periodic structure is also quasiperiodic. If $n > 3$ it is aperiodic and quasiperiodic, also called incommensurate.

Quasi-periodicity can also be seen macroscopically. The lattice periodic crystals have, generally, flat surfaces for which the orientation is governed by the law of rational indices. The facets can be characterized by normals belonging to the reciprocal lattice. Moreover, they show, generally, symmetry, for example the six-fold symmetry of a snow crystal.

Also aperiodic, quasiperiodic crystals have, in general, flat surfaces, now with normals belonging to the Fourier module of the structure (See e.g. [6]). Also in this case the facets may show the point group symmetry, but here the point group is not necessarily one of the well known 32 point groups. Crystals of AlMn may show five-fold symmetry, and the icosahedral phase of AlMnPd shows icosahedral symmetry. Such point groups (5m and $\bar{5}\bar{3}$m) cannot occur for lattice periodic structures, because it is easily shown that a lattice periodic structure may have only one of the 32 three-dimensional point groups as symmetry group.

1.2 Examples of Quasiperiodic Crystals

The earliest examples of aperiodic crystals were the incommensurately modulated structures (See [2, 11]). They show in the diffraction pattern a reciprocal lattice with main reflections and in addition satellites having irrational indices with respect to this reciprocal lattice. An example is $\gamma-Na_2CO_3$ (P.M. de Wolff in ref. [2]) with a monoclinic reciprocal lattice containing the main reflections and with a Fourier module consisting of vectors

$$\boldsymbol{k} = \sum_{i=1}^{3} h_i \boldsymbol{a}_i^* + m\boldsymbol{q}. \qquad (2)$$

The vector $\boldsymbol{q}$ can be expressed as $\alpha\boldsymbol{a}_1^* + \gamma\boldsymbol{a}_3^*$, where α and γ are temperature dependent, and therefore generally irrational. The rank (the value of n) is four in this case.

The diffraction pattern is explained by the structure which can be described as a periodic deformation of a lattice periodic structure. The positions of the atoms are

$$\boldsymbol{r}_{nj} = \boldsymbol{n} + \boldsymbol{r}_j + \boldsymbol{f}_j(\boldsymbol{q}.\boldsymbol{n}), \qquad (3)$$

where n belongs to the three-dimensional lattice, $\boldsymbol{r}_j$ is the average position of the $j-$th atom in the unit cell and $\boldsymbol{f}_j$ is a periodic function with period 1. The function is called the modulation function. Because $\boldsymbol{q}$ is incommensurate (*i.e.* has irrational indices) two atoms of type j in the unit cells $\boldsymbol{n}$ and $\hat{\boldsymbol{n}}$ only give the same argument in $\boldsymbol{f}_j$ if their difference is perpendicular to $\boldsymbol{q}$.

A second class of aperiodic crystals is formed by incommensurate composites ([9, 17]). Such composites consist of two or more subsystems which are themselves incommensurately modulated. The subsystems labeled by ν have periodic basic structures with lattice Λ_ν. The position of the j-th atom in unit cell n_ν of system ν is given by

$$\boldsymbol{x}_{\nu nj} = \boldsymbol{n}_\nu + \boldsymbol{r}_{\nu j} + \sum_{k\in M^*} \boldsymbol{f}(\boldsymbol{k}\,(\boldsymbol{n}_\nu + \boldsymbol{r}_{\nu j}))\,, \qquad (4)$$

where $\boldsymbol{n}_\nu$ is a lattice vector of lattice Λ_ν, $\boldsymbol{r}_{\nu j}$ a vector in the unit cell of that lattice, whereas the modulation function has Fourier components in the Fourier module M^* of the whole system without the reciprocal lattice Λ_ν^*. Generally, the modulations are caused by the interactions with the other subsystems. That is the reason why the modulation wave vectors are combinations of reciprocal lattice vectors of the other subsystems. The vector module is spanned by the basis vectors $\boldsymbol{a}_{\nu i}^*$ of the reciprocal lattices. If the subsystems are mutually incommensurate the full system is aperiodic and the positions of the Bragg peaks are given by

$$\boldsymbol{k} = \sum_{\nu i} m_{\nu i}\boldsymbol{a}_{\nu i}^* = \sum_{i=1}^{n} h_i \boldsymbol{a}_i^*, \qquad (5)$$

where the vectors $\boldsymbol{a}_i^*$ are linear combinations of the vectors $\boldsymbol{a}_{\nu i}^*$ such that they are a minimal set of vectors spanning the Fourier module. In addition, there may be satellites caused by other mechanisms, which increase the rank of the module.

An example is the compound $(PbS)_{1.18}(TiS_2)_2$ consisting of layers of PbS and layers of TiS_2 (Fig. 2). They are periodically stacked in the $z-$direction, but are mutually incommensurate in the x-y plane. Then the PbS layers are modulated with wave vectors belonging the the TiS_2 reciprocal lattice and vice versa.

A third class of aperiodic crystals are the quasicrystals [18, 20]. They have a rank higher than three. A precise definition is lacking, but most of

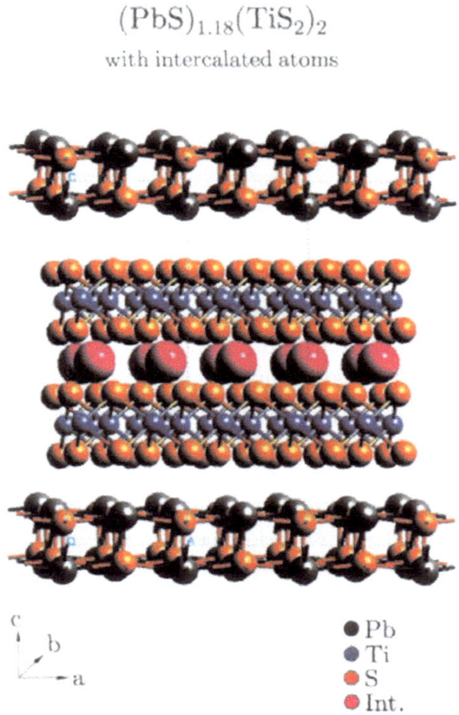

Fig. 2. An incommensurate composite: $(PbS)_x(TiS_2)_y$ with incommensurate layers

them can be considered as tilings or have a (possibly broken) point group symmetry which is non-crystallographic in three dimensions. An example is the alloy i-$Al_xMn_yPd_z$ in a certain composition range. It has as point group the non-crystallographic symmetry group of the icosahedron.

Quasicrystals often contain Al, most of them, but not all, are ternary or quaternary alloys. Building blocks often found are Mackay and Bergman clusters with icosahedral symmetry. In general, these clusters overlap. There are families with icosahedral, decagonal, dodecagonal or octagonal symmetry. However, such non-crystallographic 3D point groups are not essential.

1.3 Approximants

Modulated crystal phases have satellites next to the main reflections. We consider here the case that the rank is equal to four. Then the satellites differ from the main reflections by multiples of the modulation wave vector $\boldsymbol{k}$. In general, this wave vector is temperature dependent. For simplicity we take $\boldsymbol{k} = \gamma(T)\boldsymbol{c}^*$. As a consequence the value of γ will pass through rational values, and in that case the structure is lattice periodic. The temperature dependence may be such that there are plateau's in the function $\gamma(T)$. Moreover, at low temperature the value of γ often locks in at a rational value r/s and then the structure is an s-fold superstructure. The structures at rational values of γ may be considered as approximants: the structure for irrational values may be approximated by a series of periodic structures. This gives a means of studying physical properties of incommensurate structures from those of the approximants.

For quasicrystals the precise structure strongly depends on the precise chemical composition. If the composition differs slightly from that of a quasiperiodic quasicrystal it often has locally a similar structure as the quasicrystal, but it is periodic, although with many atoms in the unit cell. Such a structure is also called an approximant. The determination of approximants may be very helpful for the structure determination of the quasicrystal.

2 Description in Superspace

2.1 Embedding

In general, a quasiperiodic system can be considered as a density function $\rho(\boldsymbol{r})$ with Fourier decomposition

$$\rho(\boldsymbol{r}) = \sum_{\boldsymbol{k}\in M^*} \hat{\rho}(\boldsymbol{k}) \exp(i\boldsymbol{k}\cdot\boldsymbol{r}), \tag{6}$$

where the Fourier module M^* is the set of vectors

$$\boldsymbol{k} = \sum_{i=1}^{n} m_i \boldsymbol{a}_i^*. \tag{7}$$

Such a function can be seen as the restriction to the three-dimensional physical space of an n-dimensional function. Choose a basis $(\boldsymbol{a}_i^*, \boldsymbol{b}_i^*)$ in the n-dimensional space such that it projects on the basis $(\boldsymbol{a}_i^*)$ of the Fourier module. Each vector $(\boldsymbol{k}, \boldsymbol{k}_I)$ in the lattice spanned by this basis is the unique vector that projects on the vector $\boldsymbol{k}$ of the Fourier module. Then define a function on n-dimensional space of vectors $(\boldsymbol{r}, \boldsymbol{r}_I)$

$$\rho_s(\boldsymbol{r}, \boldsymbol{r}_I) = \sum_{\boldsymbol{k_s}\in\Sigma^*} \hat{\rho}(\pi\boldsymbol{k_s}) \exp\left(i(\boldsymbol{k}.\boldsymbol{r} + \boldsymbol{k_I}\cdot\boldsymbol{r_I})\right), \tag{8}$$

where Σ^* is the reciprocal lattice in n dimensions spanned by the n independent vectors $(\boldsymbol{a}_i^*, \boldsymbol{b}_i^*)$, and π the projection of Σ^* on M^*. Its restriction to the physical space $(\boldsymbol{r}_I = 0)$ then is exactly the function $\rho(\boldsymbol{r})$ and the projection of the Fourier transform in nD is exactly the Fourier transform of ρ.

Examples of embeddings of quasiperiodic systems belonging to different classes are given in Fig. 3. Modulated crystals are embedded as arrays of $(n-3)-$dimensional hypersurfaces in n dimensions. These are called the atomic surfaces. When the modulation functions are continuous the atomic surfaces stretch out until infinity. For incommensurate composites there are periodic arrays of atomic surfaces for each of the subsystems. For quasicrystals, the atomic surfaces are bounded, generally, although it is possible to construct 3-dimensional quasiperiodic patterns that could be considered as quasicrystals from a periodic array of unbounded atomic surfaces. In the figure the Fibonacci chain (a quasiperiodic array of rank two) is embedded in 2 dimensions. The array of atomic surfaces of length $1+\tau$ ($\tau = (\sqrt{5}-1)/2$) in the lattice points $(n + m\tau, -n\tau + m)$ intersect the physical space in the positions of the chain.

For modulated phases the additional coordinate can be seen from the embedding:

$$x_n = na + f(qna) \;\rightarrow\; (na + f(nqa + t),\; t\;) \qquad -\infty < t < \infty. \tag{9}$$

This is a periodic pattern with translation symmetry generated by $(a, -qa)$ and (0,1). The variable t is just the phase of the modulation function. t is the phase variable. Certain excitations in such system may be considered to be phase oscillations, for which the term phason was introduced. The excitations in quasicrystals have certain aspects in common with this. There are jumps that can be seen as jumps in superspace, and here also the term phason was used. Very often phenomena involving the additional space (for quasicrystals usually called perpendicular space) get a name with the term phason. For example, a strain in a quasicrystal can be divided into 'phonon strain' and 'phason strain'.

2.2 Symmetry

By construction the function $\rho(\boldsymbol{r}_s)$ in n dimensions is lattice periodic with lattice Σ for which Σ^* is the reciprocal lattice. Its symmetry group is an n-dimensional space group G with elements $\{R|a\}$. The orthogonal transformations R are pairs of an orthogonal transformation R_E in physical space and an orthogonal transformation R_I in internal, or perpendicular, space. The translation a is similarly a pair (a_E, a_I). The element $\{R|a\}$ is an element of the space group G if it leaves ρ_s invariant:

$$\rho_s(\boldsymbol{r}_s) \;=\; \rho_s(\{R|a\}\boldsymbol{r}_s). \tag{10}$$

In reciprocal space this corresponds to the relation

$$\hat{\rho}(\boldsymbol{k}) \;=\; \exp(i\boldsymbol{k}\cdot\boldsymbol{a}_E)\exp(i\boldsymbol{k}_I\cdot\boldsymbol{a}_I)\hat{\rho}(R\boldsymbol{k}). \tag{11}$$

If a vector $\boldsymbol{k}$ in the Fourier module is left invariant by a point group element R this relation implies that the Fourier component $\rho(\boldsymbol{k})$ is zero if there is a space group element with vector part $(\boldsymbol{a}_E, \boldsymbol{a}_I)$ such that $\boldsymbol{k}.\boldsymbol{a}_E + \boldsymbol{k}_I.\boldsymbol{a}_I \neq 0$ mod 2π. Since the static structure factor is the component of a superspace

Fig. 3. Embeddings into higher-dimensional space of a number of quasiperiodic structures. *Left*: Fibonacci chain, *middle*: IC composite (dashed: system 1, solid: system 2), *right*: sinusoidally modulated chain (dashed: lattice)

group invariant function on reciprocal space this leads to extinction rules for aperiodic crystals. They can be compared to the well known extinction rules that play an important role in crystal structure determination.

The translations in the superspace group form a lattice for which the projection on the additional space is dense. A three-dimensional structure in physical space ($t_I = 0$) is equivalent to a three-dimensional structure for arbitrary small value of t_I by a lattice translation $\sum n_i a_i$ with internal coordinate $\sum n_i \boldsymbol{a}_{Ii} \approx 0$. The sets of distances in both sections are equal. Therefore, if the $n-$dimensional structure is smooth the potential energies of the structures are equal, which means that the energy is degenerate in the internal coordinate. If the atomic surfaces are discontinuous, the energy is still the same for a dense set of three-dimensional sections, but this set is of measure zero. For smooth atomic surfaces the internal coordinate may be changed adiabatically without loss of energy. Then there are excitations with low frequency corresponding to this internal coordinate shift, and these are called phasons.

2.3 Approximants

Approximants are periodic structures which have a structure that is similar to that of a quasiperiodic structure. Because of the periodicity there is a three-dimensional lattice of translations in the physical space.

One may obtain such a periodic structure from a quasiperiodic structure by a linear strain in superspace. The lattice of translations of the quasiperiodic structure has rank less than three. It is possible to bring three independent lattice vectors into the physical space by the linear strain. Generally, the symmetry of the strained lattice is different from the quasiperiodic structure embedded in n dimensions. This is clear when the point group of the quasiperiodic structure is non-crystallographic in three dimensions.

Approximants may be useful for finding the structure, or in calculations for determining the physical properties. In the latter case it is useful to have a series of approximants converging to the quasiperiodic structure. As an example consider the Fibonacci chain. Its embedding has a lattice of translations generated by the two-dimensional vectors $(1, -\tau)$ and $(\tau, 1)$, where

$\tau = (\sqrt{5}-1)/2$. The golden ratio τ may be written in a continued fraction expansion as

$$\tau = 1/(1+1/(1+1/(1+/\ldots)))$$

Truncation of this expansion gives a series of fractions F_n/F_{n+1}, with $F_{n+1} = F_n + F_{n-1}$, $F_1 = 1, F_2 = 2$. The sequence $1/2, 2/3, 3/5, 5/8, 8/13\ldots$ converges to τ. A strain leading to a lattice with basis vectors $(1, -F_n/F_{n+1}), (\tau, 1)$ yields a lattice with an intersection with the physical space of rank one, whereas the intervals remain of length 1 and τ. Calculation of physical properties of the approximants may take place using the conventional means. The limit, if it exists, then gives the properties of the quasiperiodic chain.

3 Modulated Phases: DIFFOUR Model

3.1 The Model

A simple model in which the ground state may be aperiodic is the following. It is a linear chain with particles with one degree of freedom, for example the deviation of its position from that in an equidistant array. The potential is a non-linear function of the deviations and there is a interaction between a particle and its first and second neighbours. The Hamiltonian of the model is given by

$$H = \sum_n \left(\frac{p_n^2}{2} + V_1(x_n) + V_2(x_n - x_{n-1}) + V_3(x_n - x_{n-2}) \right). \tag{12}$$

An example is

$$H = \sum_n \left(\frac{p_n^2}{2} + Ax_n^2/2 + x_n^4/4 + Bx_nx_{n-1} + Cx_nx_{n-2} \right). \tag{13}$$

The on-site potential is here a ϕ^4 function. The terms with B and C may favour different ground states, which leads to frustration. Therefore, the model is called the Discrete Frustrated ϕ^4 (DIFFOUR) model ([13, 14]).

The ground state of eq.(13) for $T = 0$ is given by the coupled non-linear equations

$$Ax_n + x_n^3 + B(x_{n+1} + x_{n-1}) + C(x_{n+2} + x_{n-2}) = 0. \tag{14}$$

Periodic solutions with period N can be found by the solution of a finite set of coupled equations. Aperiodic solutions with wave vector $2\pi q_i$ can be found as the limit of periodic solutions with $q = L/N$ when N tends to infinity such that q tends to the irrational value q_i. The ground state is found as the lowest energy solution for all wave vectors q.

There are two types of aperiodic ground states. In the first type, there is a value q such that $qn \bmod \mathbb{Z}$ cover the whole unit interval and there is a smooth and periodic function $f(z)$ such that $x_n = f(qn)$. This corresponds to a smooth modulation function. For the other type the $qn \bmod \mathbb{Z}$ still fill the unit interval densely, but the function f is discontinuous. This ground state corresponds to a chain for which the values of x_n are given by $x_n = f(q_c n + \phi(n))$, where q_c is a rational approximant and $\phi(n)$ a phason function. If $\phi(n)$ is constant the structure is periodic, but in the non-constant case the regions where ϕ is not constant correspond to discommensurations.

The phase diagram can be constructed from the determination of the ground state for given values of the parameters. In the A/C versus B/C plane incommensurate phases are concentrated around the origin. For high values of A/C the solution $x_n = 0$ (the para-phase) is the ground state. For large absolute values of B/C the ground state is ferroic (period 1 different from the para-phase) or antiferroic (period 2). For low values of A/C ground states are commensurate. Around the origin the wave vector is incommensurate or commensurate and the phase diagram is complicated. This means that for comparable values of A/C and B/C the ground state may be quasiperiodic and the ground state is degenerate.

The typical situation is that for decreasing value of A/C the para-phase becomes unstable, a sinusoidally modulated structure sets in, the modulation gets higher harmonics and later becomes discontinuous with domains of almost constant phase, and finally these domains grow and the structure becomes commensurately modulated (a superstructure). An example is given in Fig.3.1. Here a is proportional to $-A/C$. For small values of a the modulation is smooth, for larger values it becomes discontinuous. As long as the modulation function is smooth the ground states form a continuous manifold.

3.2 The Ground State and Non-linear Mappings

The set of coupled equations (14) is equivalent with a non-linear mapping in a four-dimensional space. To that end introduce the 4D vector

$$v_n = (x_{n+1}, x_n, x_{n-1}, x_{n-2})$$

and consider the mapping $v_n \to v_{n+1}$ with components

$$x_{n+2} = -\frac{B}{C}x_{n+1} - \left(\frac{A}{C}x_n + \frac{1}{C}x_n^3\right) - \frac{B}{C}x_{n-1} - x_{n-2},\quad x_{n+1},\quad x_n,\quad x_{n-1}.$$

This is a non-linear mapping with Jacobian

$$\begin{pmatrix} -\frac{B}{C} & -\frac{A}{C} - \frac{3x_n^2}{C} & -\frac{B}{C} & -1 \\ 1 & 0 & 0 & 0 \\ 0 & 1 & 0 & 0 \\ 0 & 0 & 1 & 0 \end{pmatrix}. \tag{15}$$

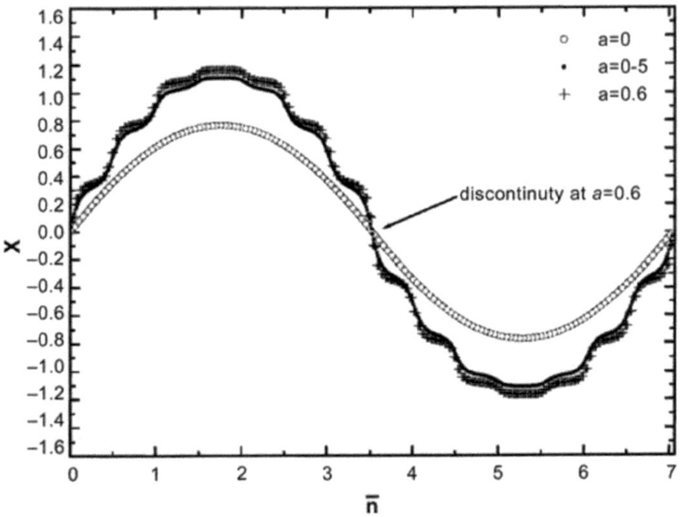

Fig. 4. The modulation function for the ground state of the DIFFOUR model determined for 3 different values of the parameters

When $C = 0$ the mapping becomes 2-dimensional with Jacobian

$$\begin{pmatrix} -\frac{A}{B} - \frac{3x_n^2}{B} & -1 \\ 1 & 0 \end{pmatrix} .$$

The origin is a fixed point for any value of the parameters, but it changes its character as a function of A, B and C. For $C = 0$ the fixed point is hyperbolic if $A^2 > 4B^2$, otherwise it is elliptic. The bifurcation point corresponds to the appearance of a zero frequency mode, because the dispersion relations for the chain corresponding to the f.p. (an equidistant array because all values of x_n are zero) are $\omega(k)^2 = -A - 2B\cos(k)$. There is a close relationship between the appearance of a zero frequency oscillation and the bifurcation where a pair of real eigenvalues meet and move as complex conjugates over the unit circle. The situation is similar for the 4-dimensional mapping. In this case the eigenvalues of the linearized mapping are $\lambda, \lambda^*, 1/\lambda$, and $1/\lambda^*$. Starting from 4 complex eigenvalues, these meet pairwise and the 2 conjugate complex eigenvalues move over the unit circle in the complex plane. In the latter case there is a zero frequency mode.

The ground state corresponds to an orbit in the 4-dimensional space. For high values of A this is the fixed point in the origin, for lower values the orbit lies on a smooth curve that is densely covered by the orbit. For still lower values the orbit corresponding to the ground state becomes discontinuous, and for the lowest values the orbit is periodic and consists of a finite number of points, It corresponds to a periodic superstructure of the chain (Fig. 5). The transition from a ground state orbit which lies on a smooth curve, to

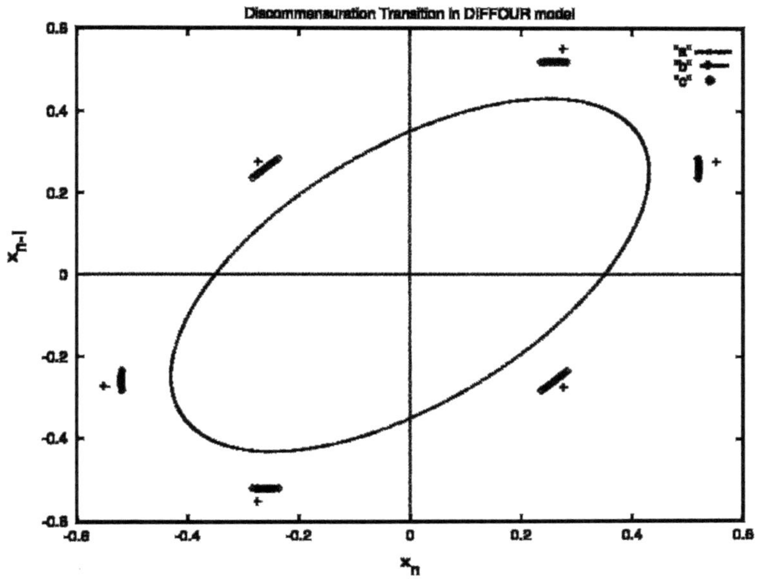

Fig. 5. Three orbits under the non-linear 4-dimensional mapping corresponding to the DIFFOUR ground state equations. The smooth line is for a just above a_i and corresponds to a smooth modulation function (a), the crosses are for a period six ground state (b), and the others correspond to a discontinuous, aperiodic modulation (c)

one with a larger fractal dimension than unity is the discommensuration transition.

4 Incommensurate Composites: Double Chain Model

4.1 The Model

For aperiodic composites we study a system consisting of two one-dimensional chains, one with atoms at positions x_n, and one with atoms at y_m. The potential energy is given by

$$V = \sum_n V_1(x_n - x_{n-1}) + \sum_m V_2(y_m - y_{m-1}) + \sum_{nm} W(x_n - y_m). \quad (16)$$

The intra-chain couplings are either harmonic ($V_1(x) = \alpha(x-a)^2/2, V_2(y) = \beta(y-b)^2/2$) or they are Lennard-Jones potentials with minima for $x = a$ and $y = b$, respectively [15, 3]. The inter-chain coupling has been chosen as Gaussian or as Lennard-Jones potential:

$$W(r) = \lambda \exp(-\gamma r^2), \quad \text{or} \quad W(r) = \lambda\left(\left(\frac{\sigma}{r}\right)^{12} - 2\left(\frac{\sigma}{r}\right)^6\right),$$

where $r^2 = (x-y)^2 + d^2$, if d is the inter-chain distance.

Generalizations consider 2-dimensional cases of alternating parallel $x-$ and $y-$chains, different masses for the atoms in the two chains, or a unit cell with more than one atom. The model is in fact a generalization of a model introduced by Dehlinger, and which was studied by Frenkel and Kontorova and by Frank and Van der Merwe [8]. It consists of a linear chain on a fixed substrate, with Hamiltonian

$$H = \sum_n \left(\frac{p_n^2}{2} + \alpha(x_n - x_{n-1} - a)^2/2 + \lambda \cos(2\pi x_n/b + \phi) \right) . \qquad (17)$$

Usually it is called the Frenkel-Kontorova model, also when the lattice constant a is incommensurate with the periodicity b of the substrate potential. In the DCM also the substrate is deformable.

4.2 The Ground State

The ground state is determined from the coupled non-linear equations

$$V_1'(x_n - x_{n-1} - a) - V_1'(x_{n+1} - x_n - a) + \sum_m W'(x_n - y_m) = 0 \qquad (18)$$

$$V_2'(y_m - y_{m-1} - b) - V_2'(y_{m+1} - y_m - b) - \sum_n W'(x_n - y_m) = 0 .$$

In general, the solutions can be written as

$$x_n = x_0 + na + f(x_0 + na), \qquad f(x) = f(x+b) \qquad (19)$$

$$y_m = y_0 + mb + g(y_0 + mb), \qquad g(y) = g(y+a) . \qquad (20)$$

The modulation functions can be determined from

$$f(na \bmod b) = x_n - na, \quad g(mb \bmod a) = y_m - mb$$

when x_n, y_m are found determining the minimal energy configuration numerically. For small interaction parameter λ the functions f and g are continuous, for larger value of λ they become simultaneously discontinuous. In the latter region the modulation functions are approximately piecewise linear. This means that locally the lattice parameter of each chain is changed. Because the density of the particles is fixed, the discontinuities provide an overall incommensurability. In this sense the transition from smooth to discontinuous can be called a discommensuration transition as well.

A phase diagram shows here in the plane of the relevant parameters (λ/α and λ/β) the transition from smooth modulation functions to discontinuous modulation functions (cf. figure 6). These have been obtained by keeping λ/α fixed and varying λ/β monitoring the value of the discontinuity. In the real calculations, which were based on approximants, this means that the largest gap exceeds a threshold value. If one increases the size of the approximant

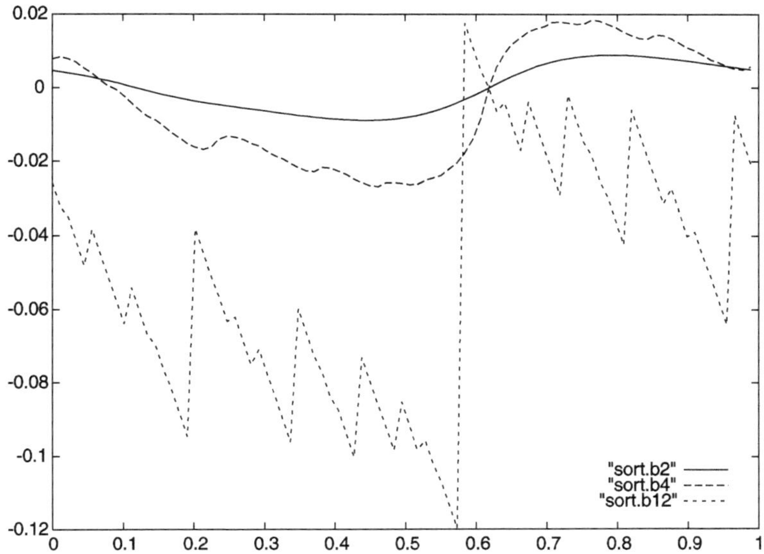

Fig. 6. The modulation function for one of the chains in the ground state of the DCM. There is a transition from smooth to discontinuous functions

the transition becomes more pronounced. For Gaussian inter-chain potential a convex line in the plane is found, for Lennard-Jones potentials the line is no longer convex.

5 Phonons and Phasons

5.1 Phonons in Aperiodic Crystals

Phonons are the collective vibrational excitations around the equilibrium positions. These displacements with Cartesian coordinates $u_{n\alpha}(t)$ satisfy the equations of motion

$$m_n \ddot{u}_{n\alpha} = -\sum_{m\beta} U(n,m)_{\alpha\beta} u_{m\beta}$$

in the harmonic approximation. For lattice periodic crystals the solutions can be written as

$$u(nj)_\alpha \;=\; \sum_{k\nu} Q_{k\nu} \epsilon(k\nu|j\alpha) \exp(ikn) \;+\; c.c., \tag{21}$$

where the summation is over all k-vectors in the Brillouin zone and over all branches ν. The index j numbers the particles in the unit cell, and $\epsilon(k\nu)$ is the eigenvector of the dynamical matrix with eigenvalue $\omega_{k\nu}^2$.

For aperiodic crystals there is no Brillouin zone and the eigenmodes can no longer be labeled with $k\nu$. For the numerical calculation of these eigenmodes one uses approximants. The incommensurate structure is approximated by a series, usually based on a continued fraction expansion of the irrational numbers, of lattice periodic structures for which the usual techniques can be used. For these approximants quantities like the (partial or full) density of states, the integrated density of states and the dynamical structure for neutron scattering can be calculated. The corresponding quantities for the incommensurate structure then are obtained by extrapolation. Examples are the Fibonacci chain and three-dimensional Penrose tilings. Both are quasi-periodic, of rank two, and six, respectively. In both the golden number $\tau = (\sqrt{5}-1)/2$ determines the aperiodicity. For this number the relation

$$\tau = \lim_{n\to\infty} \frac{F_n}{F_{n+1}}, \qquad (F_{n+1} = F_n + F_{n-1},\ F_0 = F_1 = 1), \tag{22}$$

holds. Substituting one of the approximants of τ into the internal coordinates gives an approximant structure, for which the dynamical matrix is finite.

There is an important qualitative difference between lattice periodic and incommensurate phases. In the former an eigenmode is characterized by a wave vector k, which implies that going from one unit cell to another the function $u_{n\nu}$ gets a simple phase factor. Therefore the absolute value in each unit cell is the same, which means that for an ideal crystal all modes are extended. For incommensurate structure, however, three types of eigenmodes may occur: extended states not decaying towards infinity, localized states falling off exponentially, and so-called critical states. The latter are neither extended nor localized and usually fall off with a power law or they have scaling behaviour. This character may be recognized in the dynamical structure factor. For low frequencies the dispersion curves are sharp and the states are extended, but for higher frequencies the states are critical and the dispersion curves broaden.

5.2 Phason Excitations

The transition from a lattice periodic to an incommensurate displacively modulated structure can be modeled by the DIFFOUR model. In a mean field approximation the equations of motion have the same form but the coefficient A becomes temperature dependent. This means that for a critical temperature T_i the critical value $A_i = A(T_i)$ is reached and the periodic array becomes unstable. This is seen in the dynamics from the dispersion relation for the periodic array ($x_n = 0$):

$$\omega(q)^2 = A + 2B\cos(q) + 2C\cos(2q), \qquad -\pi < q \leq \pi. \tag{23}$$

The minimum of the curve is at q_c with $\cos(q_c) = -B/4C$, if $|B/4C| \leq 1$. (Otherwise at $q = 0$ or $q = \pi$) The frequency goes to zero if A decreases to $A_c = 2C + B^2/4C$. This is called a soft mode.

For $A < A_c$ the structure with $x_n = 0$ is unstable. The ground state just below $A = A_c$ is a modulated structure with modulation wave vector q_c. The two degenerate modes at $\pm q_c$ are coupled by the modulation. Near the transition point the modulation wave is given by $x_n = \Delta \cos(q_c n)$. The equations for the normal coordinates Q_k in $u_n = \sum_k Q_k \exp(ikn)$ are

$$\omega^2(k)Q_k = (A + 2B\cos(2k) + 2C\cos(2k))Q_k + \frac{3}{4}\Delta^2(Q_{k+2q_c} + 2Q_k + 2Q_{k-2q_c}). \quad (24)$$

This gives a non-linear coupling between eigenmodes which differ by $2q_c$.

In the neighbourhood of $\pm q_c$ the dispersion curves are linear:

$$\omega(q_c + k)^2 = [4C - B^2/4C]\, k^2 \quad (25)$$

for $A = A_c$. The coupled modes give new modes which are the symmetric and anti-symmetric combinations of the original modes. One is proportional to the sinusoidal modulation function, the other differs by a phase of $\pi/2$ and corresponds to the derivative of the modulation function. This means that the first changes the amplitude, the other the phase of the modulation. These modes are called the amplitude and phase modes, or amplitudon and phason, respectively. The branch starting from this zero frequency mode is called the phason branch. Because the phase of the modulation corresponds to the internal coordinate in superspace, when one embeds the aperiodic chain in higher dimensions, a phason may be described as an oscillation in superspace with a polarisation pointing out of the physical space.

Because the potential energy of the crystal is invariant under a phase shift of the incommensurate modulation, it is to be expected that the phason with $k = 0$ ($q = q_c$) has frequency zero, also for $A < A_c$. Numerical calculations of the dispersion curves in the modulated phase show that this is true in the neighbourhood of $A = A_c$, but not generally for any value of A. The typical situation is that the frequency remains zero in the interval $A_d < A < A_c$ for some value of A_d that is larger than the value of A for which the ground state becomes commensurate. Below A_d the frequency of the lowest phason is non-zero, a phason gap opens. A careful analysis shows that at the transition point the modulation function is no longer smooth, but shows discontinuities. It is the discommensuration transition, also found in similar systems under the name ‘transition by breaking of analyticity’ [1].

Excitations with zero or low frequency and with eigenvectors which correspond to motions that can be interpreted as motions in the additional space (and therefor can be called phasons) have been found in the DIFFOUR model, the DCM and in the Frenkel-Kontorova model (see section 9). In a certain parameter or temperature range the minimal frequency of these modes is zero. Let us summarize the results for the three models at zero temperature, as function of the parameters. Fixing B/C in the DIFFOUR model, there is a zero frequency phason mode in the range from A_d/C to A_i/C. Above

A_i/C the paraphase is stable, and there is no phason. Below A_d/C there is a phason gap. For the DCM there is a line in the λ/α-λ/β-plane separating the region with zero frequency phason from that with a phason gap. This happens both for the Gaussian and for the Lennard-Jones potential. In the Frenkel-Kontorova model, another model for composites which will be discussed in section 9, there is a critical value of the chain-substrate interaction λ above which there is a phason gap, and below which the gap is zero.

In all these cases the character of the modulation functions has been studied. The line (or point) in parameter space where the phason gap opens coincides always exactly with the appearance of discontinuities in the modulation function. Therefore, the conclusion is that the discommensuration transition is connected to the softening of the phason. If the modulation functions are smooth, the phason gap is zero. For a zero frequency acoustic phonon (a displacement of the crystal) increasing the amplitude does not change the distances between the particles.

6 Solitons in Incommensurate Phases

A quasiperiodic ground state may show discommensurations, corresponding to discontinuities in the modulation function. They may also appear as static or dynamic excitations in the chain. If they are static they correspond to metastable states, if they are dynamic they are solutions to the equations of motion. We study these first in a continuum approximation. For particles with unit mass the equations of motion in the DIFFOUR model are

$$\ddot{x_n} = -Ax_n - x_n^3 - B(x_{n+1} + x_{n-1}) - C(x_{n+2} + x_{n-2}). \tag{26}$$

There are simple periodic stationary solutions, such as $N = 1$ with x_n=0, or $N = 1$ with $x_n = \pm\sqrt{-(A+2B+2C)}$, or $N = 2$ with $x_n = \pm(-1)^n \cdot \sqrt{-(A-2B+2C)}$. If the wave vector of the modulation is not very different from a simple rational value, the equations of motion may be considered in a continuum approximation by writing x_n as the product of a rapidly and a slowly varying function [16]. For example, if the wave vector is close to 1/2, one writes $x_n = (-1)^n Q_n$, where now Q varies slowly. The differential equations of motion for the function $Q(\xi, t)$ become

$$-\ddot{Q} + 2(B - 2C)Q'' = (A - 2B + 2C)Q + Q^3, \tag{27}$$

where Q'' is the second derivative of Q with respect to ξ. This equation has the solitary wave solution

$$Q(\xi, t) = \sqrt{-(A-2B+2C)}\tanh\left((\xi - \xi_0 - vt)\sqrt{\frac{-A+2B-2C}{4B-8C-v^2}}\right).$$

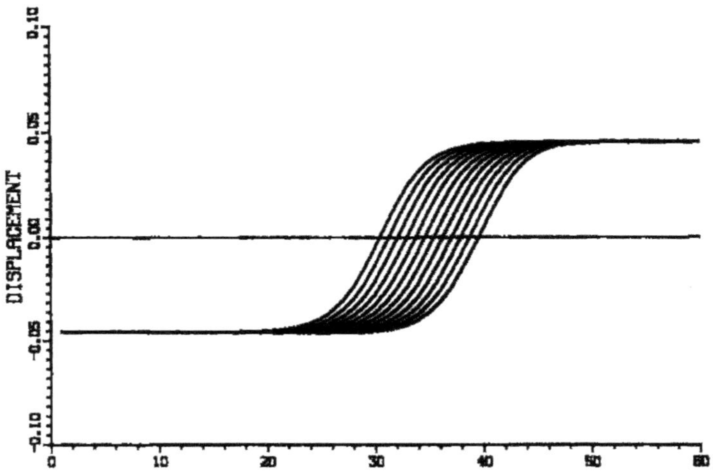

Fig. 7. A solitary wave with low initial speed; almost friction-less motion

This is of the form of a moving kink. For the general case where the wave vector is close to $q = 2\pi L/N$ the normal coordinates Q are complex and it is necessary to derive equations of motion for Q defined by

$$x_n(t) = e^{iqn}Q(n,t) + e^{-iqn}Q^*(n,t). \tag{28}$$

For arbitrary values of L/N these continuum equations have been analyzed by Slot and Janssen [16].

The exact solutions of the continuum approximation can be used as initial configurations for a numerical integration of the discrete equations of motion, for which they are no longer exact solutions.

$$x_n(0) = e^{iqn}Q(n) + e^{-iqn}Q^*(n) \tag{29}$$
$$\dot{x}_n(0) = -\zeta v \left(e^{iqn}Q'(n) + e^{-iqn}Q^{*\prime}(n)\right) . \tag{30}$$

Here ζ is a parameter allowing to vary the ratio between the initial speed and the speed in the solution.

The solitary waves found in the continuum approximation survive for the discrete system provided the propagation speed is not too high. For low speeds the kink keeps its shape, and very little phonons are created (Fig. 7). For higher initial speeds the kink remains, but its speed decreases and energy is radiated in the form of phonons (Fig. 8). For very high initial speeds the kink is no longer stable. A kink may give rise to a kink and a kink-antikink pair. The kink and the antikink move in opposite directions (Fig. 9). The loss of energy and the creation of phonons sets in if the initial speed approaches the phason speed. The kink becomes unstable when the kinetic energy is larger than the creation energy of two kinks. For initial speed equal to zero the kink is a non-linear static excitation.

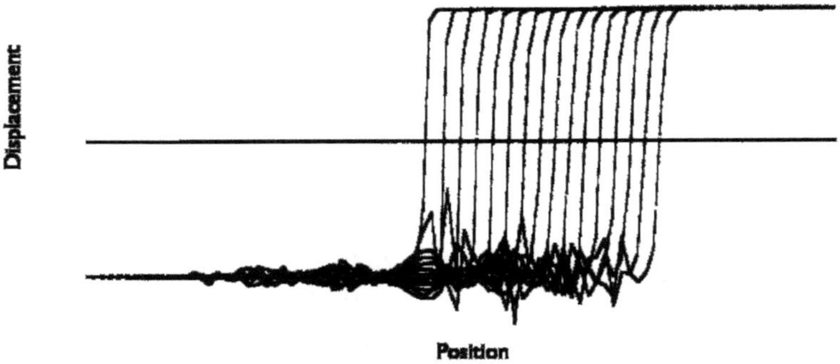

Fig. 8. Solitary wave with speed approximately equal to the phason velocity: creation of phonons

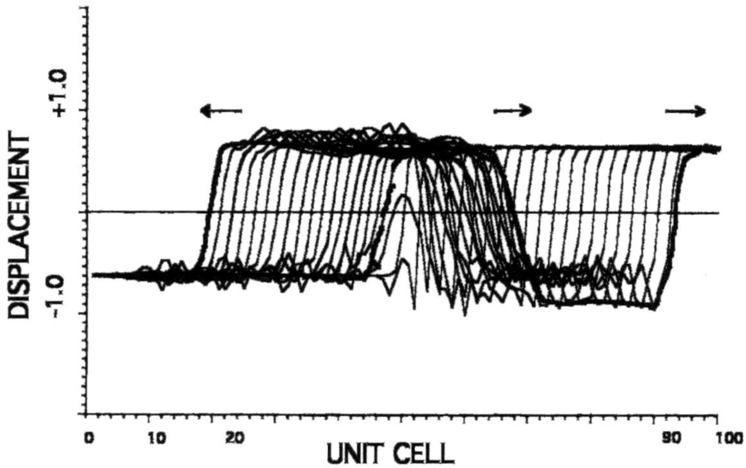

Fig. 9. Solitary wave with initial speed higher than the phason velocity: creation of kink-antikink pair

7 Non-linear Phason Dynamics in Modulated Phases

The eigenmodes of a (periodic or aperiodic) crystal are solutions of the linearized dynamical problem. Among them are the phasons in modulated phases which are phonons with a character that can be described as a shift of the modulation function. The frequency of the homogeneous shift, the pure phason with wave vector zero, is zero if the modulation function is smooth, due to the degeneracy of the potential energy. The latter argument can be used to show that also an arbitrary static shift will not cost energy. However, in this case non-linear terms will come in. In principle, these could give rise to a coupling between the phason and other phonons.

We consider this question in the frame of the DIFFOUR model [12]. We first introduce a new parameterization of the model such that the Hamiltonian is given by

$$H = \sum_n \left(\frac{\dot{x}_n^2}{2} - \frac{bx_n^2}{2} + \frac{x_n^4}{4} + (x_n - x_{n-1})^2 + d(x_n - x_{n-2})^2 \right). \tag{31}$$

Notice that the parameter change from A, B and C to b and d is such that A and b have different sign. If the ground state is given by the modulation function $x_n = f(kna)$ a shift of the modulation function gives $x_n(t) = f(k(na - vt))$ or an initial speed

$$\dot{x}_n(0) = -kvf'(kna) = \epsilon u_n, \qquad \sum_n |u_n|^2 = 1.$$

For a sinusoidal modulation $x_n = U\cos(kna)$ the speed is given by $v = \epsilon\sqrt{2a/L}/kU$, where L is the length of the normalization domain. The speed should be compared with the phason velocity, the slope of the phason branch, which is equal to $v_{\mathrm{ph}} = \sqrt{1/2d - 8d}$.

We look for solitary wave solutions of the equations of motion of the form

$$x_n(t) = f(na - vt) \tag{32}$$

and start with a continuum approximation. The Lagrange function has the form

$$\mathcal{L} = \sum_n \left(\frac{\dot{x}_n^2}{2} + \frac{bx_n^2}{2} - \frac{x_n^4}{4} - (x_n - x_{n-1})^2 - d(x_n - x_{n-2})^2 \right). \tag{33}$$

The solution $f(z)$ is periodic with period p just as the modulation function. Then the Lagrangian function in the continuum approximation is an integral over the unit cell p.

$$\mathcal{L} = \frac{N}{p} \int_0^p \left(\frac{mv^2 f'(z)^2}{2} - \frac{f(z)^4}{2} \right. \tag{34}$$

$$\left. + \frac{bf(z)^2}{2} - (f(z) - f(z+a))^2 - d(f(z) - f(z+2a))^2 \right) \mathrm{d}z. \tag{35}$$

Here N is the number of particles.

For $v = 0$ the action is extremal. Here it is a maximum. The function f which depends still on v should maximize the Lagrangian. For a trial function $f(z) = f_0 \sin kz$ this means that

$$\left(\frac{mk^2v^2}{4} + \frac{b}{4} - 2\sin(ka/2)^2 - 2\sin(ka)^2 \right) A^2 - \frac{3}{16} A^4$$

is maximal, which is an equation for k. This transcendental equation has non-trivial solutions provided

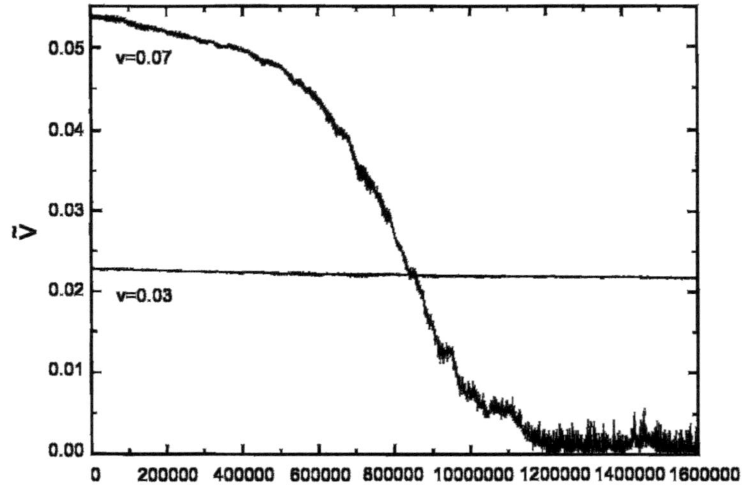

Fig. 10. The speed of a non-linear excitation in the DIFFOUR model, when the modulation function is smooth, for 2 different initial speeds. For the lower speed the motion is almost dissipation-less

$$mv^2 \; < \; -8d - 2. \tag{36}$$

If v satisfies this condition and k_0 maximizes the Lagrangian, then the equation for A has a non-trivial solution if

$$\frac{mk_0^2v^2}{4} + \frac{b}{4} - 2\sin(k_0a/2)^2 - 2\sin(k_0a)^2 > 0 \; .$$

For speeds higher than $\sqrt{(-8d-2)/m}$ the solution is unstable. This leads to the conjecture that if the modulation function is smooth (which is required for using the continuum approximation) there is a solitary wave solution moving through the crystal without energy loss provided its speed remains below the threshold value. In the discrete system there will nevertheless be some energy loss due to the coupling to phonons, but it may be expected to be small.

To check these expectations the equations of motion were numerically solved for the DIFFOUR model with as initial positions the positions of the ground state configurations, and as initial velocities a factor ϵ times the eigenvector coordinates of the phason, i.e. proportional to the derivative of the modulation function. After many integration steps the shape of the modulation function has not changed for small initial speed $v = 0.03$. The speed itself remains practically constant. There is only a very small energy loss to internal vibrations because of the discreteness of the system. However, for an initial speed $v = 0.07$ the speed decreases immediately and goes to zero. The energy then is completely transferred to the phonons (Fig. 10).

8 Sliding and Friction in IC Composites

The ground state configuration in the DCM is given by a solution of eqs. (19). The vibrations around the equilibrium positions are u_n for the first and v_m for the second chain. These quantities satisfy the equations

$$\begin{aligned} M_n\omega^2 u_n = & \left(V_1''(x_n^0 - x_{n-1}^0) + V_1''(x_{n+1}^0 - x_n^0)\right) u_n \\ & -V_1''(x_n^0 - x_{n-1}^0)u_{n-1} - V_1''(x_{n+1}^0 - x_n^0)u_{n+1} \\ & +\sum_\ell W''(x_n^0 - y_\ell^0)u_n - \sum_\ell W''(x_n^0 - y_\ell^0)v_\ell \quad , \end{aligned} \tag{37}$$

and an analogous equation for v_m. The superscript '0' denotes the positions in the ground state. Determination of the dispersion curves for approximants shows that there are two zero frequency modes if the modulation function is smooth, one corresponding to an acoustic mode and one with phason character. However, when the modulation function is discontinuous, there is a phason gap. For low frequency modes in the continuous case the amplitude may become large. The question is whether there is also in this case a sliding mode with large amplitude.

The vibrations around the equilibrium positions, described in terms of phonons, are harmonic. The harmonic approximation is valid only for small displacements. For larger displacements the equations become non-linear. They are

$$m_1\ddot{x}_n = -V_1'(x_n - x_{n-1}) - V_1'(x_n - x_{n+1}) - \lambda\sum_m W'(x_n - y_m) \tag{38}$$

$$m_2\ddot{y}_m = -V_2'(y_m - y_{m-1}) - V_2'(y_m - y_{m+1}) + \lambda\sum_n W'(x_n - y_m). \tag{39}$$

We suppose that the displacements u_n and v_m remain small in the moving frame.

$$V_1'(x_n - x_{n-1}) + V_1'(x_n - x_{n+1}) = \alpha(2u_n - u_{n-1} - u_{n+1})$$

and a similar expression for V_2'. The phonons then are non-linearly coupled by the W terms.

The solutions $x_m(t)$ and $y_m(t)$ determine also the motion of the centre of mass, and the motion of the internal coordinate t_I. The equilibrium state is embedded in superspace as

$$(na + Z_1t_I + f(na + Zt_I), t_I),\ (mb - Z_2t_I + g(mb - Zt_I), t_I),\ (Z = Z_1 + Z_2).$$

This embedding is invariant under the lattice translations $(Z_1b/Z, b/Z)$ and $(Z_2a/Z, -a/Z)$. The centre of mass shift in physical space and the shift in the internal coordinate t_I are given by

$$\Delta_{C.M.} = \frac{Z_2Q_0^{(1)} + Z_1Q_0^{(2)}}{Z},\quad \Delta t_I = \frac{Q_0^{(1)} - Q_0^{(2)}}{Z}, \tag{40}$$

where

$$Q_0^{(1)} = v_1 t + \frac{1}{N}\sum_n u_n, \qquad Q_0^{(2)} = v_2 t + \frac{1}{L}\sum_m v_m. \tag{41}$$

The equations of motion for these variables are

$$\ddot{Q}_0^{(1)} = -\frac{\lambda}{m_1}\sum_{nm} W'(x_n - y_m) \tag{42}$$
$$\ddot{Q}_0^{(2)} = \frac{\lambda}{m_2}\sum_{nm} W'(x_n - y_m) \ .$$

The right hand sides can be developed in powers of the normal coordinates $Q_k^{(j)}$ according to

$$\sum_{nm} W'(x_n - y_m) = \sum_{nm}\sum_j \frac{1}{j!} W^{(j+1)}(na - vt - mb)(u_n - v_m)^j \ . \tag{43}$$

The first order terms are given by

$$\sum_{nm} W''(na - vt - mb)u_n = \sum_{K_2 s} \hat{f}_{K_2 s} Q_{K_2}^{(1)} \exp(-is\Omega_1 t)$$
$$\sum_{nm} W''(na - vt - mb)v_m = \sum_{K_1 s} \hat{g}_{K_1 s} Q_{K_1}^{(2)} \exp(-is\Omega_2 t)$$

where K_1 is a multiple of $2\pi/a$. K_2 a multiple of $2\pi/b$, $\Omega_1 = 2\pi v/b$ and $\Omega_2 = 2\pi v/a$. The latter frequencies correspond to the frequencies with which the particles of one chain move over the particles of the other chain. Furthermore, $\hat{f}_{ks}$ and $\hat{g}_{ks}$ are the Fourier transforms of $\sum_m W'(na - vt - mb)$ and $\sum_n W'(na - vt - mb)$, respectively.

By the expressions in terms of the normal coordinates the modes with wave vector k in chain one are coupled to modes at $k + K_2$ in the same, and to modes at $k + K_1$ in the other chain, and vice versa. The centre of mass motions are in first approximation coupled to modes with wave vectors in one of the two reciprocal lattices. Then the equations of motion become

$$\ddot{Q}_0^{(1)} = -\frac{\lambda}{m_1}\left[\sum_{K_2} \hat{f}_{K_2 1} Q_{K_2}^{(1)} \exp(-i\Omega_1 t) + \sum_{K1} \hat{g}_{K_1 1} Q_{K_1}^{(2)} \exp(-i\Omega_2 t)\right]$$
$$\ddot{Q}_0^{(2)} = \frac{\lambda}{m_2}\left[\sum_{K_2} \hat{f}_{K_2 1} Q_{K_2}^{(1)} \exp(-i\Omega_1 t) + \sum_{K1} \hat{g}_{K_1 1} Q_{K_1}^{(2)} \exp(-i\Omega_2 t)\right]$$
$$\ddot{Q}_{K_2}^{1} = -\omega_{1K_2} Q_{K_2}^{(1)}$$
$$\ddot{Q}_{K_1}^{2} = -\omega_{2K_1} Q_{K_1}^{(2)} \quad .$$

The solutions give for the change in the internal coordinate

$$Z_1 - Z_2 \approx vt + A\exp(-i\Omega_1 t \pm i\omega_{1K_2} t) + B\exp(-i\Omega_2 t \pm i\omega_{2K_1} t). \tag{44}$$

The result is a quasiperiodic oscillation of the internal coordinate around a mean value v. Through the coupling to other modes energy flows from this centre of mass motion to the phonon bath. The flow is most important in the regions where one of the frequencies $s\Omega_i$ becomes equal to a frequency ω_{1K_2} or ω_{2K_1}. The effect is even more pronounced at frequencies where the participation of the chains is comparable in size (*i.e.* approximately 0.5).

The analysis given above may be illustrated by numerical calculations [10]. This allows to explore the region that is not accessible to analytical treatment. We consider the DCM with truncated Lennard-Jones potentials:

$$V_i(x) = \left((a_i/x)^{12} - 2(a_i/x)^6\right)\exp(-rx^2).$$

Chain lengths up to N=89 and L=144 were considered, with periodic boundary conditions. The equations of motion were integrated with a four-step Runge-Kutta procedure. For various values of the lengths L and N the equations were integrated with the equilibrium positions as positional initial conditions, zero velocity for the particles of chain one, and a uniform initial velocity of the particles of chain two. The monitored properties were the momenta of the two chains, and their kinetic energies as function of time.

In the first simulations the inter-chain coupling was taken to be so small that the modulation functions were smooth. In Fig. 11 the momentum of chain two is plotted as function of time for a number of initial velocities. For momentum smaller than a critical value $v = 1.1$ it remains practically constant for a very long time. For $v \approx 0.65$ the coupling becomes stronger, the energy is lost faster and there are stronger oscillations due to resonance of Ω_1 and ω_{1K_2}. The resonance disappears for higher values of v. Above the critical value the energy loss is much stronger. There is no longer a sliding mode. If the relative motion vanishes the momentum of the second chain goes to $L/(L+N)$ of its original value because of conservation of total momentum. Fig. 12 shows the kinetic energy in both chains as a function of time, when there is a strong dissipation ($v = 1.5$). The chain 2 looses quickly its kinetic energy to chain 1, until the point that the energy is evenly distributed over the modes of both chains. The cross-over from almost dissipation-less to strong dissipative behaviour is very similar to that in the Frenkel-Kontorova model for weak coupling [4]. The calculations show that for low velocities the energy loss, and therefore also the damping of harmonic modes is very small. In experiments the phason and sliding modes have been found usually as strongly damped. This would be then not an intrinsic property of the dynamics of incommensurate phases, but probably due to other effects, such as the coupling to defects and pinning. In [5] the dynamics of incommensurate phases has been studied with a phenomenological approach to the damping.

When the coupling between the chains becomes stronger the modulation functions are no longer continuous, and the analysis in terms of normal coordinates of the two chains is no longer valid. The two chains are still mutually incommensurate, which means that the ground state remains infinitely degenerate. However, in this case there are barriers between the various ground

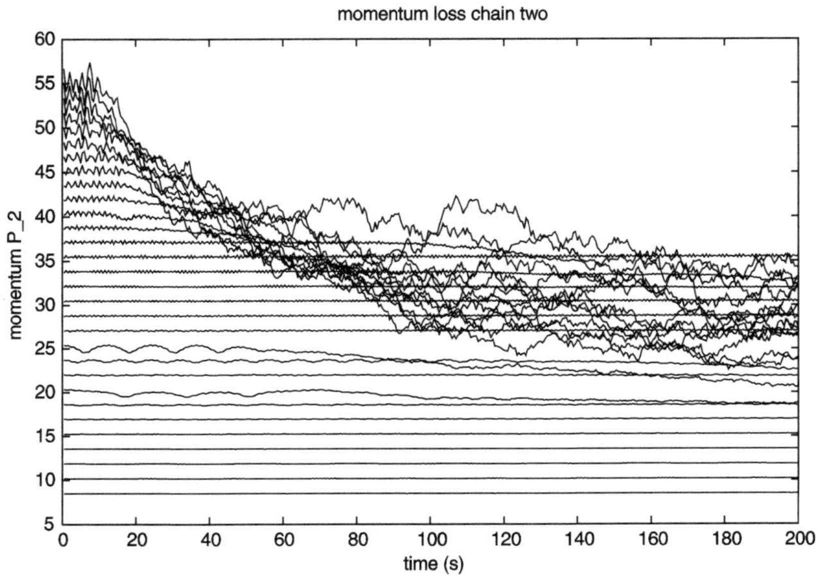

Fig. 11. Time dependence of the momentum of one chain in the DCM for initial speeds ranging from 0.5 to 2.0, with intervals of 0.05

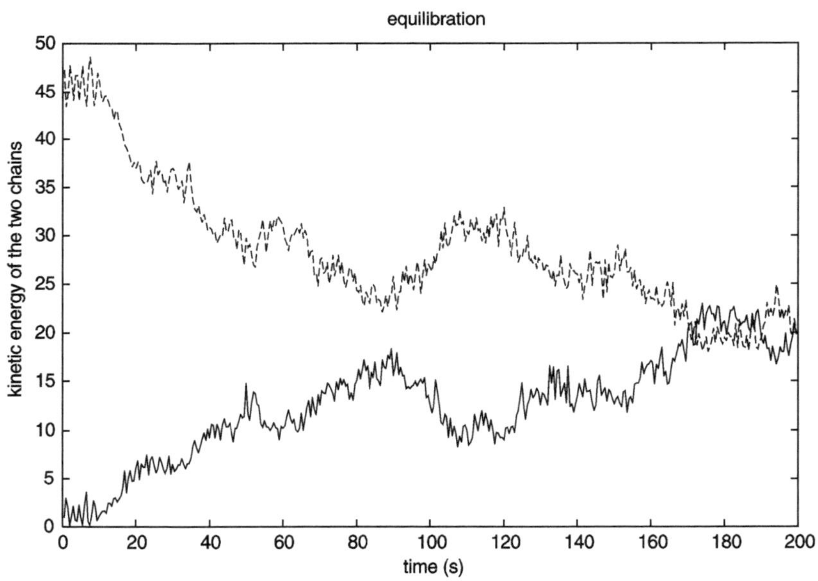

Fig. 12. Time dependence of the kinetic energies of the two chains in the DCM, for initial speed $v = 1.5$

states involving finite jumps of the particles. Therefore, the displacements are no longer harmonic. A numerical integration of the equations of motion gives another cross-over behaviour. For low relative momenta the kinetic energy is not sufficient to cause the particles to move over the barriers. Then the kinetic energy is exchanged between the two subsystems and the center of mass oscillates. For higher momenta the two chains may slide over each other, and the kinetic energy is quickly transferred to the phonon degrees of freedom.

9 Sliding on a Quasiperiodic Substrate

To study the motion of a crystal on the surface of another (periodic or aperiodic) crystal, a simple model can be used that has its origin in the study of incommensurate composites, the double chain model (DCM) [15]. However, still simpler is the model that usually is called the Frenkel-Kontorova model, or a generalization of it (GFKM).

We consider a linear chain of atoms with mass m with harmonic interatomic interaction, on a rigid substrate. In the standard FK model this is a simple sinusoidal potential.

$$H = \sum_n \left(\frac{p_n^2}{2} + \frac{\alpha}{2}(x_n - x_{n-1} - a)^2 + \frac{\lambda}{2\pi} \cos(2\pi(\frac{x_n}{b} + \phi)\right) . \qquad (45)$$

The ground state and the dispersion curves can be determined in a similar way as done for the DIFFOUR model. Numerically and rigorously ([1]) it has been shown that a critical value λ_c exists such that for lower values the ground state is a modulated chain with smooth modulation function, and with a closed phason gap. For values above λ_c there is a phason gap and the modulation function is discontinuous. The character of the phason (with $q = 0$, i.e. a uniform shift of the modulation function) can be seen from the participation ratio

$$P = \frac{1}{N} \frac{|\sum_n u_n^2|^2}{\sum_n u_n^4} \qquad (46)$$

which ranges from 0 (localized) to 1 (extended). The phason eigenvector is extended for all values of λ not in the neighbourhood of λ_c [7]. The phason gap, the jump in the modulation function and the participation ratio of the phason mode are plotted in Fig.13, where they are compared with the situation for an aperiodic substrate.

For the aperiodic case we consider the chain with Hamiltonian (GFK: generalised Frenkel-Kontorova)

$$H = \sum_n \left(\frac{p_n^2}{2m} + \frac{\alpha}{2}(x_n - x_{n-1} - a)^2 + \lambda V(x_n)\right) . \qquad (47)$$

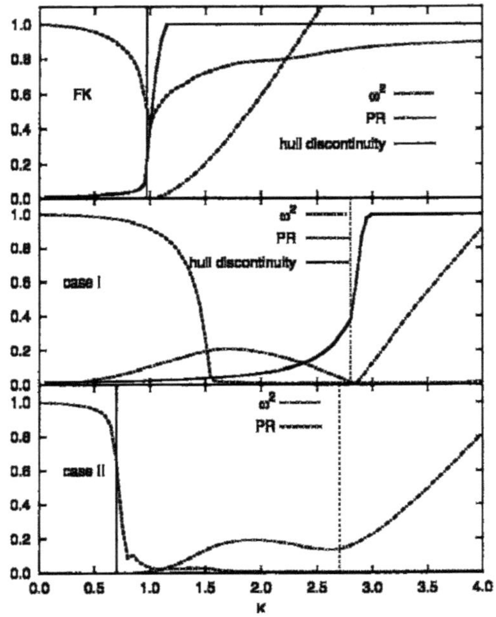

Fig. 13. Phason gap, modulation function jump and participation ratio for (from top to bottom) The Frenkel-Kontorova model, the GFK with rank 2 and the GFK with rank 3

The substrate potential is quasiperiodic and can be written as

$$V(x) \;=\; \sum_{K \in M^*} V_K \exp(iKx), \tag{48}$$

where M^* is a Fourier module of rank 1 or 2. For $\lambda = 0$ the equilibrium positions of the atoms are $x_n = x_0 + na$, and the particles may oscillate around these equilibrium positions

$$x_n \;=\; x_0 + na + u_n \;=\; x_0 + na + \sum_k Q_k(t) \exp(ikna). \tag{49}$$

The normal coordinates vary as $Q_k(t) = Q_k(0) \exp(i\omega_k t)$ with

$$\omega_k^2 = 4\alpha \sin(ka/2)^2/m.$$

For general value of λ we suppose that the deviations of the inter-atomic distances from a remain small. If Z is the centre of mass the positions of the particles are

$$x_n \;=\; Z + na + u_n \;=\; Z + na + \sum_k Q_k(t) \exp(ikna). \tag{50}$$

The equations of motion are

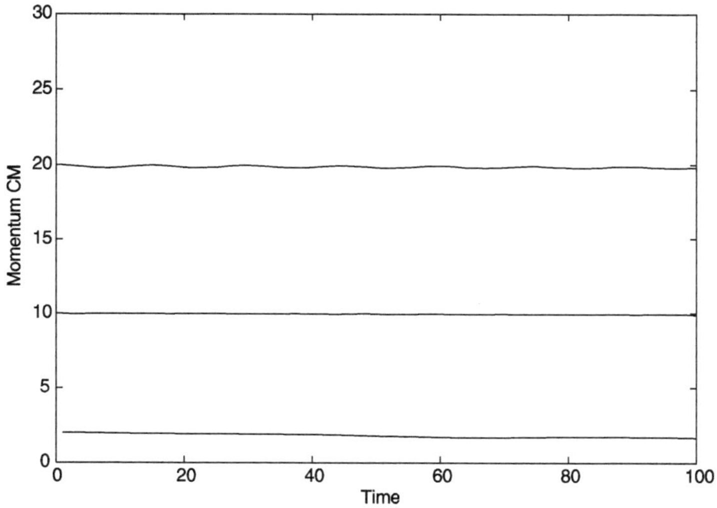

Fig. 14. The Centre of Mass velocity as function of time for 3 different initial velocities for weak chain-substrate interaction. The dissipation is low

$$m\ddot{Q}_k = -\omega_k^2 Q_k - \frac{i\lambda}{m}\sum_{K\in M^*} V_K \exp(i\Omega_K t)\delta(k-K) \tag{51}$$
$$-\frac{\lambda}{m}\sum_{K\in M^*} K^2 V_K \exp(i\Omega_K t) Q_{k-K}.$$

Here $\Omega_K = Kv$ and v the centre of mass velocity. The equations of motion for the center of mass are

$$\ddot{Q}_0 = -\frac{\lambda}{mN}\sum_K V_K K^2 \exp(i\Omega_k t) Q_{-K}. \tag{52}$$

There is strong coupling, and a resulting resonance, if Ω_K is close to the frequency of the mode K, and this depends on the velocity v of the centre of mass. For certain velocities resonances occur between the 'wash board frequency' Ω_K and the mode at K. This leads to oscillations in the CM velocity. Eventually this leads to dissipation of the CM kinetic energy.

If the CM velocity is not in a resonance region, the dissipation is low, and the motion is almost frictionless. This is a direct consequence of the aperiodicity. For still higher speeds the coupling of the CM motion with the phonons in the chain becomes larger. A dynamic phase transition occurs from low friction to high friction.

We check these results numerically. For the substrate potential we choose a quasiperiodic function

$$V(x) = s\,\cos(2\pi x/b) + (1-s)\,\cos(2\pi x/c), \quad (0 \le s \le 1) \tag{53}$$

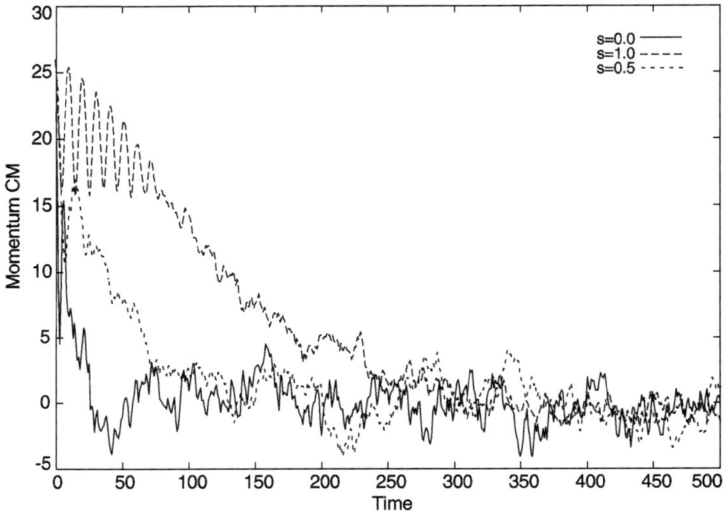

Fig. 15. CM energy dissipation for strong chain-substrate coupling for 3 values of the parameter s

where a and b are incommensurate numbers, *i.e.* a/b is irrational. As shown in [7] the situation depends on the rank of the module generated by $1/a$, $1/b$ and $1/c$. If b/c is irrational the rank may be two or three. For $s = 0$ or $s = 1$ the substrate is periodic, and the system is the usual FK model. There is an pronounced difference between the case I (rank equal to 2) and case II (rank 3). For rank 3 the embedding is in a 3-dimensional space and the modulation function is a function of two variables. In Fig. 13 the corresponding phason gap, maximal modulation function jump and participation ratio are given.

For the non-linear dynamics we shall consider the cases of rank 2 and 3. In particular, for the numerical integration the values $a = 1$, $b = (\sqrt{5}-1)/2$, and $c = \sqrt{2} - 1$ were used. Then the rank is 2 for $s = 0$ or $s = 1$, and it is 3 otherwise. Because the chain is in the potential of the substrate, there is no translation invariance. Nevertheless, there may be a vibration mode with very low frequency. The chain is quasiperiodically modulated. Because of the incommensurability the ground state is infinitely degenerated, because there is a dense set of phases of the modulation function with the same energy. If the modulation function is smooth this yields a zero frequency vibration mode. This has been found in the FK model [1] and in the DCM [15, 12, 3]. This same situation occurs if the substrate is quasiperiodic of rank 2. Then the modulated chain is of rank three, and there are two independent phases. The solutions of the equations of motion with an external force and with an explicit damping term were studied in [19, 22] for the periodic FK substrate, and in [21] for an aperiodic substrate. Here the energy loss is studied with such a damping. Energy transfer to the heat bath is determined by the non-linear coupling to the phonons. It is a Hamiltonian system.

In Fig. 14 the CM momentum is given for the case of small coupling λ. The modulation function is smooth and there is a zero frequency sliding mode. For three initial velocities in the figure the CM momentum remains practically constant, which means a low friction. However, for higher initial velocities there is energy loss to the phonon bath, and the behaviour of the CM is similar to that for strong coupling in the DCM. (See Fig. 15)

If there is a zero frequency sliding mode, i.e. a vibration mode that can be interpreted as an oscillation in the two-dimensional internal space, an adiabatic motion connects two ground states without loss of energy. In the present model one finds, just as in the DCM, that for sufficiently low relative velocities of chain and substrate the motion is almost frictionless.

In principle, there are more resonances for a system of rank three than for one with rank two. Nevertheless, there is little difference between the behaviour when varying the parameter s. For $s = 1$ or $s = 0$ the substrate is periodic, and the rank is two. Three curves are presented in Fig. 15 for various values of s. The initial velocities are the same for the 3 cases. For $s = 1$ the system is near a resonance for this initial speed, but soon the dissipation takes over. There is no clear difference between the case of rank 3 ($s = 0.5$) and of rank 2 (s =0 or 1).

10 Summary

Non-linearity plays a role in the theory of aperiodic crystals in several places. First the presence of a soft mode leading to a phase transition is inherently due to non-linear terms.

Second, the equations for the ground state configuration may be interpreted as non-linear discrete mappings in a space of a dimension dependent on the range of the interactions.

The presence of a phason gap in the vibrational spectrum is directly related with the breaking of the continuity of the modulation functions. If there are low frequency phason modes the amplitudes may become large. Then non-linearity comes into play in the third place, when phase excitations move through the system. Here a dynamic transition takes place from a practically dissipation-less motion to a dissipative motion.

The same dynamic transition has been found in the sliding of subsystems in an aperiodic composite (internal friction) or of the sliding of one crystal on top of another (normal friction).

Acknowledgement

Many of the results presented here have been obtained in collaboration with several people. I would like to thank Linda Brussaard, Luca Consoli, Titus van Erp, Annalisa Fasolino, Ovidiu Radulescu, Alexei Rubtsov, Han Slot and John Tjon.

References

1. S. Aubry, Physica D **7**,240 (1983).
2. R. Blinc, A.P. Levanyuk (eds.), *Incommensurate phases in dielectrics* (2 vols.), North-Holland , Amsterdam (1986).
3. L.A. Brussaard, A. Fasolino, T. Janssen: Phys. Rev. B **63**, 214302, (2001)
4. L. Consoli, A. Fasolino, H. Knops, T. Janssen: Ferroelectrics, **250**, 111 (2001).
5. R. Currat, E. Kats, I. Luk'yanchuk, Eur.Phys.J. B**26**, 339 (2002).
6. B. Dam, A. Janner, J.D.H. Donnay: Phys. Rev. Lett. **21**, 2301 (1985)
7. T. van Erp, A. Fasolino, O. Radulescu, T. Janssen: Phys. Rev. B**60**, 6522 (1999).
8. F.C. Frank, J.H. van der Merwe, Proc. R.Soc. London, Ser. A **198**, 205 (1949).
9. A. Janner, T. Janssen, Acta Cryst. A **36**, 399, 408 (1980).
10. T. Janssen: J.Phys. Cond. Matt., **14**, 12411 (2002).
11. T. Janssen, A. Janner, Adv. Phys. **36**, 519-624 (1987)
12. T. Janssen, O. Radulescu, A.N. Rubtsov: Eur.J.Phys. B **29**, 85 (2002).
13. T. Janssen, J.A. Tjon, Phys. Rev. B **25**, 3767 (1982).
14. T. Janssen, J.A. Tjon: J.Phys. C **16**, 4789 (1983).
15. O. Radulescu, T. Janssen: Phys. Rev. B **60**, 12737 (1999).
16. J.J.M. Slot, T. Janssen: Physica D **32**, 27 (1988).
17. S. van Smaalen, Cryst. Rev. **4**, 79-202 (1994).
18. Z.M. Stadnik (ed.), Physical properties of quasicrystals, Springer, Berlin (1999).
19. T. Strunz, F.-J. Elmer, Phys. Rev. E**58**, 1601, 1612 (1998).
20. J.B. Suck, M. Schreiber, P. Haeussler (eds.), Quasicrystals: An introduction to structure, physical properties and applications, Springer, Berlin (2002).
21. A. Vanossi, J. Röder, A.R. Bishop, V. Bortolani, Phys. Rev. E**63**, 17203 (2001).
22. M. Weiss, F.-J. Elmer, Z.Phys. B**104**, 55 (1997).

Part III

Complex and Chaotic Behaviour

Introduction to Part III

The nonlinear dynamics of interacting particles or systems is generally accepted as one of the corner-stones of equilibrium and non-equilibrium statistical mechanics. Although it is clear that unstable and chaotic behaviour on a microscopic scale is of fundamental importance in this field, there exists, in contrast to systems with few degrees of freedom, still a considerable lack of understanding. One reason is the appearance of qualitatively new, collective effects in the thermodynamic limit. Also in the more general context of interacting entities in biological, social, or economic systems, the understanding of emerging complex behaviour is still in its infancy. This reasoning explains why in the following contributions the interplay between Nonlinear Dynamics and Statistical Physics plays a major role.

The first contribution by Günter Radons is naturally connected to the previous part on disordered systems: It is concerned with nonlinear dynamics in the presence of quenched disorder. He first points out that even quite simple disordered dynamical systems with one chaotic degree of freedom can naturally give rise to non-equilibrium phenomena such as aging, anomalous transport, and the dynamical phase transitions usually observed in complex many-particle systems. The second half of the paper is devoted to large ensembles of interacting limit cycle oscillators, i.e. prototype elements describing periodically active subsystems such as neurons or cells. The randomness lies in the individual frequencies and possibly in the interaction terms. In the first case, which is known as Kuramoto model a famous collective effect, the synchronization transition is explained in detail and the corresponding Lyapunov spectrum is calculated analytically. For random interactions the non-equilibrium counterpart of a spin-glass transition is identified and its manifestation in the Lyapunov spectrum is investigated.

The Lyapunov spectra of many particle systems is also one of the central objects of interest in the second paper by Harald Posch and Christina Forster. The corresponding perturbation dynamics is introduced and investigated in detail for fluids in equilibrium and in stationary non-equilibrium situations. As models, hard sphere gases or many-particle systems with soft interaction potential are investigated by molecular dynamics methods. The authors present an interesting recent finding in the equilibrium situation, namely the occurrence of steps in the Lyapunov spectrum of hard sphere systems and the observation of wave-like structures in the associated Lyapunov vectors. These hydrodynamic Lyapunov modes provide an example of a surprising new collective effect in chaotic many particle systems. The second part of this contribution is devoted to the non-equilibrium molecular dynamics (NEMD) of such systems. In systems which are to be kept in a stationary state far from equilibrium one needs thermostats to get rid of the excess heat produced by the external mechanical or thermal forces. The authors discuss how such states are reflected in the Lyapunov spectra of these systems, and how the latter are connected to macroscopic transport coefficients. Some aspects of this approach, e.g. the appearance of multifractal phase space structures, are subject of an ongoing scientific discussion to which the present article is a profound contribution.

The link between Nonlinear Dynamics and Equilibrium Statistical Physics is also the subject of the contribution of Wolfram Just. Here, however, the connection is more on a formal level where the analogy between time series of symbols and, in the simplest case, spin-chains is exploited. This subject is usually referred to as symbolic dynamics or the thermodynamic formalism. In this article first an elementary introduction to this topic is given. The concepts of Markov partitions and Markov maps are explained with the aid of simple models, which are helpful also for the understanding of the more complicated dynamics of the maps introduced in the first contribution by Günter Radons in this part. Subsequently these concepts are extended to deal with questions of space-time chaos, a subject of great current interest. For this purpose so-called coupled map lattices are introduced, which are constructed such that the Markov property is preserved. The time series generated by such an infinite one-dimensional array of coupled maps can be understood in terms of the equilibrium properties of the 2-dimensional Ising model, which is known to display a phase transition. This implies the occurrence of a new kind of dynamical phase transition for coupled maps, a macroscopic collective effect due to the extended nature of the system. This phenomenon is discussed as a special case of similar effects in more general coupled maps, where due to the typically greater complexity of the symbolic dynamics also transitions to spin-glass phases etc. can be expected.

The last contribution to this part by Heinz Georg Schuster introduces a quite different and novel aspect. The concept of adaptivity opens a whole realm of applications of nonlinear dynamics for instance in the biological sciences and in economy. As an example, agents modify their actions in response to reactions of other agents with the goal of optimising some environmental influence on them. Such problems define coupled nonlinear systems, which can be investigated with the tools of Nonlinear Dynamics. The approach is explained with two examples. The first is a self-recognition problem consisting in the determination of the status of a member in a hierarchy, and the second is the solution of the minority game with the aid of adaptive probabilistic Boolean functions.

Disordered Dynamical Systems

Günter Radons

Institute of Physics, Chemnitz University of Technology,
09107 Chemnitz, Germany
radons@physik.tu-chemnitz.de

1 Introduction

Disordered dynamical systems constitute a research field, which, despite its importance, is still not well explored. The interplay between dynamical and static, quenched disorder can lead to a wealth of phenomena, which are only partially understood. The difficulty with this sort of complex systems comes from the need to combine concepts from dynamical systems theory and from the theory of disordered systems. Each of them constitutes a formidable research field by its own and is usually treated in different scientific communities. In this sense disordered dynamical systems constitute an interdisciplinary research topic. The theory of dynamical systems without disorder is already well established [1, 2]. It has applications in all branches of physics, e.g. atomic physics, hydrodynamics, quantum optics, or solid state physics. The key concepts for the characterization of such systems are Lyapunov exponents, various dynamical entropies, and fractal dimensions of strange attractors in dissipative systems. Many investigations considered systems with few effective degrees of freedom living in a compact phase space. Cases where phase space is infinite were often treated in the presence of a discrete translational symmetry, which allows a reduction to a unit cell, again leading to compact phase spaces. The presence of a continuous translational symmetry usually leads to some simplifications, too. An example is the presence of hydrodynamic modes in extended systems. Further examples can be found in Part I and also here in Part III of this book. Clearly these are limiting cases, and one wonders what happens for systems with broken translational invariance, or for systems with many non-equivalent degrees of freedom. An idealization of this situation is captured by the concept of disorder. Disordered systems constitute a branch of statistical physics or solid state physics [3] with concepts and methods very different from the ones in dynamical systems theory. Part II of this book introduces some of the important ideas in this field. It is clear that there exist many ways for including disorder into dynamical systems depending on the level of description. For instance, one can include disorder in partial differential equations, in the coupling between ordinary differential equations, or between iterated maps leading to coupled map lattices with disorder. In the following we treat two model classes of disordered dynamical systems, which are at least in limiting cases rather well understood. These examples suggest that spin-glass phases , aging , or disorder induced anomalous transport are phenomena that arise very frequently also in disordered dynamical systems . We will also see how the above mentioned characteristics of dynamical systems reflect these disorder phenomena. This article is organized as follows. In Sect 2 we introduce simple disordered extended maps with quenched disorder. Despite their simplicity these systems exhibit a kind of collective behavior for ensembles of states, although there is no mutual interaction. Furthermore aging, anomalous transport and other interesting phenomena are found. In Sect. 3 we consider limit cycle oscillators with random frequencies and with random interactions. The Kuramoto model, a special case, is treated in some detail because it is the paradigm for another collective phenomenon, the occurrence of mutual synchronization, and because it allows for an exact calculation of its Lyapunov spectrum in the thermodynamic limit. For random interactions chaotic behavior is observed in the "paramagnetic" phase, and in addition in the non-chaotic regime a spin-glass phase is shown to exist.

2 Disordered Maps

The simplest disordered dynamical systems are those with one or two degrees of freedom with disorder in the environment. An example is the Hamiltonian motion of a point particle in a two-dimensional disordered potential consisting of hard discs (Lorentz gas [4]) or smoothed version thereof. The periodic counterparts of these models have recently received much experimental and theoretical attention in connection with mesoscopic transport in quantum dot lattices [5, 6]. There the importance of classical phase space structures and chaotic diffusion was realized [6]. Here we will concentrate first on dissipative systems which may also show chaotic transport. A simple much studied example is the damped motion of a periodically driven particle in a periodic potential. This provides e.g. a model for superionic conductors in an external field and currently finds renewed interest in the context of ratchet physics. Simplified versions which are assumed to capture essential aspects of these systems are one-dimensional iterated maps [7, 8, 9]. While the periodicity in the equations of motion allows for the application of advanced methods such as periodic orbit theory, the thermodynamic formalism, or Levy flight statistics [10, 11, 12], it is clearly of much greater importance to understand the effects of static disorder in such systems.

2.1 Anomalous Transport

From the physics of disordered systems, it is known that static or quenched randomness may drastically alter macroscopic quantities such as transport coefficients. In the following we will treat such an effect for dynamical systems, namely the total *suppression of normal or anomalous chaotic diffusion by quenched randomness* in the equations of motion. This turns out to be a non-trivial effect, since the mean-square displacement will remain finite, although chaotic transport is *not* inhibited locally. This was observed for the first time in [13]. We will also show that these systems exhibit the phenomenon of *aging*. Generalizations of these systems which include a global bias show various *dynamical phase transitions* characterized by *anomalous chaotic transport* properties. Interestingly all these phenomena can occur in both, dissipative and Hamiltonian dynamical systems.

We will first concentrate on one-dimensional non-invertible maps of the type studied in [7, 8, 9, 10, 11, 12, 13, 14]. They have the general form $x_{t+1} = f(x_t) = x_t + F(x_t)$, with $F(x)$ periodic in x. The periodicity interval, which we set equal to one, i.e. $F(x) = F(x+1)$, defines cells or half open intervals $A_i = [i, i+1)$, $i \in \mathbb{Z}$, on the real axis. We will modify these dynamical systems by randomly changing $F(x)$ in each cell A_i to a function $F^{(i)}(x)$ resulting in

$$x_{t+1} = x_t + F^{(i)}(x_t) \tag{1}$$

for $x_t \in A_i$. As a possible origin of this modification one could imagine a random variation in space of some driving force felt by a particle. A natural choice for $F^{(i)}(x)$ consists of random shifts of F

$$F^{(i)}(x) = F(x) + \varepsilon(i). \tag{2}$$

In order to avoid complications connected with a global bias we assume for the moment the symmetry $F(-x) = -F(x)$, and further that the $\varepsilon(i)$ are independent, identically distributed random variables with a symmetric distribution function $p(\varepsilon) = p(-\varepsilon)$ implying $\overline{\varepsilon(i)\varepsilon(j)} \propto \delta_{ij}$ and $\overline{\varepsilon(i)} = 0$. Through the cell index i, defined as $i = [x]$, the largest integer smaller than x, the term $\varepsilon(i)$ is recognized as piecewise constant random function of x. In contrast to other studies [7], where time-dependent noise was added to the deterministic dynamics, the random term $\varepsilon(i)$ remains constant in time, Eq. (1) is still deterministic, it describes a dynamical system with *quenched* disorder.

Let us now investigate the effect of this static randomness first for the simplest maps which in the absence of disorder ($\varepsilon(i) = 0$) exhibit chaotic diffusion. These are systems where $F(x)$ varies linearly in each cell, i.e. $F(x) = a\{x\} - a/2$ with $\{x\} = x - [x]$. Since the slope of $f(x)$ is $a+1$ these maps are chaotic for $a > 0$ and show chaotic diffusion for $a > 1$. The dashed graph in Fig. 1(a) is an example with $a = 3$. The diffusive motion for this ordered case is verified by the linear increase of the mean-square displacement $\sigma^2(t) = \langle x_t^2 \rangle - \langle x_t \rangle^2 = 2Dt$ with the correct diffusion constant $D = 1/4$ [8] (dashed line in Fig. 1(b)). This and the following results for $\sigma^2(t)$ were obtained numerically by iterating ensembles of 2×10^4 points (initially distributed homogeneously or inhomogeneously in one cell) for 10^6 (occasionally 10^7) time steps. An example of a map with binary disorder, $\varepsilon(i) = \pm 1/2$ in Eq. (2), is shown as full line in Fig. 1(a). Now, with disorder $\varepsilon(i) \neq 0$,

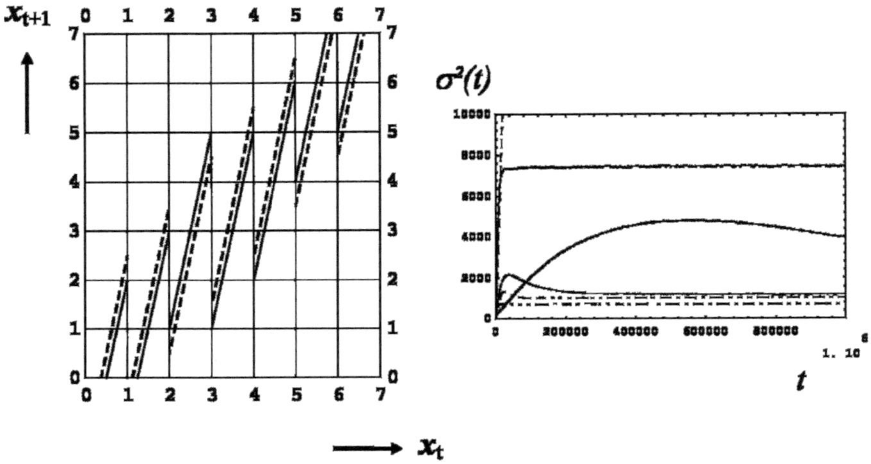

(a) (b)

Fig. 1. (**a**) Simple piecewise linear maps corresponding to a periodic (*dashed*) and a random driving 'force' (*bold*). The mean-square displacement (**b**) increases linearly for the former (*dashed line*) and saturates in the latter case as shown for several disorder realizations (*full lines*: $\sigma^2(t)$ for $\varepsilon(i) = \pm 1/2$,*dot-dashed graphs*: $\varepsilon(i)$ equally distributed in $(-1/2, +1/2)$). There exist environments where a first constant level is observed only after more than $t = 10^6$ iterations ($0.5 \ldots 1.0 \times 10^7$ for the second graph from the top)

a very different behavior is observed: $\sigma^2(t)$ saturates and remains bounded for large times. This is a dynamical localization effect, which is, however, of totally different origin than quantum mechanical localization in disordered solids. As is seen from Fig. 1(b) this localization occurs for discrete random variations $\varepsilon(i)$ as well as for continuously distributed random variables $\varepsilon(i)$. We emphasize that for both cases there exists no obvious reason why the spreading of the distribution $\rho_t(x)$ should be limited because the *a priori* probability for reaching one of the neighboring cells is always finite. More explicitly, *independent* of the chosen sequence $\varepsilon(i)$, a fraction $p = 1/4$ of a homogeneous distribution in some cell A_i is always transferred to the right neighboring cell A_{i+1}, the same fraction to the left cell A_{i-1}, and one quarter remains within the cell A_i. From this point of view there is no difference between the homogeneous situation ($\varepsilon(i) = 0$) and the inhomogeneous case ($\varepsilon(i) \neq 0$)! The randomness affects only the last quarter, which is mapped into one or both of the next-nearest cells $A_{i\pm 2}$. Note also that the strength of chaos as measured by the Lyapunov exponent (Fig. 1: $\lambda = \ln 4$) is not altered by the random shifts because the slope of $f(x)$ remains at its value $f'(x) = 4$. For reasons to become clear below we call the disordered map of Fig. 1(a) a *map with topological disorder*.

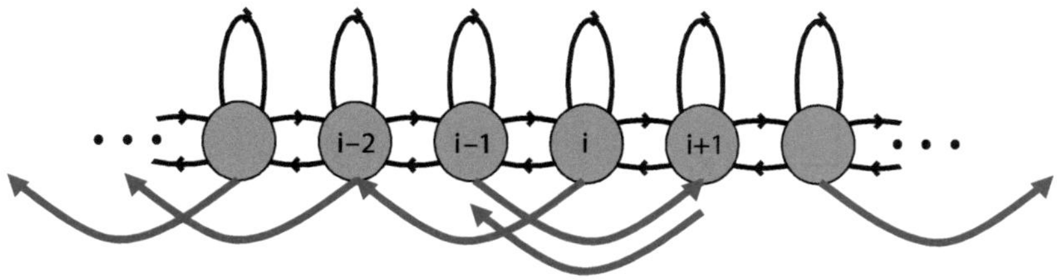

Fig. 2. The Markov model corresponding to the disordered map of Fig. 1(a). All non-zero transition probabilities are equal to $p = 1/4$ and the disorder is present only in the transitions to the next-nearest neighbors. The saturation of $\sigma^2(t)$ of Fig. 1(b) for $\varepsilon(i) = \pm 1/2$ were also checked by iterating Eq. (3) with transitions probabilities as defined by this graph

The explanation of this localization effect in the case of quenched randomness follows from the following connection. For the map $f(x)$ with discrete random shifts $\varepsilon(i) = \pm 1/2$ as in Fig.1 the cells A_i define a (generating) Markov partition [15, 16] and the corresponding map is called a Markov map. This implies that the evolution of piecewise constant distributions $\rho_t(x)$ (constant in the cells A_i) is fully equivalent to a Markov process, i.e. the content $\pi_i(t) = \int_i^{i+1} \rho_t(x)\, dx$ of cell A_i at time t is iterated according to

$$\pi_j(t+1) = \sum_i \pi_i(t) p_{ij}. \qquad (3)$$

For the above piecewise linear map with $\varepsilon(i) = \pm 1/2$ (Fig. 1) the only non-zero transition probabilities p_{ij} are given by $p_{ii} = p_{i,i\pm 1} = 1/4$ and $p_{i,i\pm 2} = (1/2 \pm \varepsilon(i))/4$. The corresponding Markov model is shown in Fig. 2. Such a model defines a discrete random walk in a locally asymmetric random environment. In the case of Fig. 2 the randomness is fully defined by the connectivity of the Markov model, or in technical terms by the adjacency matrix, because all non-zero transition probabilities are equal. This explains why we denote the corresponding maps as maps with topological disorder.

The above localization effect, i.e. $\sigma^2(t)$ remaining finite for $t \to \infty$, is known as *Golosov phenomenon* in the random walk literature [17]. Inspired by Sinai's work [18] it was proven rigorously for one-dimensional random walks with random nearest neighbor transitions by Golosov [19]. Reversing the above arguments which led us from iterated maps to random walks, it is obvious that also for the latter systems there exist realizations in terms of dynamical systems. These consist of piecewise linear chaotic maps of the form Eq. (1), with a typical example shown in Fig. 3. Again the cells A_i provide a Markov partition for this system and thus define the Markov model also shown in the figure. The segments of length p_{ii} and $p_{i,i\pm 1}$ in each unit cell, where the map $f(x)$ is linear, correspond to the non-zero transition probabilities p_{ii} and $p_{i,i\pm 1}$ of the associated Markov chain.

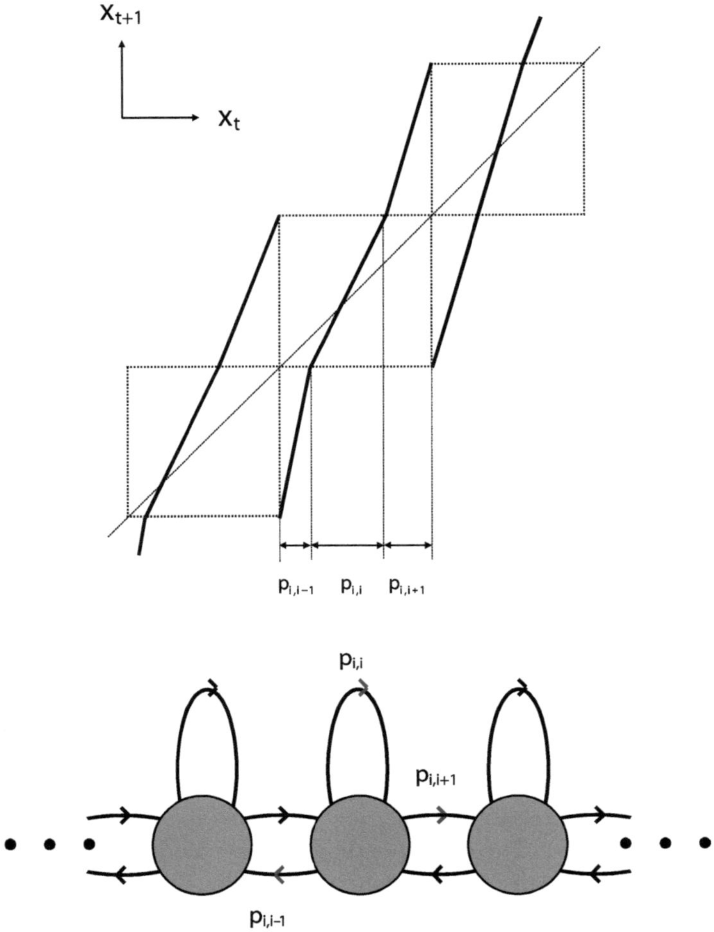

Fig. 3. An example from the class of iterated maps for which the asymptotically finite mean-square displacement follows rigorously due the work of Sinai and Golosov [18, 19]. Also shown by dashed lines are the unit squares of the integer grid (see Fig. 1(a)) along the bisectrix. The indicated intervals p_{ii} and $p_{i,i\pm1}$ mediate the transitions from the i-th cell A_i to itself and its neighbors respectively. They vary randomly from cell to cell

The asymptotically finite mean-square displacement was proven in [19] for independent random sequences p_{ii} and $p_{i,i-1}/p_{i,i+1}$ with $\overline{\ln(p_{i,i-1}/p_{i,i+1})} = 0$. Maps with these properties will be called *maps with Sinai disorder*. The latter condition means that there is no global bias in the system and that one observes recurrent behavior with probability one.

So far we have given only a formal argument for the observed localization phenomenon by mapping the dynamical systems defined in Fig. 1(a) and Fig. 3 to random walk models with (locally) asymmetric random transitions probabilities to next-nearest respectively nearest neighbors. An intuitive picture for the relevant physical processes is obtained from the continuum limit of the nearest neighbor discrete random walk model, which is Brownian motion in a spatially random force field $\tilde{F}(x)$ [17]. In this limit the dynamics is governed by the Langevin equation

$$\dot{x}(t) = -\frac{\partial U}{\partial x}(x(t)) + \xi(t) \tag{4}$$

with Gaussian white noise $\xi(t)$. The important point is that the graph of the associated potential $U(x) = -\int^x \tilde{F}(x')dx'$ itself can be thought of as a realization of a Brownian path as shown in Fig. 4. The resulting statistical self-similarity of the random potential $U(Lx) \simeq L^{1/2}U(x)$ implies the occurrence of deeper and deeper potential wells as the particle proceeds. The work of Sinai and Golosov shows that an ensemble of initially close particles moves in a coherent fashion from one deep minimum to the next deeper potential well as indicated schematically in Fig. 4.

In this stepwise process it is *typically*[1] one minimum which dominates and therefore determines the (finite) width $\sigma^2(t)$ of the ensemble[2]. Since the random environment in the neighborhood of these minima is the same only in a statistical sense one observes for a fixed environment still fluctuations in $\sigma^2(t)$. These fluctuations become extremely rare for large times t as follows from an Arrhenius argument [17] which says that the typical time to overcome the ever increasing relevant potential barriers increases exponentially with the barrier height, i.e. it takes a time of the order $\exp(b\sqrt{x})$ for the state to travel a distance x. Solving this relation for x says that the typical distance reached in time t increases only as $\ln^2 t$. Indeed it has been shown rigorously that the mean displacement grows anomalously as

$$\langle x(t) \rangle = \xi(t) \ln^2 t \tag{5}$$

with $\xi(t)$ a random function of $O(1)$ [18, 19, 17]. By averaging this law for the anomalous drift over the environment it is often expressed as

$$\overline{\langle x(t) \rangle^2} \sim \ln^4 t \tag{6}$$

sometimes called "Sinai diffusion". A more rigorous review of Sinai's and related findings for random walks in random environments can be found in the monograph of Hughes [22].

[1] Although $\sigma^2(t)$ typically remains finite for all times t, averaging $\sigma^2(t)$ over the random environments leads to a divergence for $t \to \infty$, i.e. $\lim_{t\to\infty} \overline{\sigma^2(t)} = \infty$, due to rare contributions from atypical configurations of the environment [17, 20, 21]. This is an example for the difference between typical values and their mean encountered often in disordered systems.

[2] The initial increase and subsequent decrease of $\sigma^2(t)$ as seen in Fig. 1(b) can also be understood in terms of a relaxation process of the initial distribution into a first local trap of the system. In the case of the Sinai systems this process has been visualized in Fig. 4 of Ref. [20].

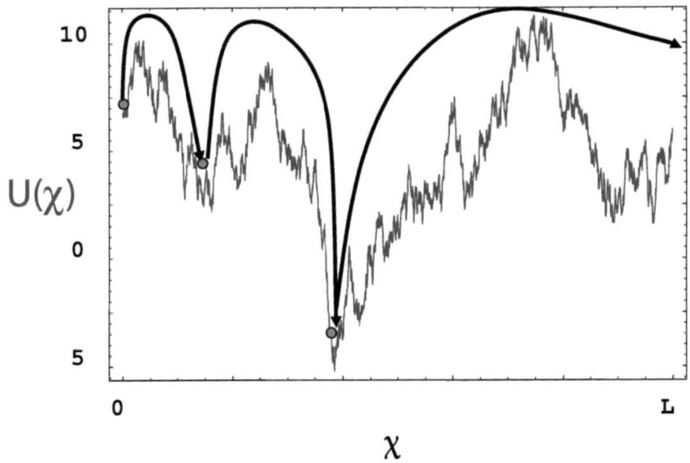

Fig. 4. The random potential associated in the continuum limit with the Sinai model and the map with Sinai disorder Fig. 3. The motion of the state variable $x(t)$ corresponds to an activated process in this potential. A localized initial distribution of states moves in a *coherent* fashion from one minimum to the next deeper minimum

2.2 Aging

Disordered systems like those treated in Part II of this book exhibit the phenomenon of aging. It is a well-known experimental fact from glasses, spin glasses, and other complex materials [23, 24]. Theoretical investigations are also mainly concerned with these systems (see e.g. [25, 26, 27], and refs. therein). Recently aging was found also in much simpler model systems [28, 29]. Aging can be defined as an anomalous behavior of response and correlation functions (see also the contribution [30] to this book). Consider e.g. the correlation function $C_{AB}(t, t_w) \equiv \langle A(t_w)B(t + t_w)\rangle$ of two variables A and B, where t_w is the waiting time after the preparation of the initial state at time $t = 0$. For $t \ll t_w$ the correlation function $C_{AB}(t, t_w)$ is independent of t_w and a fluctuation-dissipation theorem is supposed to hold. Aging is present if $C_{AB}(t, t_w)$ depends strongly on t_w when t is of the same order as t_w. Furthermore for t and t_w large one often assumes a scaling behavior of the form

$$C_{AB}(t, t_w) = t^{-\nu} F\left(\frac{t}{t_w}\right). \tag{7}$$

First numerical investigations by Marinari and Parisi [28] for random walks in random environments of Sinai type indicated a dependence $C_{AB}(t, t_w) = \ln^2(\frac{t}{t_w})$, i.e. $\nu = 0$ in Eq. (7) for various correlation functions. Recently, however, improved numerical simulations revealed a more complex behavior i.e. one finds asymptotically different logarithmic scalings depending on the ratio $\ln t / \ln t_w$ as t and t_w are sent to infinity [31]. The latter kind of scaling behavior was subsequently confirmed also analytically by an exact real space renormalization group (RSRG) calculation [32, 33]. In any case non-trivial aging is found in random walks in random environments of Sinai type. As pointed out above such systems can be implemented by dynamical systems as introduced in Fig. 3. This implies that aging occurs also in these simple disordered dynamical systems and its generalizations treated in Sect. 2.3. This fact has been realized only recently [34]. We should remark that a behavior as in Eq. (7) is observed also in much simpler non-equilibrium systems, even without disorder: It was pointed out in [35] that already the simple homogeneous, one-dimensional random walk $\dot{x}(t) = \xi(t)$, i.e. the system governed by the Langevin equation, Eq. (4), without drift term, gives rise to a violation of the fluctuation-dissipation theorem and a scaling behavior for $C_{xx}(t, t_w)$ as in Eq. (7). The scaling, however, is trivial in this case, since $C_{xx}(t, t_w) = t_w$ implying $\nu = -1$ and $F(x) = 1/x$ in Eq. (7). It follows that the same simple scaling behavior applies also to dynamical systems, which lead to simple diffusion, e.g. the iterated maps in [7, 8, 9] and the ordered map of Fig. 1(b). Note, however, that already in these simple iterated maps the scaling function depends in a non-trivial (fractal) manner on the parameters of the map. This a consequence of the fractal dependence of the diffusion coefficient on the system parameters [12]. A less trivial aging behavior was found very recently [36] for periodic iterated maps leading to anomalous diffusion, especially in the systems introduced in [11]. Similar to the disordered case one finds in such systems a broad distribution of trapping times which can be related to the aging property. In periodic iterated maps, however, the trapping is associated with marginally stable fixed points of the dynamics. Generalizations of this class of maps can exhibit in addition an extremely slow, i.e. a logarithmic increase in time of the mean square displacement [37], which is reminiscent of "Sinai diffusion", Eq. (6). Also for these maps one expects non-trivial aging. We close this section with a remark on a problem arising in the context of aging, namely the question of the validity of linear response theory and the fluctuation-dissipation theorem. This is well understood for stochastic systems as described by Langevin or Fokker-Planck equations and thus also for the continuum limit of the maps of Fig. 3 and systems with Sinai disorder (for the latter see [33]). Its applicability to general nonlinear dynamical systems, however, is still an active field of research despite the fact that first works in that direction appeared two decades ago [38]. For the current status of this question see e.g. [39] and refs. therein. In summary, the recent finding that simple ordered or disordered extended dynamical systems lead to aging, raises important questions for future research, especially whether and how the different kinds of scaling behavior are related to different classes of dynamical systems.

2.3 Generalizations and Perspectives

In Sect. 2.1 and 2.2 we have exploited the relation between certain dynamical systems and Markov models, especially those defining random walks in random environments. This connection has further consequences. An important aspect follows if one releases the constraints $\overline{\epsilon(i)} = 0$ for the system of Fig. 1 or $\overline{\ln(p_{i,i-1}/p_{i,i+1})} = 0$ for that of Fig. 3. This leads us from systems without global bias, $\mu = 0$, to such with a bias, $\mu \neq 0$. Pictorially this means that the potential $U(x)$ of Fig. 4 gets an additional contribution, $U(x) \rightarrow U(x) + \mu x$, i.e. the potential gets tilted. This picture is exact in the continuum limit of the systems with Sinai disorder, Fig. 3.

Biased Systems and Dynamical Phase Transitions

A bias may simulate an external static field or may be attributed to systematic asymmetries of the underlying potential in more realistic models, such as driven damped particles in some potential landscape. Such aspects are clearly of great importance in the context of transport in ratchets [40, 41]. Although the connection between simple one-dimensional maps and more realistic models is complicated and not fully understood, the investigation of the former with bias will provide some insight into the possible scenarios in the latter.

Exploiting again the connection between maps and discrete time Markov chains, allows us to apply known results from the literature to disordered iterated maps. Analytical results by Derrida and Pomeau [42] for the biased case $\overline{\ln(p_{i,i-1}/p_{i,i+1})} \neq 0$ lead to transport properties for the corresponding dynamical systems, which we summarize in Fig. 5 for a special example of the type of Fig. 3.

Our example consists of setting $p_{ii} = 0$ and choosing a binary distribution for the transition probabilities $p_{i,i\pm 1}$ of Fig. 3. More explicitly, we construct a chain of maps consisting of two sorts of cells (see Fig. 3), which we denote as A_+ and A_-. For the type A_+ we choose $p_{i,i+1} = a$ and correspondingly $p_{i,i-1} = 1 - a$, while for type A_- we reverse the assignments, i.e. we set $p_{i,i+1} = 1 - a$ and $p_{i,i-1} = a$. These cells are concatenated randomly and independently so that type A_+ is present in a concentration c and correspondingly a fraction $1 - c$ of cells are of type A_-. For Fig. 5 we have chosen $a = 1/3$. For $c = 0$ we have an ordered system consisting only of cells A_-, which map points with larger probability, namely with probability 2/3, to the right, and with probability 1/3 to the left. In this limit one gets chaotic diffusion with a mean drift to the right, where diffusion constant D and drift velocity V are given by $D = D_0 = 2a(1 - a)$ and $V = V_0 = 1 - 2a$, respectively. The behavior of the transport coefficients, normalized by its bare values D_0 and V_0, is shown in Fig. 5 as function of the concentration c. We need to discuss only the case $0 \leq c \leq 1/2$, since the rest follows by symmetry. For the other extreme of full disorder $c = 1/2$ we get the unbiased situation

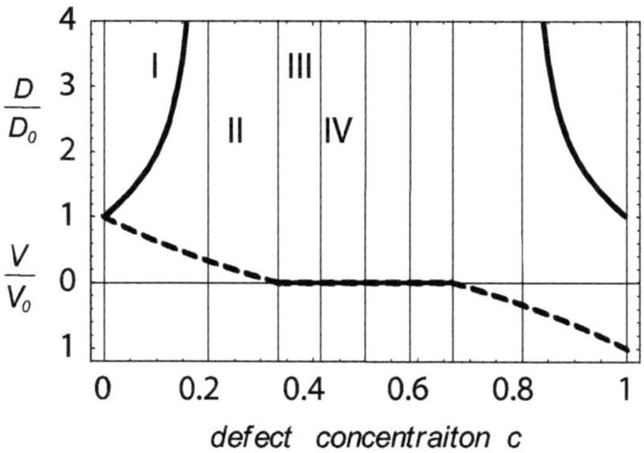

Fig. 5. The dependence of the normalized transport coefficients D/D_0 (*bold*) and V/V_0 (*dashed*) is shown as function of the defect concentration c. In the regimes where these are finite they are self-averaging quantities and therefore these values are observed with probability one independent of the disorder realization

$\overline{\ln(p_{i,i-1}/p_{i,i+1})} = 0$, where the anomalous Sinai and Golosov results of section 2.1 hold. This means we get in this case $D = 0$ and $V = 0$. Between these extremes various transitions between dynamically different phases occur. In Fig. 5 these different regimes are numbered as I-IV. In phase I both D and V are finite and non-zero, i.e. one has normal chaotic transport as in the ordered limit. The transition to phase II occurs at concentrations where conditions $\overline{(p_{i,i-1}/p_{i,i+1})^{\pm 2}} = 1$ are fulfilled (two symmetric solutions in c) . This transition is accompanied by D becoming infinite, which holds in regimes II an III. In these phases one has therefore anomalously enhanced diffusion, i.e. the mean-square displacement grows superlinear. The transition between II an III is signaled by a vanishing drift velocity, i.e. the mean displacement grows slower than linearly in time. The latter transition points are given by $\overline{(p_{i,i-1}/p_{i,i+1})^{\pm 1}} = 1$. The anomalously slow growth of the displacement holds up to the value $c = 1/2$, i.e. also in regime IV. The transition from III to the latter is characterized by a crossover from superdiffusive to subdiffusive chaotic transport implying that in addition to V also D vanishes. The last transition is observed for $\overline{(p_{i,i-1}/p_{i,i+1})^{\pm 1/2}} = 1$. The fact that the qualitative changes in the drift and diffusion properties occur at different values of the concentration is quite common in disordered systems and is expected to hold also more generally. A more extensive description and a discussion of the different regimes in terms of activated Brownian motion in tilted Brownian potentials can be found in the review article [17]. Interestingly the same conclusions were reached very recently in the context of disordered ratchets by a totally different reasoning within a continuous time model [41]. Note that dynamical phase transitions as a consequence of an applied bias may also arise in ordered dynamical systems [43].

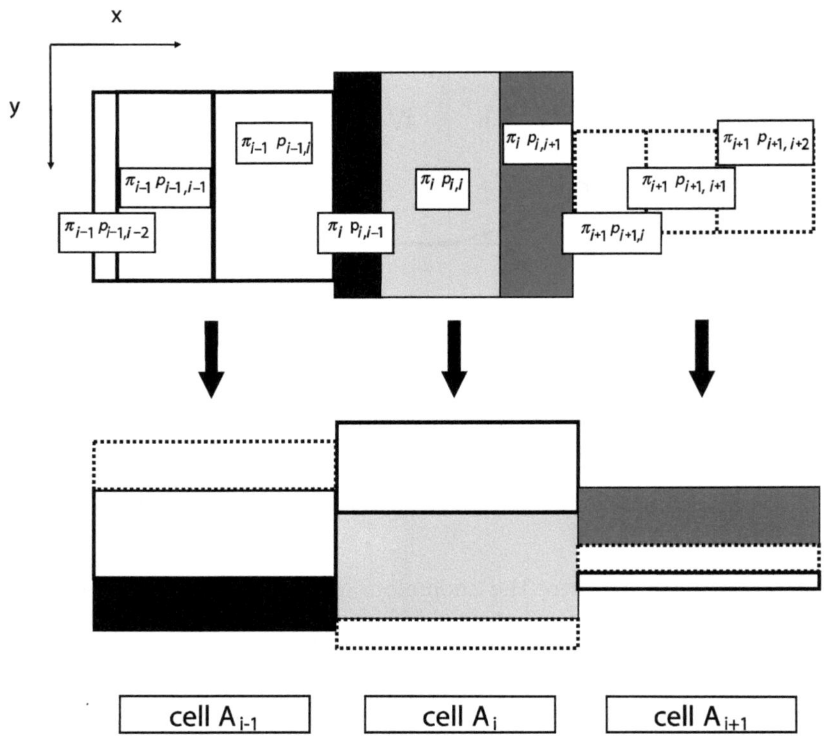

Fig. 6. The construction of an area-preserving inhomogeneous chain of baker maps, which in the projection on the x-axis reveals exactly the dynamics of the one-dimensional iterated map of Fig. 3

We close this section by noting that very recently many more exact results have been obtained for the Sinai model and its aging properties, especially in the presence of a bias [20, 21, 44, 45, 46]. These results are valid also for the associated dynamical systems of Sinai type, but its applicability to more general dynamical systems is an open question.

Area-preserving Maps

One may wonder whether dynamical localization and aging as discussed in sections 2.1 and 2.2 can be observed also in Hamiltonian systems or area preserving maps. This question is answered affirmatively by the explicit construction of an area preserving map of the baker type, which shows the same behavior [34]. The resulting inhomogeneous chain of baker maps is a generalization of a class of dynamical systems (homogeneous chains of baker maps), which recently became very popular (see [47], and refs. therein) and which were originally introduced by Hopf in [48]. The construction is shown in Fig. 6 below.

It consists of an array of L rectangles called also cells $A_i, i = 0, \ldots, L-1$, of unit width in the $x-$direction and heights π_i. During one iteration step the inscribed rectangles within one cell A_i are mapped to the neighboring cells and into the present cell as follows: Take e.g. the black rectangle of height π_i and width $p_{i,i-1}$, which is labeled by its area $\pi_i p_{i,i-1}$ in Fig. 6. This rectangle is in one iteration (from top to bottom in Fig. 6) squeezed in the $y-$direction, stretched in the $x-$direction, and transferred to the left neighboring cell A_{i-1} in an area preserving manner. In the same way the grey-shaded rectangle $\pi_i p_{i,i+1}$ is transferred to the right, and the light-grey rectangle gets squeezed and stretched, but remains in the cell A_i as shown in the figure. For a finite chain we may impose periodic boundary conditions $A_{i+L} = A_i$ ($L = 3$ in Fig. 6), or, alternatively one could confine transport within the array by appropriate modifications of the boundary maps ($p_{0,-1} = p_{L-1,L} = 0$). Observe that the $x-$coordinate x_t of a point (x_t, y_t) in this two-dimensional area preserving map is iterated exactly as a point x_t of the one-dimensional map of Fig. 3 (with slopes $p_{i,j}^{-1}$).

Note, however, that in order to get an area preserving map, the heights π_i have to be adjusted appropriately. The condition that in one iteration the outflow of area of a given cell, say A_i, has to be equal to its inflow, results in

$$\pi_{i-1} p_{i-1,i} + \pi_{i+1} p_{i+1,i} = \pi_i p_{i,i-1} + \pi_i p_{i,i+1}. \tag{8}$$

Adding to both sides of this equation the term $\pi_i p_{i,i}$ and using the normalization $\sum_j p_{i,j} = 1$, one finds that the π_i fulfill the equation for the stationary probability distribution of a Markov chain with transition probabilities $p_{i,j}$. The latter exists for random $p_{i,j}$ only if the system is finite. For reflecting boundary conditions this stationary distribution can be found exactly as

$$\pi_i = \pi_0 \prod_{k=0}^{i-1} \frac{p_{k,k+1}}{p_{k+1,k}}. \tag{9}$$

A characterization of such stationary distributions in terms of its cumulants was given recently in [49]. For periodic boundary conditions a result similar to Eq. (9) is obtained [42]. For systems with nearest-neighbor transitions, however, there is no need to restrict one-self to finite chains. Imposing for the infinite system the condition that the flow of area in one iteration from cell A_i to cell A_{i+1} is equal to the backflow from A_{i+1} to A_i implying a vanishing net current through the system, one obtains as condition for the heights π_i

$$\pi_i p_{i,i+1} = \pi_{i+1} p_{i+1,i} \tag{10}$$

which has to hold for all i. The solution of this infinite system of equations is for $i > 0$ given again by Eq. (9) and an analogous expression holds for $i < 0$. The cell label A_0 can be attached to some arbitrarily chosen cell, and the height π_0 is an arbitrary constant. Only for finite systems one can assign to the heights π_i the meaning of a stationary distribution, and Eq.

(10) then becomes the condition for detailed balance, which is automatically fulfilled in the case with reflecting boundary conditions. For quenched random transition probabilities, as in the models treated here, Eq. (10) means that the π_i follow a random multiplicative process as given by Eq. (9), which occur naturally in many branches of physics. Of course it would be interesting to see whether for some class of area preserving maps or corresponding continuous time Hamiltonian systems, such a distribution of chaotic areas occurs naturally. In such cases one expects that one finds the same anomalous transport properties as in the simple maps introduced here. This will be observable until the boundaries of the system of size L are reached, which for extended systems occurs only at exponentially large times, i.e. at times of the order $O(\exp b\sqrt{L})$. Finally we note that the construction shown in this subsection can be extended to biased systems with the aid of the stationary distribution given in [42] for periodic boundary conditions. This implies that in addition to dynamical localization and aging, also dynamical phase transitions as introduced in Sect. 2.3 can be observed in area-preserving maps and Hamiltonian systems.

Escape Rates and Spectral Properties

Instead of observing anomalous transport in terms of drift and mean-square displacement it is rewarding to investigate certain spectral properties of the systems of interest. For the maps with topological and with Sinai disorder this has been done in [50]. Without going too much into details, it is worth noting that these investigations revealed clear differences between maps with Sinai disorder on one hand, and maps with topological disorder on the other hand. Consider for instance the dependence on the length L of the escape rate γ. For that purpose one takes a system consisting of L cells or states (see Figs. (1(a) and (3)), places an absorbing state on one end, say at site $L+1$, and a reflecting state on the other end, and asks with which rate an initial distribution in the segment of length L decays, i.e. is absorbed by the absorbing state. This rate can be determined by calculating the logarithm of the largest eigenvalue of a $(L \times L)$-submatrix of the transition matrix P with elements $(P)_{ij} = p_{ij}$ (see also [47]). For ordinary diffusive systems one finds that for large L the escape rate behaves generally as $\gamma(L) \sim D(\chi/L)^2$ with D being the diffusion constant and χ a geometric factor. For details see "escape rate formalism" in [47]. In the absence of a bias we find for both kinds of disordered maps that the disorder averaged escape rate $\overline{\gamma(L)}$ decreases with the system size as $\overline{\gamma(L)} = a\exp(-bL^{\beta})$ with some constants a and b, and a characteristic exponent β. For the maps with Sinai disorder one finds $\beta = 1/2$ in accordance with the picture developed in Fig. 4. For maps with topological disorder, however, the exponent β turns out to be $\beta \simeq 0.34$, and therefore is clearly different from the Sinai value.

A related difference is observed if one considers simply the spectrum of the transition matrix P for large systems and e.g. reflecting boundary conditions.

The long-term behavior of the system dynamics is determined by the spectral properties near the maximal eigenvalue $\lambda = 1$. It turns out that the eigenvalues cluster near $\lambda = 1$. In order to quantify this clustering one introduces the distance ϵ from the value $\lambda = 1$, i.e. $\epsilon \equiv 1 - \lambda$. The integrated density of states $N(\epsilon) = \int_0^\epsilon d\epsilon' \rho(\epsilon')$, with $\rho(\epsilon)$ being the spectral density, shows a logarithmic singularity near $\epsilon = 0$. For Sinai disorder one finds asymptotically for $\epsilon \to 0$ a behavior of the form $N(\epsilon) \sim c/ |\ln \epsilon|^\delta$ with $\delta = 2$. For maps with topologically disorder, however, one finds again a different exponent, namely $\delta \simeq 2.6$. The relation to the characteristic exponent for the escape rate $\beta = 1/\delta$, which is valid for the Sinai case, is only roughly fulfilled ($\beta = 0.34$ vs. $1/\delta \simeq 0.38$), but this may be due numerical inaccuracies in determining the exponents. In any case, with respect to these spectral properties one finds a pronounced difference between systems with Sinai and with topological disorder. This is attributed mainly to a difference in the anomalous drift behavior, which, however, is hard to observe in direct numerical simulations of the mean displacement. This difference seems to be related to the fact that both systems behave different already in equilibrium i.e. for finite systems. The equilibrium distribution of the Sinai systems obeys detailed balance, whereas the topologically disordered system does not, due to the transitions to the next-nearest neighbors. The latter allow for permanent probability currents in the stationary state. As a consequence, the picture of a random landscape as developed in Fig. 4 is strictly valid only for the Sinai system, but not for systems with transitions to next-nearest neighbors or neighbors even further apart. The surmise that these probability currents are responsible for the modified drift behavior in non-equilibrium situations deserves further investigations.

Open Problems

In the previous sections we exploited very much the existence of Markov partitions in certain dynamical systems. Such partitions exists only in rare cases and one wonders, how general the effects are which we have described. In Sect. 2.1 we have already seen that the dynamical localization phenomenon was not bound to the existence of Markov partitions, because in systems with continuously distributed shifts $\varepsilon(i)$ localization was found, but there Markov partitions do not exist. In [13] it was numerically found in addition that the piecewise linear behavior of the maps is not necessary for localization to occur. This means that such a localization in disordered dynamical systems seems to be an universal phenomenon. The question whether or not it can be observed in systems described by ordinary differential equations is still open. The problem there is that other localization mechanisms, e.g. attracting regions, typically interfere with the mechanisms described above. The question of anomalous drift is obviously more subtle. We have seen that it depends on the range of the transitions in systems where Markov partitions exist. In general maps, where no Markov partition exists, it is totally unclear how the drifting behavior depends on the features of the maps considered. Even the simple question, under what condition the bias-free situation analogous to the Sinai disorder is attained, is not answered. The answer appears to be not simple: already in systems with next-nearest neighbor transitions one has to determine the Lyapunov spectrum of random products of certain (4×4)-matrices to decide this question [51] (see also [50]). Even less is known for general diffusive disordered maps with regard to aging properties as treated in Sect. 2.2, or dynamical phase transitions as function of a bias mentioned in Sect. 2.3.

3 Coupled Phase Oscillators

In this section we consider the other extreme of disordered dynamical systems, namely that of many degrees of freedom, each confined to a compact set (a circle) and coupled in a random manner. The results below are obtained for a model defined by N interacting phase oscillators. Interest in such models comes from biological phenomena like cell cycles, pacemaker cells, the coherent flashing of fireflies [52, 53, 54], or coherent oscillations in the brain [55, 56], and from laser physics [57, 58], and other branches of physics [59]. They are a good approximation for many dynamical systems where limit cycle oscillators are the basic entities [54, 59]. The model is defined by the equations of motion

$$\dot{\phi}_j(t) = \omega_j + \sum_{i=1}^{N} J_{ij} \sin(\phi_i(t) - \phi_j(t)), \tag{11}$$

where ϕ_i are angles, $\phi_i \in [-\pi, \pi)$, the ω_i are the unperturbed oscillator frequencies, and the J_{ij} are the couplings between the different limit cycle oscillators. This model has mostly been investigated for the special case of random, uncorrelated frequencies ω_i and uniform all-to-all interactions $J_{ij} = K/N$ [59, 60, 61, 62, 63, 64, 65, 66]. For a recent review see [67]. In this case, which is usually referred to as the Kuramoto model, it is well known that one observes a transition from completely incoherent motion of the oscillators for small interaction strength $K < K_c$, to partially coherent motion (phase locking) above a critical interaction strength K_c. Because of its importance we will first, in Sect. 3.1, give an account on the essence of the Kuramoto model and then in Sect. 3.2 we will treat the more complicated case where the interaction strengths J_{ij} are random variables, too.

3.1 Kuramoto Model

The importance of the Kuramoto model comes from the fact that it is an exactly solvable dynamical model, which exhibits a synchronization transition. The latter is an elementary collective effect in disordered dynamical systems, with relevance for all the fields mentioned above. The phenomenon of synchronization itself is known at least since Huygens and found renewed interest quite recently. For an overview and many more applications we refer to [68]. We will show below that the Kuramoto model is also one of the few models, where the full spectrum of Lyapunov exponents can be calculated analytically.

For constant couplings $J_{ij} = K/N$ the equations of motion Eq. (11) can be rewritten exactly as

$$\dot{\phi}_j(t) = \omega_j - K\,|Z|\sin(\phi_j - \Psi), \tag{12}$$

where we have introduced the quantity

$$Z = |Z|\exp(i\Psi) \equiv \frac{1}{N}\sum_{i=1}^{N}\exp(i\phi_i). \tag{13}$$

The latter is in general a time dependent complex quantity. From Eq. (12) one sees that it describes a global mean field which acts equally on all oscillators. The problem has reduced to solving the dynamics of N independent one-oscillator equations subject to a yet unknown field Z. The assumption is now that the distribution of frequencies

$$g(\omega) = \frac{1}{N}\sum_{i=1}^{N}\delta(\omega - \omega_i) \tag{14}$$

is symmetric with respect to the mean frequency $\overline{\omega} = \int \omega\ g(\omega)\ d\omega$ and that it becomes continuous in the limit $N \to \infty$. Below we will give some explicit results for a Gaussian $g(\omega)$. By a transformation to a rotating frame $\phi_j(t) = \overline{\omega}t + \widetilde{\phi}_j(t)$ in Eq. (11) we see that without loss of generality we can set $\overline{\omega} = 0$ by considering the equations of motion for $\widetilde{\phi}_j(t)$. In the following we do not formally distinguish between $\phi_j(t)$ and $\widetilde{\phi}_j(t)$, but one should keep in mind that for $\overline{\omega} \neq 0$, ϕ is interpreted as an angle relative to the frame rotating with frequency $\overline{\omega}$. In order to solve Eq. (11), one has to determine Z self-consistently from Eqs. (12) and (13). We will see below that a time-independent $Z = const$ is a self-consistent solution. Actually, it will turn that Z can be chosen to be real with $Z \geq 0$. In the following derivation we will first assume $Z = const$, and afterwards one sees that this assumption is indeed self-consistent. Note first that for constant Z, the one-oscillator equation (12) has depending on the strength of $K\,|Z|$ two types of solutions: for $|\omega_j| \leq |KZ|$ Eq. (12) has stationary points $\dot{\phi}_j = 0$, whereas for $|\omega_j| \geq |KZ|$ it has none. This can be visualized by regarding ϕ as an extended variable $-\infty < \phi < \infty$, and Eq. (12) as the relaxation dynamics $\dot{\phi} = -\frac{d}{d\phi}V(\phi)$ in the potential $V(\phi) = -\omega\phi + K\,|Z|\cos(\phi - \Psi)$, where in the notation we have now for convenience suppressed the index j. Correspondingly $V(\phi)$ has local minima and maxima in the first case and $\phi(t)$ converges to a fixed point $\phi(t) \to \phi^*$

for $t \to \infty$. The corresponding oscillator gets locked at ϕ^*. In contrast, for $|\omega| > |KZ|$ the potential $V(\phi)$ is monotonously decreasing or increasing and $\phi(t)$ is continuously increasing or decreasing, or in other words, the phase oscillator keeps rotating for all times. Actually the equation of motion Eq. (12) can be solved explicitly. We get as an exact result

$$\phi(t) = \begin{cases} 2\text{arccot}[\frac{\Lambda}{\omega}\coth(\frac{\Lambda}{2}t) + \frac{K|Z|}{\omega}] & \text{if } |\omega| < |KZ| \\ 2\text{arccot}[\frac{\Omega}{\omega}\cot(\frac{\Omega}{2}t) + \frac{K|Z|}{\omega}] & \text{if } |\omega| \geq |KZ| \end{cases} \tag{15}$$

where we have introduced $\Omega = \sqrt{\omega^2 - |KZ|^2}$ and $\Lambda = \sqrt{|KZ|^2 - \omega^2}$. We have set for convenience $\Psi = 0$ and the initial condition is $\phi(0) = 0$. The solution for other initial conditions and $\Psi \neq 0$ is easily deduced from Eq. (15). For completeness and further reference we give also simplified formulas for the phase velocity

$$\dot{\phi}(t) = \begin{cases} \dfrac{-\omega\Lambda^2}{\omega^2 - |KZ|^2\cosh(\Lambda t) + |KZ|\,\Lambda\sinh(\Lambda t)} & \text{if } |\omega| < |KZ| \\ \dfrac{\omega\Omega^2}{\omega^2 - |KZ|^2\cos(\Omega t) - |KZ|\,\Omega\sin(\Omega t)} & \text{if } |\omega| \geq |KZ| \end{cases} \tag{16}$$

In the given form Eq. (15) is valid for ϕ on the circle $\phi \in [-\pi, \pi)$, with the consequence that $\phi(t)$ jumps as the angle $\pm\pi$ is crossed. To get the continuous solutions in the extended space $-\infty < \phi < \infty$ appropriate jumps by $\pm 2\pi$ have to be added. Note that in the case of continuous rotation $|\omega| \geq |KZ|$ the oscillator rotates on average with the effective frequency $\pm\Omega$.

Let us now return to the calculation of Z. Since the latter is assumed to be constant, one can also consider its time-average $\langle Z \rangle_t$ without affecting the result

$$Z = \langle Z \rangle_t = \lim_{t\to\infty} \frac{1}{t} \int_0^t dt' \, \frac{1}{N} \sum_{i=1}^{N} \exp[i\phi_i(t')]. \tag{17}$$

In the evaluation of this expression one has to distinguish between the two types of solutions. For $|\omega_i| < |KZ|$ each of the oscillator gets locked and contributes to Z an amount

$$\exp[i(\phi_i^* + \Psi)] = \exp[i\Psi](\text{sgn}(K)\sqrt{1 - \left|\frac{\omega_i}{KZ}\right|^2} + i\frac{\omega_i}{K|Z|}) \, ,$$

where we have used that

$$\phi_i^* = \lim_{t\to\infty} \phi(t) = 2\ \text{arccot}[\frac{1}{\omega_i}(\sqrt{\omega_i^2 - |KZ|^2} + K|Z|)]$$

is the limiting stationary point of (15) for $|\omega_i| < |KZ|$. Below we will show that the rotating oscillators do not contribute to $\langle Z \rangle_t$ in Eq. (17). Thus $|Z|$ is determined by the following self-consistency equation

$$|Z| = \text{sgn}(K) \int_{-|KZ|}^{|KZ|} \sqrt{1 - \left|\frac{\omega}{KZ}\right|^2} \, g(\omega) \, d\omega, \tag{18}$$

where we have used the definition Eq. (14) of $g(\omega)$ and exploited the symmetry $g(\omega) = g(-\omega)$. The phase Ψ is an arbitrary gauge field which can be chosen arbitrarily because it cancels from the self-consistency condition for Z. The vanishing of contributions of oscillators with $|\omega_i| > |KZ|$ is most easily seen in the following way. Note first that the long-term average in (17) reduces to an average over one period $T = 2\pi/\Omega$ because $\phi(t) = \phi(t+T)$, i.e. $\lim_{t\to\infty} \frac{1}{t}\int_0^t dt' \, \exp[i\phi_i(t)] = \frac{1}{T}\int_0^T dt' \, \exp[i\phi_i(t')]$. We now employ that $\frac{1}{T}\int_0^T dt' \, \exp[i\phi_i(t')] = \int_{-\pi}^{\pi} d\phi \, \exp[i\phi] \frac{1}{T}\int_0^T dt' \, \delta[\phi - \phi_i(t')]$, which gives by using the properties of the Dirac δ-function $\int_{-\pi}^{\pi} d\phi \, \exp[i\phi] \frac{1}{T}\frac{1}{|\dot\phi_i(\phi)|} = \frac{1}{T}\int_{-\pi}^{\pi} d\phi \, \frac{\cos\phi + i\sin\phi}{\omega_i - K|Z|\sin\phi}$. The real part of the latter integral vanishes identically. This can also be seen by evaluating the time average of $\cos\phi_i(t)$ directly with the aid of Eq. (15). The integral $I(\omega_i) = \int_{-\pi}^{\pi} d\phi \, \frac{\sin\phi}{\omega_i - K|Z|\sin\phi}$ does not vanish, but it can be shown that $I(-\omega_i) = -I(\omega_i)$ as is also clear from the symmetry $\phi(t; -\omega_i) = -\phi(t; \omega_i)$ of the rotating solution of (15). Thus for a symmetric frequency distribution, i.e. for $g(\omega) = g(-\omega)$, the contributions from frequencies with opposite signs cancel exactly. This shows that Eq. (18) is already the correct self-consistency equation, which will be discussed now in more detail. Obviously $|Z| = 0$ is always a solution, and for $K < 0$ it is the only solution. Therefore we consider in the following the more interesting case $K > 0$. That there may exist a second solution with $|Z| \neq 0$ is seen by substituting $x = \frac{\omega}{K|Z|}$ and dividing Eq. (18) by $|Z|$ yielding

$$\begin{aligned}
1 &= K \int_{-1}^{1} dx \sqrt{1-x^2} \, g(xK\,|Z|) \\
&= K[g(0) \int_{-1}^{1} dx \sqrt{1-x^2} + \frac{1}{2}\,|KZ|^2 \, g''(0) \int_{-1}^{1} dx \, x^2 \sqrt{1-x^2} + O(|KZ|^4)] \\
&= K[g(0)\frac{\pi}{2} + \frac{1}{2}\,|KZ|^2 \, g''(0)\frac{\pi}{8} + O(|KZ|^4)],
\end{aligned} \tag{19}$$

where we have applied a Taylor expansion around $|Z| = 0$. Neglecting the higher order terms one can solve for $|KZ|^2$ to obtain $|KZ|^2 = (1 - Kg(0)\frac{\pi}{2})/(K\frac{\pi}{16}g''(0))$. Assuming that $g(\omega)$ has a local maximum at $\omega = 0$, i.e. $g''(0) < 0$, one finds that a solution with non-zero Z exists only for $K > K_c = 2/(\pi g(0))$. It is also seen that the non-zero solution branches off like $|Z| \sim \sqrt{\frac{K-K_c}{K_c}}$, similar to a second order phase transition in equilibrium statistical mechanics. That this solution with non-zero $|Z|$ is indeed attained for $K > K_c$, instead of the solution with $|Z| = 0$, is more difficult to show. The stability problem has essentially been solved after initial attempts [61] by Strogatz and Mirollo in [62] (see also [64]). For still open questions see [67]. It turns out that the solution $|Z| = 0$ is indeed unstable

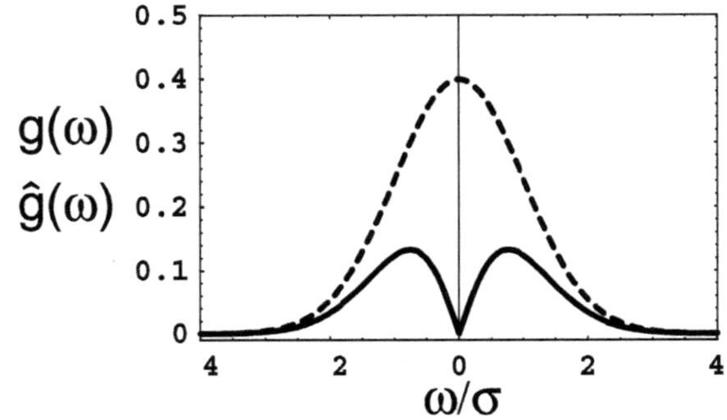

Fig. 7. The distribution of effective frequencies in the incoherent state (*dashed* $g(\omega)$) coincides for $K < K_c$ with the distribution of Eq. (14) for the uncoupled oscillators (shown for $\sigma = 1$). In the partially synchronized state which exists for $K/\sigma > K_c/\sigma = \sqrt{8/\pi} \simeq 1.6$ it is modified to $\widehat{g}(\omega)$ (*full line*) as shown here for $K/\sigma = 1.75$. The distribution $\widehat{g}(\omega)$ contains a singular part $\sim \delta(\omega)$ as indicated by the vertical line at $\omega = 0$, which corresponds to oscillators which are synchronized to the mean frequency $\overline{\omega}$

for $K > K_c$ and neutrally stable for $K < K_c$. The latter fact is somewhat surprising and leads to an unusual relaxation behavior [64]. It is also in these stability considerations where the nature of the frequency distribution $g(\omega)$ enters [64, 65, 66, 67]. For a continuous, symmetric, and one-humped $g(\omega)$ we have now the following picture. A) For $K < K_c$ the order parameter $|Z|$ vanishes, and all oscillators rotate incoherently as if there were no coupling between them. B) For K above K_c there exist two sub-populations: the oscillators with frequency $|\omega_i| > K|Z|$ run incoherently, but with modified frequencies $\Omega_i = \overline{\omega} \pm \sqrt{\omega_i^2 - |KZ|^2}$, and the oscillators with $|\omega_i| < K|Z|$ run coherently with a fixed phase relation and with the mean frequency $\Omega_i = \overline{\omega}$. This phenomenon is called "entrainment" or "phase locking", and is a collective effect as a result of mutual synchronization. In Fig. 7 this behavior is visualized by plotting the distribution $\widehat{g}(\omega)$ of the effective frequencies Ω_i in the rotating frame

$$\begin{aligned}\widehat{g}(\omega) &= \langle \delta(\omega - \Omega_i \rangle = \int \delta(\omega - \Omega(\omega'))\, g(\omega')\, d\omega' \qquad (20)\\ &= 2\int_{K|Z|}^{\infty} \delta(\omega - \sqrt{\omega'^2 - |KZ|^2})\, g(\omega')\, d\omega' + \delta(\omega)\int_{-K|Z|}^{K|Z|} g(\omega')\, d\omega' \\ &= g(\sqrt{\omega^2 + |KZ|^2}) \frac{|\omega|}{\sqrt{\omega^2 + |KZ|^2}} + \delta(\omega)\int_{-K|Z|}^{K|Z|} g(\omega')\, d\omega'.\end{aligned}$$

The first, continuous part stems from the rotating oscillators, and the second, singular part are the locked oscillators. In the rest frame the effective frequency distribution is given by $\widehat{g}(\omega - \overline{\omega})$.

For the figure and for the following we have used a Gaussian $g(\omega) = (2\pi\sigma^2)^{-1/2}\exp[-\omega^2/(2\sigma^2)]$. The order parameter $|Z|$ has to be determined numerically by solving Eq. (18). For a Gaussian $g(\omega)$ Eq. (18) or (19) reduces to the transcendental equation $1 = \sqrt{\frac{\pi}{2}}\frac{K}{2\sigma}\exp[-(\frac{K}{2\sigma}|Z|)^2]\{I_0[(\frac{K}{2\sigma}|Z|)^2] + I_1[(\frac{K}{2\sigma}|Z|)^2]\}$ (I_0 and I_1 are the modified Bessel functions [69]). The solution is seen to depend only on the fraction K/σ, which we have chosen as $K/\sigma = 1.75$ resulting in $|Z| \simeq 0.5$. K/σ is the only relevant parameter of the Kuramoto model. This becomes clear by dividing Eq. (11) with σ. The parameter σ just sets the time scale, e.g. for ω and for the Lyapunov exponents (see below), and is taken to be $\sigma = 1$. The fraction of the locked oscillators is obtained for the Gaussian as $\int_{-K|Z|}^{K|Z|} g(\omega)\ d\omega = \mathrm{erf}(K|Z|/\sqrt{2\sigma^2})$, which gives $\simeq 0.62$ for our parameters.

The Kuramoto model is one of the rare cases where the Lyapunov spectrum can be calculated analytically (see [70] for more details on the numerical determination of Lyapunov spectra for other many-particle systems). Below we will give an explicit formula for the case of a Gaussian $g(\omega)$. The Lyapunov spectrum describes the possible expansion or contraction rates in the neighborhood of some reference trajectory. For a general dynamical system $\dot{x} = F(x)$ with solution $x(t) \in \mathbb{R}^N$ the behavior of infinitesimal displacements $\delta x(t)$ from $x(t)$ is described by the linearized equations of motion in tangent space

$$\delta\dot{x} = \frac{dF}{dx}(x(t)) \cdot \delta x, \tag{21}$$

where $\frac{dF}{dx}(x(t))$ is the Jacobian matrix evaluated at $x(t)$. With the formal solution $\delta x(t) = M(t,0) \cdot \delta x(0)$ of Eq. (21) the spectrum of Lyapunov exponents is the set of N numbers $\{\lambda_i\}, i = 1, \ldots, N$ which can be assumed by the limit

$$\lambda = \lim_{t\to\infty} \frac{1}{t} \log \frac{\|\delta x(t)\|}{\|\delta x(0)\|} = \lim_{t\to\infty} \frac{1}{t} \log \frac{\|M(t,0)\cdot\delta x(0)\|}{\|\delta x(0)\|} \tag{22}$$

as the initial displacement vector $\delta x(0)$ varies ($\|\ldots\|$ is the usual vector norm). Equivalently the $\{\lambda_i\}$ are the logarithms of the eigenvalues of the so-called Oseledec matrix $\lim_{t\to\infty}[M(t,0)^+ \cdot M(t,0)]^{1/(2t)}$ where M^+ is the transpose of the matrix M. Obviously, in the mean-field limit of the Kuromoto model also the tangent dynamics Eq. (21) decomposes into N independent one-oscillator equations, i.e. the Jacobian $\frac{dF}{dx}$ is diagonal, and therefore also the matrix M. This can be shown also formally by introducing an appropriate order parameter analogous to Z, which however vanishes in the stationary mean field limit [71, 72]. In other words, the spectrum of Lyapunov exponents is just the union of the individual Lyapunov exponents, each obtained from the effective one-particle dynamics of the corresponding oscillator in the ensemble. For an individual oscillator i the Lyapunov exponent is easily obtained as

$$\lambda_i = \lim_{t\to\infty} \frac{1}{t} \log \left| \frac{\delta\phi_i(t)}{\delta\phi_i(0)} \right| = \lim_{t\to\infty} \frac{1}{t} \log \left| \frac{\dot{\phi}_i(t)}{\dot{\phi}_i(0)} \right| = \lim_{t\to\infty} \frac{1}{t} \log \left| \dot{\phi}_i(t) \right| , \tag{23}$$

where the one-dimensional nature of the problem was exploited. With the results of Eq. (16) the last expression is simply evaluated as

$$\lambda_i = \begin{cases} -\sqrt{|KZ|^2 - \omega_i^2} & \text{if } |\omega_i| < |KZ| \\ 0 & \text{if } |\omega_i| \geq |KZ| \end{cases} . \tag{24}$$

This result expresses simply the fact the velocity $\dot{\phi}$ of a locked oscillator vanishes exponentially with the rate λ as $\phi(t)$ approaches its limiting value ϕ^*. The vanishing of the Lyapunov exponent for a continuously rotating, unlocked oscillator comes from the fact that $\left|\dot{\phi}\right|$ remains bounded away from zero, i.e. it cannot increase or decrease exponentially. Usually for the Lyapunov spectrum the $\{\lambda_i\}$ are ordered from the largest to the smallest, i.e. $\lambda_1 \geq \lambda_2 \geq \ldots \geq \lambda_N$, and plotted against i or better against $X = i/N$. The latter is preferred because it can be used in the thermodynamic limit $N \to \infty$. Its only in this limit where mean-field theory and the result of Eq. (24) becomes exact. In the thermodynamic limit the Lyapunov spectrum becomes a function $\lambda(X)$ of the now continuous variable X with $0 < X \leq 1$. The variable X is just the fraction of Lyapunov exponents with values λ' greater than $\lambda(X)$. To obtain $\lambda(X)$ one needs to know the density of Lyapunov exponents

$$\rho(\lambda) = \langle \delta(\lambda - \lambda_i) \rangle = \lim_{N\to\infty} \frac{1}{N} \sum_{i=1}^{N} \delta(\lambda - \lambda_i), \tag{25}$$

then one calculates the fraction

$$X(\lambda) = \int_{\lambda}^{\infty} \rho(\lambda')\, d\lambda', \tag{26}$$

which by inversion gives $\lambda(X)$. For the Kuramoto model one obtains $\rho(\lambda)$ with the aid of Eq. (24) as

$$\rho(\lambda) = \int_{-\infty}^{\infty} \delta(\lambda - \lambda(\omega)) g(\omega)\, d\omega \tag{27}$$

$$= \int_{-K|Z|}^{K|Z|} \delta(\lambda + \sqrt{|KZ|^2 - \omega^2}) g(\omega)\, d\omega + 2\delta(\lambda) \int_{K|Z|}^{\infty} g(\omega)\, d\omega. \tag{28}$$

Thus $\rho(\lambda)$ is seen to be the sum of two contributions $\rho(\lambda) = \rho_c(\lambda) + \rho_d(\lambda)$, where the second, discrete part $\rho_d(\lambda)$ is due to the non-locked oscillators. The non-singular part $\rho_c(\lambda)$ which comes from the locked oscillators can be further evaluated to yield $\rho_c(\lambda) = g(\sqrt{|KZ|^2 - \lambda^2}) \frac{2|\lambda|}{\sqrt{|KZ|^2 - \lambda^2}}$ for $-K|Z| < \lambda < 0$,

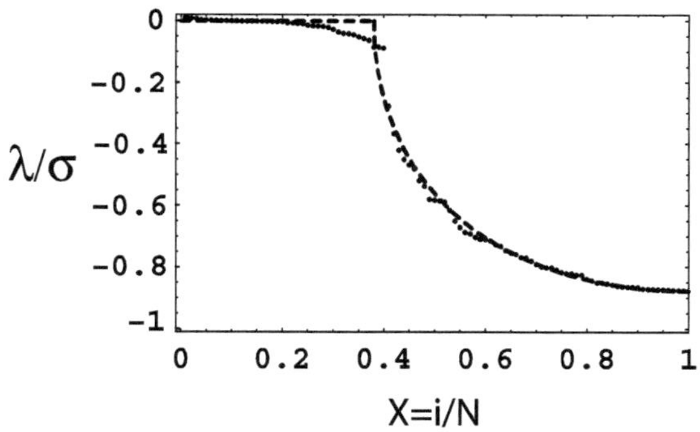

Fig. 8. The spectrum of Lyapunov exponents for the Kuramoto model in the partially synchronized phase. The dashed line corresponds to the analytical expression of Eq. (30), the dots are the result of a simulation with $N = 100$ coupled oscillators. The parameters are those of Fig. 7. The zero exponents in the left part of the spectrum correspond to unlocked phase oscillators, the negative part stems from phase locked, synchronized oscillators

and zero elsewhere. With these results the integral in (26) can be calculated explicitly for a Gaussian $g(\omega)$ resulting in $X(\lambda) = 0$ for $\lambda > 0$, $X(\lambda) = 1$ for $\lambda < -|KZ|$, and

$$X(\lambda) = \text{erf}(\frac{|KZ|}{\sqrt{2\sigma^2}}) - \text{erf}(\frac{\sqrt{|KZ|^2 - \lambda^2}}{\sqrt{2\sigma^2}}) + \theta(-\lambda)[1 - \text{erf}(\frac{|KZ|}{\sqrt{2\sigma^2}})] \tag{29}$$

for $-|KZ| < \lambda \leq 0$, where $\theta(x)$ denotes the Heaviside step function. Inverting this yields for the Lyapunov spectrum the exact result

$$\lambda(X) = \begin{cases} 0 & \text{if } 0 < X \leq 1 - \text{erf}(\frac{|KZ|}{\sqrt{2\sigma^2}}) \\ -\sqrt{|KZ|^2 - 2\sigma^2[\text{erf}^{-1}(1-x)]^2} & \text{if } 1 - \text{erf}(\frac{|KZ|}{\sqrt{2\sigma^2}}) < X \leq 1 \end{cases}, \tag{30}$$

where $y = \text{erf}^{-1}(x)$ denotes the inverse error function, i.e. the solution of $x = \text{erf}(y)$. In Fig. 8 this analytical result is compared with the Lyapunov spectrum, which was determined numerically with the standard algorithm (see e.g. [2]) for N=100 coupled phase oscillators. Apart from finite size effects near the transition from unlocked ($\lambda(X) = 0$) to locked oscillators ($\lambda(X) < 0$) the correspondence is very good.

We close this subsection with some remarks on various generalizations of the Kuramoto model. The replacement of the sinusoidal interaction term by higher order Fourier components or the inclusion of cosine interactions was treated in [73]. The main result is that the solution with $Z \neq 0$ no longer branches off with a square root law but with a linear dependence on

$K - K_c$. Many papers treat the influence of uncorrelated white noise in the equations of motion [62, 64, 65, 66, 67]. A remarkable finding is that the stability of the $Z = 0$ state for $K < K_c$ is affected quite drastically [62, 64, 67]. The replacement of the one-humped frequency distribution $g(\omega)$ by a bimodal distribution leads to a richer phase diagram, especially the order parameter Z is no longer constant in general [65, 66, 67]. Finally we mention that generalizations of the Kuramoto model such as the Winfree model shows a wealth of collective phenomena, which can also be treated analytically [74].

3.2 Randomly Coupled Oscillators

Here we present results for an interesting generalization, where also the couplings are quenched random variables. We assume that in addition to the frequencies also J_{ij} in Eq. (11) are Gaussian distributed variables with mean zero $\overline{J_{ij}} = 0$ and covariances

$$\overline{J_{ij} J_{kl}} = \frac{J^2}{N}(\delta_{ik}\delta_{jl} + \eta\delta_{il}\delta_{jk}). \tag{31}$$

The condition $\overline{J_{ij}} = 0$ means that the couplings are on the average neither ferromagnetic nor antiferromagnetic, but since the J_{ij} are random with positive and negative values one gets competing interactions and possibly frustration for a given oscillator. The system is characterized by two independent parameters in the limit $N \to \infty$: J with $0 < J < \infty$ is a measure for the interaction strength, and $-1 \leq \eta \leq 1$ is the so-called symmetry parameter. As in the Kuramoto model σ just provides the units for the couplings and for time, respectively. From Eq. (31) follows that for $\eta = 0$ all coupling parameters are uncorrelated, while $\eta = 1$ imposes the constraint $J_{ij} = J_{ji}$ (symmetric interactions) and $\eta = -1$ implies $J_{ij} = -J_{ji}$ (antisymmetric interactions). The case of symmetric interactions has been treated previously in [75]. Our main results for the general case are summarized in the phase diagram of Fig. 9.

As function of the parameters η and J one finds two different phases which can be regarded as a paramagnetic phase and a spin glass phase (SG in Fig. 9). The latter is characterized by a non-vanishing Edwards-Anderson (EA) parameter [3], which in our case may be defined as

$$q = \lim_{t\to\infty} \mathrm{Re}\,\langle \exp(i\phi_j(t_0)\exp(-i\phi_j(t+t_0)\rangle\,. \tag{32}$$

The spin glass phase, which exists above the full, bold line $J_c(\eta)$ in the region $0.8 \lesssim \eta \leq 1$ and for $J \geq J_c \equiv J_c(\eta = 1) \simeq 24.3$, is characterized by a vanishing "magnetisation" $|Z| = 0$ and an EA parameter $q = 1$. In addition we find in this phase that the maximal Lyapunov exponent is zero and that all other Lyapunov exponents are negative. This means that all oscillators

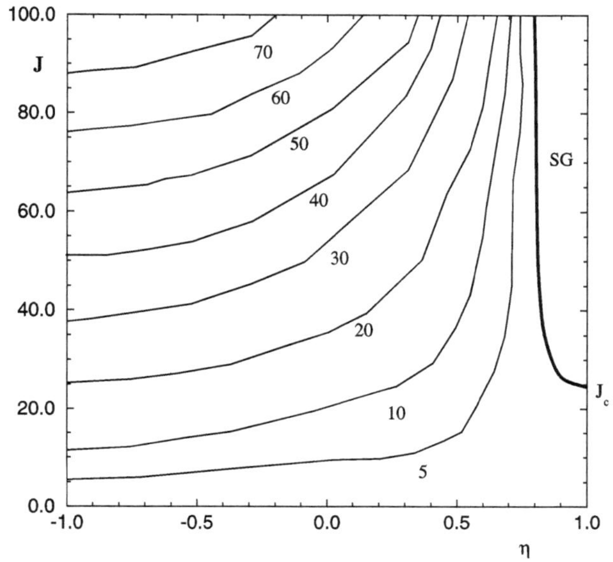

Fig. 9. Value of the largest Lyapunov exponent $\lambda_{\max}$ in dependence on symmetry parameter η and interaction strength J represented by contour lines at levels $\lambda_{\max} = 5, 10, 20, 30, 40, 50, 60, 70$ ($N = 100$). The part of the parameter space (η, J) above $J_c(\eta)$ (bold line) where the oscillators freeze in random positions, is denoted by "SG" In this regime $\lambda_{\max} = 0$. For $\eta < 0.8$ there is no spin glass order. The largest Lyapunov exponent decreases monotonically with η and for $\eta < 0.8$ grows monotonically with J

are frozen into random directions, which is indeed a characteristic feature of a spin glass state at temperature $T = 0$. A similar freezing transition at a symmetry parameter $\eta_c \simeq 0.83$ has also been found in an asymmetric random neural network model [76], which may be related to our model at $J = \infty$. Note that in the symmetric case $\eta = 1$ the maximal Lyapunov exponent is zero for all values of J. Below $J_c \simeq 24.3$, however, not only one, but a finite fraction of zero Lyapunov exponents appears to exist (similar to Fig. (8) for the Kuramoto model), which indicates that in this case a non-chaotic motion of the system on a lower dimensional manifold in phase space is present.

The situation is very different in the "paramagnetic" phase. There in addition to $|Z|$ also q vanishes, meaning that we are in a situation of dynamical disorder. This is also reflected in a positive maximal Lyapunov exponent implying chaotic motion, as can be seen from the level lines for the maximal Lyapunov exponent in Fig. 9. Thus we have here a certain correspondence between order parameters for disordered systems and characteristic quantities for dynamical systems.

A final point of interest, which shall be presented here concerns the nature of the dynamical phase transition at the transition line $J_c(\eta)$. Other interesting aspects of the Lyapunov spectra, behavior of correlation functions, the

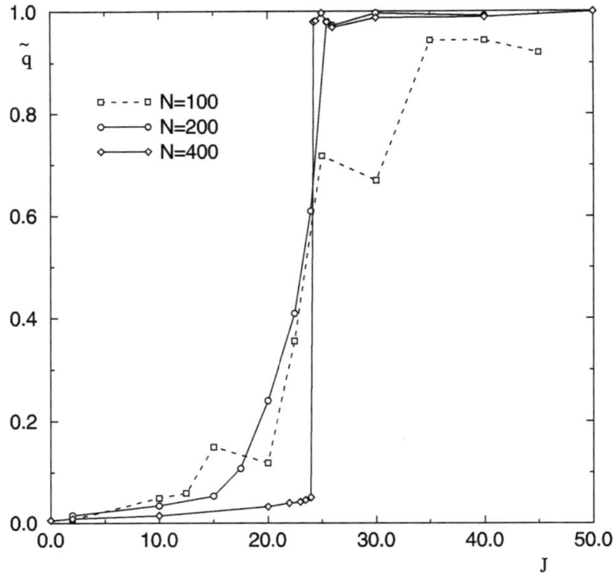

Fig. 10. Order parameter q in dependence on interaction strength J for $N = 100, 200, 400$ oscillators. The transition from $q \approx 0$ for $J < J_c$ to $q \approx 1$ for $J > J_c$ becomes sharper with growing system size N. Between $J = 24$ and $J = 24.3$ the order parameter q changes from 0.046 to 0.982

time-dependence of the magnetisation, etc. can be found in [77]. The phase transition appears to be of first order as is demonstrated for $\eta = 1$ in Fig. 10. There the dependence of q on the coupling strength J is plotted for increasing system size. These simulations clearly suggests that the spin glass order parameter q jumps from zero to one in the thermodynamic limit $N \to \infty$ as J crosses $J_c \simeq 24.3$. The first order character of the transition is also confirmed for $0.8 \lesssim \eta \leq 1$, where interestingly we find for finite N strong intermittency phenomena in the neighborhood of the transition line [77]. We should also mention that advanced techniques such as dynamical mean field theory (see [30]) can be developed also for this model [77] resulting in an effective one-particle equation. The latter, however, can be solved only numerically with the drawback that the asymptotic long time behavior is difficult to reach. A last speculative remark on the spin glass phase of this model is that one expects also aging phenomena to occur here.

4 Summary

We have shown that disordered dynamical systems exhibit complex and interesting behavior. One-dimensional maps can show collective behavior for ensembles, although there is no mutual interaction. In addition, aging, anomalous transport, and dynamical phase transitions occur. In Sect. 3 we considered random limit cycle oscillators including the Kuramoto model. Collective phenomena such as the occurrence of mutual synchronization, chaotic behavior in "paramagnetic" phases, and the occurrence of spin-glass phases has been considered. The relation to characteristic quantities of dynamical systems, such as Lyapunov exponents and Lyapunov spectra, have been discussed.

References

1. H.G. Schuster, *Deterministic chaos*, 3rd augm. ed. (VCH, Weinheim, 1995).
2. E. Ott, *Chaos in Dynamical Systems.* (Cambridge University Press, Cambridge, 1993).
3. M. Mézard, G. Parisi, M.A.Virasoro, *Spin glass theory and beyond* (World Scientific, Singapore, 1987).
4. H. van Beijeren, J.R. Dorfman, Phys. Rev. Lett. **74**, 4412 (1995); and refs. therein.
5. D. Weiss et al.: Phys. Rev. Lett. **66**, 2790 (1991).
6. R. Fleischmann, T. Geisel, R. Ketzmerick, Phys. Rev. Lett. **68**, 1367 (1992).
7. T. Geisel, J. Nierwetberg, Phys. Rev. Lett. **48**, 7 (1982).
8. S. Grossmann, H. Fujisaka, Phys. Rev. A **26**, 1779 (1982); H. Fujisaka, S. Grossmann, Z. Phys. B **48**, 261 (1982).
9. M. Schell, S. Fraser, R. Kapral, Phys. Rev. A **26**, 504 (1982).
10. R. Artuso, Phys. Lett. A **202**, 195 (1992); G. Zumofen, J. Klafter, Phys. Rev. E **47**, 851 (1993); X.-J. Wang, C.-K. Hu, Phys. Rev. E **48**, 728 (1993); R. Artuso, G. Casati, R. Lombardi, Phys. Rev. Lett. **71**, 62 (1993); R. Stoop, Phys. Rev. E **49**, 4913 (1994); G. Radons, Phys. Rep. **290**, 67 (1997).
11. T. Geisel, S. Thomae, Phys. Rev. Lett. **52**, 1936 (1984).
12. R. Klages, J.R. Dorfman, Phys. Rev. Lett. **74**, 387 (1995).
13. G. Radons, Phys. Rev. Lett. **77**, 4748 (1996).
14. T. Geisel, J. Nierwetberg, A. Zacherl, Phys. Rev. Lett. **54**, 616 (1985).
15. C. Beck, F. Schlögl, *Thermodynamics of Chaotic Systems* (Cambridge University Press, Cambridge, 1993).
16. W. Just, On symbolic dynamics of spae–time chaotic models, in this book.
17. J.-P. Bouchaud, A. Georges, Phys. Rep. **195**, 127 (1990).
18. Ya. G. Sinai, Theor. Prob. Appl. **27**, 247 (1982).
19. A. Golosov, Commun. Math. Phys. **92**, 491 (1984).
20. J. Chave, E. Guitter, J. Phys. A **32**, 445 (1999).
21. C. Monthus, P. Le Doussal, Phys. Rev. E **65**, 066129 (2002).
22. B.D. Hughes, *Random Walks and Random Environments*, Vol.2: Random Environments, Chap.6 (Clarendon Press, Oxford, 1996).
23. L. Lundgren et al., Phys. Rev. Lett. **51**, 911 (1983).
24. L.C.E. Struik, *Physical Aging in Amorphous Polymers and other Materials* (Elsevier, Amsterdam, 1978).
25. M. Rubí, C. Pérez-Vicente (eds.), *Complex Behaviour in Glassy Systems* (Springer, Berlin, 1997).
26. J.P. Bouchaud, L.F. Cugliandulo, J. Kurchan, M. Mézard, in: A.P. Young (ed.), *Spin Glasses and Random Fields* (World Scientific, Singapore, 1998), cond-mat/9702070.

27. E. Vincent, et al., in: [25], p.184, cond-mat/9607224.
28. E. Marinari, G. Parisi, J. Phys. A **26**, L1149 (1993).
29. F. Ritort, Phys. Rev. Lett. **75**, 1190 (1995).
30. H. Horner, Glassy dynamics and aging in disordered systems, in this book.
31. L. Laloux, P. Le Doussal, Phys. Rev. E **57**, 6296 (1998).
32. D.S. Fisher, P. Le Doussal, C. Monthus, Phys. Rev. Lett. **80**, 3539 (1998).
33. P. Le Doussal, C. Monthus, D.S. Fisher, Phys. Rev. E **59**, 4795 (1999).
34. G. Radons, Adv. Solid State Physics / Festkörperprobleme **38**, 439 (1999).
35. L.F. Cugliandulo, J. Kurchan, G. Parisi: J.Phys. I France **4,** 1641 (1994).
36. E. Barkai, Phys. Rev. Lett. **90**, 104101 (2003).
37. J. Dräger, J. Klafter, Phys. Rev. Lett. **84**, 5998 (2000).
38. J. Heldstab, H. Thomas, T. Geisel, G. Radons, Z. Phys. B **50**, 141 (1983), S. Grossmann, Z. Phys. B **57**, 77 (1984).
39. M. Bianucci, R. Manella, X. Fan, P. Grigolino, B.J. West, Phys. Rev. E **47**, 1510 (1993), M. Bianucci, R. Manella, B.J. West, P. Grigolino, Phys. Rev. E **50**, 2693 (1994).
40. P. Jung, J.G. Kissner, P. Hänggi, Phys. Rev. Lett. **76**, 3436 (1996).
41. T. Harms, R. Lipowsky, Phys. Rev. Lett. **79**, 2895 (1997).
42. B. Derrida, Y. Pomeau, Phys. Rev. Lett. **48**, 627 (1982), B. Derrida, J. Stat. Phys. **31**, 433 (1983).
43. E. Barkai, J. Klafter, Phys. Rev. Lett. **79**, 2245 (1997).
44. S.N. Majumdar, A. Comtet, Phys. Rev. E **66**, 061105 (2002).
45. C. Monthus, Localization properties of the anomalous diffusion phase $x \sim t^{\mu}$ in the directed trap model and in the Sinai diffusion with bias. cond-mat/0212212.
46. C. Monthus, P. Le Doussal, Energy dynamics in the Sinai model. cond-mat/0206035.
47. P. Gaspard, *Chaos, Scattering and Statistical Mechanics* (Cambridge University Press, Cambridge, 1998).
48. E. Hopf, *Ergodentheorie* (Springer, Berlin, 1937).
49. G. Radons, J. Phys. A **31**, 4141 (1998).
50. G. Radons, Physica D **187**, 3 (2004).
51. E.S. Key, Ann. Probab. **12**, 529 (1984).
52. J. Buck, E. Buck, Scientific American **234**, 74 (1976).
53. S.H. Strogatz, I. Stewart, Scientific American **269,** 102 (1993).
54. A.T. Winfree, J. Theoret. Biol. **16**, 15 (1967).
55. R. Eckhorn, et al., Biol. Cybern. **60**, 121 (1988).
56. A.K. Engel, et al., Eur. J. Neurosci. **2**, 588 (1990).
57. K. Otsuka, Phys. Rev. Lett. **67**, 1090 (1991).
58. K. Wiesenfeld, et al., Phys. Rev. Lett. **65**, 1749 (1990).
59. Y. Kuramoto, *Chemical oscillations, waves and turbulence* (Springer, Berlin, 1984).
60. H. Daido, J. Phys. A **20**, L629 (1987).
61. Y. Kuramoto, I. Nishikawa, J. Stat. Phys. **49,** 569 (1987).
62. S.H. Strogatz, E. Mirollo, J. Stat. Phys. **63**, 613 (1992).
63. W.F. Wreszinski, J.L. van Hemmen, J. Stat. Phys. **72**, 145 (1993).
64. S.H. Strogatz, E. Mirollo, P.C. Matthews, Phys. Rev Lett. **68**, 2730 (1992).
65. J.D. Crawford, J. Stat. Phys. **74**, 1047 (1994), Phys. Rev Lett. **74**, 4341 (1995).
66. L.L. Bonilla, C.J. Pérez Vicente, R. Spigler, Physica D **113**, 79 (1998).
67. S.H. Strogatz, Physica D **143**, 1 (2000).

68. A. Pikovsky, M. Rosenblum, J. Kurths, *Synchronization* (Cambridge University Press, Cambridge, 2001).
69. M. Abramowitz, I.A. Stegun (Eds.), *Handbook of Mathematical Functions* (Dover, New York, 1965).
70. H. Posch, C. Forster: Lyapunov nstability of fluids, in this book.
71. J.C. Stiller, G. Radons, Lyapunov Spectrum of the Kuramoto Model of Nonlinear Oscillators, preprint.
72. J.C. Stiller, *Dynamik von Phasenoszillatoren*, Diploma thesis, Kiel (1994).
73. H. Daido, Phys. Rev. Lett. **73**, 760 (1994), Physica D **91**, 24 (1996).
74. J.T. Ariaratnam, S.H. Strogatz, Phys. Rev Lett. **86**, 4278 (2001).
75. H. Daido, Phys. Rev. Lett. **68**,1073 (1992).
76. P. Spitzner, W. Kinzel, Z. Phys. B. **77**, 511 (1989).
77. J.C. Stiller, G. Radons, Phys. Phys. Rev. E 58, 1789 (1998), J. C. Stiller, G. Radons, Phys. Rev. E 61, 2148 (2000).

Lyapunov Instability of Fluids

Harald A. Posch and Christina Forster

Institut für Experimentalphysik, Universität Wien,
Boltzmanngasse 5, 1090 Wien, Austria
Harald.Posch@univie.ac.at; Christina.Forster@univie.ac.at

Summary. The key assumptions of classical statistical mechanics, ergodicity and mixing, are facilitated by the instability of a phase-space trajectory with respect to small perturbations of the initial conditions. Such perturbations typically grow or shrink exponentially with time, which is described by a set of rate constants, the Lyapunov exponents. The set of all exponents is referred to as the Lyapunov spectrum. Here, we summarize current ideas and methods for the computation and interpretation of the Lyapunov spectra of simple fluids. Systems in equilibrium and in stationary nonequilibrium states are considered. Emphasis is given to hard-particle systems, for which the *equilibrium* phase-space perturbations associated with *small* Lyapunov exponents display periodic patterns in space reminiscent of the modes of fluctuating hydrodynamics, the so-called Lyapunov modes. *Stationary nonequilibrium states* are generated by forcing the system away from equilibrium and, simultaneously, by removing the irreversibly-produced excessive heat with a "thermostat". Dynamical and stochastic thermostats are considered. The phase-space probability function of dynamically thermostated systems is a multifractal distribution with an information dimension smaller than the phase-space dimension. This is related to the irreversible transport in such systems, and to the Second Law of Thermodynamics. A possible extension of these ideas to stochastically-thermostated nonequilibrium flows is also attempted.

1 Introduction

Our theoretical understanding of macroscopic systems in classical statistical mechanics is based on two assumptions, the thermodynamic limit and ergodicity. In the former the number of particles, N, is taken to infinity while keeping all densities, such as the particle density, N/V, constant and finite. The latter means that any allowed state of such a dynamical system will be reached, eventually, by dynamical evolution, if it starts out from an arbitrary initial state (up to states of measure zero). Generically, such systems have the property that the finite state very sensitively depends on small perturbations of the initial condition, which, on average, grow or shrink exponentially with time. This so-called Lyapunov instability is described by a set of rate constants, the Lyapunov exponents, and the whole set of exponents is referred to as the Lyapunov spectrum. Positive Lyapunov exponents are taken as a measure for the chaotic evolution of a dynamical system. It appears that the

Lyapunov instability may be viewed as the ultimate instability responsible for the statistical properties of classical particle systems both in and out of thermodynamic equilibrium. Here we attempt a review of recent concepts and results concerning this instability, where the emphasis is on many-particle systems resembling simple fluids.

Dense fluids, and liquids in particular, have long been viewed as belonging to the most difficult problems for statistical mechanics. To some extent this is still the case. Although powerful mathematical tools such as perturbation theories for equilibrium properties, and projection-operator techniques and modern kinetic theory for the dynamics, have become available, our understanding is still largely based on computer-simulated phase diagrams and correlation functions. And the basic questions of irreversible behavior out of equilibrium are to a large extent unsolved. In such a situation any new tool is welcome, and dynamical-systems theory offers just that: The Lyapunov spectrum. It is relevant, since all real systems are Lyapunov unstable due to the dispersing nature of the convex atomic surfaces. And it offers new insights even for equilibrium systems, since the evolving perturbed states belonging to the Lyapunov exponents provide an unbiased separation of the phase-space dynamics into various projections onto submanifolds with different exponential rates for perturbation growth.[1] To be useful, the Lyapunov spectrum should exist in the (thermodynamic) limit of (infinitely) many particles.

The Lyapunov spectrum is even more useful for stationary processes far from thermodynamic equilibrium. It offers an explanation for the macroscopic irreversibility of such processes. Stationary nonequilibrium systems are very common; one might think of the heat conducted by water placed between two large heat reservoirs, or, more ambitiously, of a living organism in a nutrient solution, acting as a chemical reactor and generating irreversibly dissipated heat in the process. However, the explanation the Lyapunov spectrum offers in this case is based on a trick, which reduces the infinitely-many degrees of freedom of the heat reservoirs to a small - even vanishing - number. The time-reversible thermostats invented by Hoover and Evans and discussed in Sect. 5 are artificial in the sense that they project the system dynamics onto manifolds such as the hypersurface of constant kinetic energy in the Gaussian isokinetic case. "Cheating" it is called by David Ruelle [1], but it is successful cheating, since it recaptures the essential properties one expects for stationary irreversible systems far from thermodynamic equilibrium - even for a very small number of particles. The isokinetically-driven Lorentz gas, for example, has a three-dimensional reduced phase space, and the one-dimensional conductivity model treated in Sect. 6 is of similar complexity. The treatment links irreversibility with multifractal phase-space probabilities and provides a geo-

[1] One could ask the question, which degree of freedom of a complex molecule is most susceptible to small perturbations? In some cases it is found that a particular degree of freedom - for example, the rotation around a particular bond - accounts for the maximum exponent.

metrical explanation for the Second Law of Thermodynamics. It is hoped that the gross simplification provided by the dynamical thermostats is a decisive stepping stone towards a satisfactory understanding of irreversible processes.

In this paper we review the main ideas and methods for studying the Lyapunov instability of fluids and solids and provide a few results for grossly-idealized models connected with the general problem introduced above. Section 2 is devoted to a general introduction into the Lyapunov instability of fluids and the properties of the Lyapunov spectrum. Since the theory of transport in fluids draws heavily on the understanding of equilibrium properties, we concentrate in Sect. 3 and Sect. 4 on systems in thermodynamic equilibrium. We restrict our considerations to simple classical fluids. Since hard-disk and hard-sphere systems are good reference models for successful perturbation theories of atomic fluids, we devote Section 3 to this case. There we discuss also the recently-discovered Lyapunov modes and the degeneracy of the Lyapunov exponents. Fluids interacting with soft, continuous pair potentials are treated in Sect. 4. It is shown that the choice of the potential significantly affects the magnitude and shape of the spectrum.

In the remaining Sects. 5 to 9 we discuss fluids in nonequilibrium stationary states. An introduction into nonequilibrium molecular dynamics (NEMD) is given in Sect. 5. This method relies heavily on the dynamical thermostats already mentioned above. The nonequilibrium properties one computes - transport coefficients, Lyapunov spectrum, dimensionality reduction of the multifractal attractor - are all completely insensitive to the particular choice of the thermostat. This strengthens our hope that the results have conceptual meaning also for the physical systems studied in the laboratory. In this sense one may view the choice of a particular thermostat to be analogous to the choice of a Gibbs ensemble in equilibrium statistical mechanics.

The numerical examples include in Sect. 6 a one-dimensional conductivity problem of a single charged particle interacting with a periodic potential and driven by a constant external field. The probability distribution in the three-dimensional phase space is demonstrated to be multifractal. The two-dimensional sheared flow studied in Sect. 7 provides results which are typical for all NEMD simulations carried out with stationary nonequilibrium many-particle systems. For completeness, we extend in Sect. 8 the discussion of hard-disk systems in Sect. 3 to nonequilibrium color-conductivity simulations with this model. In the final section, Sect. 9, we argue that it should be possible to carry over some of the concepts developed for dynamical thermostats to systems interacting with a stochastic heat bath.

Some of our work has been reviewed before [2, 3]. The monographs by Wm. G. Hoover [4, 5], D. J. Evans and G. P. Morriss [6], P. Gaspard [7], and J. R. Dorfman [8] offer an excellent introduction to nonequilibrium molecular dynamics and to various aspects of the statistical mechanics for thermostated systems far from equilibrium. The current status of this theory, and some open problems, have been summarized by D. Ruelle [9]. Finally, a very useful survey of hard-disk and hard-sphere systems is provided in a recent book edited by D. Szasz [10].

2 Dynamics in Phase Space and in Tangent Space

We consider a general classical system characterized by a state vector $\mathbf{\Gamma}(t)$ in an L-dimensional phase space. It consists of N particles in d-dimensions, $L = 2dN$, and $\mathbf{\Gamma} = \{\mathbf{q}_i, \mathbf{p}_i; i = 1, \ldots, N\}$, where $\mathbf{q}_i$ and $\mathbf{p}_i$ denote the respective d-dimensional position and linear momentum vectors of particle i.

The particles interact with each other and, possibly, with boundaries and external fields. First, we assume that these interactions are smooth and continuous, such that $\mathbf{\Gamma}(t)$ evolves according to time-reversible motion equations,

$$\dot{\mathbf{\Gamma}} = \mathbf{F}(\mathbf{\Gamma}), \tag{1}$$

written as a system of first-order differential equations. The (formal) solution,

$$\mathbf{\Gamma}\left(t\right) = \Phi^t\left(\mathbf{\Gamma}\left(0\right)\right), \tag{2}$$

is called the reference trajectory, and Φ^t is the flux. The phase space is assumed to be bounded. The stability of a reference trajectory with respect to small perturbations of the initial conditions is tested by following the evolution of (infinitesimally displaced) perturbed (offset) trajectories starting from $\mathbf{\Gamma}_s(0)$, which is connected to $\mathbf{\Gamma}(0)$ by a parametric path with parameter s such that $\lim_{s\to 0} \mathbf{\Gamma}_s(0) = \mathbf{\Gamma}(0)$. Similarly, for later times, $t > 0$, we have $\lim_{s\to t} \mathbf{\Gamma}_s(t) = \mathbf{\Gamma}(t)$. The vectors $\delta\mathbf{\Gamma}(t) \equiv \lim_{s\to 0}\left(\mathbf{\Gamma}_s\left(t\right) - \mathbf{\Gamma}\left(t\right)\right)/s$ are vectors in the tangent space at time t, and evolve according to the *linearized* equations of motion,

$$\delta\dot{\mathbf{\Gamma}} = \partial\mathbf{F}/\partial\mathbf{\Gamma}\cdot\delta\mathbf{\Gamma} \equiv \mathbf{D}(\mathbf{\Gamma}(t))\cdot\delta\mathbf{\Gamma}. \tag{3}$$

$\mathbf{D}(\mathbf{\Gamma}(t))$ is the Jacobian matrix, which varies with the phase point and, hence, with time. The formal solution of this linearized equation is written as

$$\delta\mathbf{\Gamma}(t) = \mathbf{L}^t\cdot\delta\mathbf{\Gamma}(0), \tag{4}$$

where the propagator matrix $\mathbf{L}^t$ is a time-ordered exponential of the integrated Jacobian.

To be more general, we allow also instantaneous collisional events to take place at times $\{\tau_1, \tau_2, \tau_3, \ldots\}$. In each encounter, an initial state, $\mathbf{\Gamma}^i$, is instantaneously mapped into a final state, $\mathbf{\Gamma}^f$, by a collision map

$$\mathbf{\Gamma}^f = \mathbf{M}\left(\mathbf{\Gamma}^i\right), \tag{5}$$

which is assumed to be reversible and differentiable with respect to the independent variables. Any perturbation vector prior to the collision, $\delta\mathbf{\Gamma}^i$, is similarly mapped into

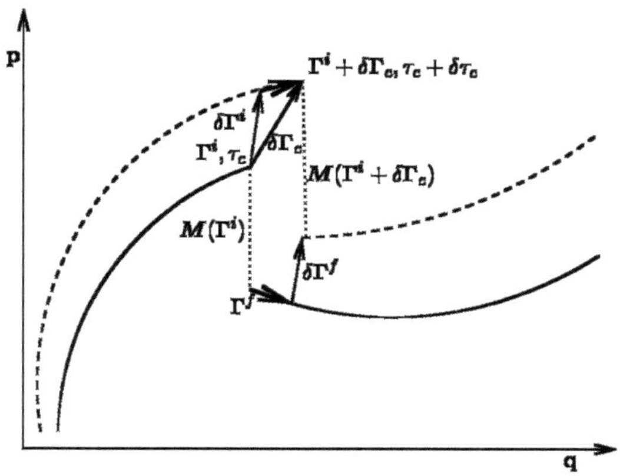

Fig. 1. Schematic representation of a collision for a reference trajectory (*smooth*) and an offset trajectory (*dashed*) in phase space

$$\delta\mathbf{\Gamma}^f = (\partial\mathbf{M}/\partial\mathbf{\Gamma}^i)\cdot\delta\mathbf{\Gamma}^i \equiv \mathbf{S}(\mathbf{\Gamma}^i)\cdot\delta\mathbf{\Gamma}^i, \tag{6}$$

where $\mathbf{S}$ is the linearized collision map. In Fig. 1 the collisions of a reference trajectory (smooth) and of an offset trajectory (dashed) are schematically depicted. By strictly keeping only terms linear in $\delta\mathbf{\Gamma}^i$, the linearized map (6) becomes

$$\delta\mathbf{\Gamma}^f = \frac{\partial\mathbf{M}}{\partial\mathbf{\Gamma}^i}\cdot\delta\mathbf{\Gamma}^i + \left[\frac{\partial\mathbf{M}}{\partial\mathbf{\Gamma}^i}\cdot\mathbf{F}\left(\mathbf{\Gamma}^i\right) - \mathbf{F}\left(\mathbf{M}\left(\mathbf{\Gamma}_i\right)\right)\right]\delta\tau_c\left(\mathbf{\Gamma}^i,\delta\mathbf{\Gamma}^i\right), \tag{7}$$

where $\delta\tau_c$ accounts for the time delay between the offset-trajectory collision and that of the reference trajectory. $\delta\tau_c$ may be positive or negative and depends on $\mathbf{\Gamma}^i$ and - linearly - on $\delta\mathbf{\Gamma}^i$ (see Fig. 1). In the applications considered below, it is determined from the spatial offset component orthogonal to the collision plane and the reference-particle velocity also perpendicular to that plane, immediately prior to the collision.

The reference trajectory for the combined phase-space motion - continuous streaming, interrupted by collisions - formally follows from Eqs. (2) and (5),

$$\mathbf{\Gamma}(t) = \Phi^{t-\tau_n}\circ\mathbf{M}\circ\Phi^{\tau_n-\tau_{n-1}}\circ\ldots\circ\Phi^{\tau_2-\tau_1}\circ\mathbf{M}\circ\Phi^{\tau_1}\;\left(\mathbf{\Gamma}(0)\right). \tag{8}$$

Similarly, the evolution of a corresponding perturbation vector in tangent space is constructed by combining the streaming solution (4) between collisions with the linearized collision map (6), applied at times τ_k,

$$\delta\mathbf{\Gamma}(t) = \mathbf{L}^{t-\tau_n}\cdot\mathbf{S}\cdot\mathbf{L}^{\tau_n-\tau_{n-1}}\ldots\mathbf{L}^{\tau_2-\tau_1}\cdot\mathbf{S}\cdot\mathbf{L}^{\tau_1}\;\cdot\delta\mathbf{\Gamma}(0). \tag{9}$$

For the computation of $\delta\mathbf{\Gamma}(t)$ the reference trajectory $\mathbf{\Gamma}(t)$ must be known at any instant of time, and the motions in phase space and tangent space have to be solved simultaneously.

For Lyapunov-unstable systems the norm of the tangent vectors $\boldsymbol{\Gamma}(t)$ grows exponentially with time. The multiplicative ergodic theorem by Oseledec [11, 12] states that for ergodic systems, for which the domain is an L-dimensional manifold, there exist L Lyapunov exponents,

$$\lambda_l = \lim_{t\to\infty} \frac{1}{t} \ln \frac{|\delta\boldsymbol{\Gamma}_l(t)|}{|\delta\boldsymbol{\Gamma}_l(0)|} \quad , \quad l = 1, \dots, 2dN, \tag{10}$$

which are independent of the initial state, $\boldsymbol{\Gamma}(0)$, and of the metric used in phase space. In principle, the initial perturbations, $\delta\boldsymbol{\Gamma}_l(0)$, must be carefully chosen to point into directions in tangent space which follow from the solution of an eigenvalue problem and are difficult to compute [12, 13]. In practice, this difficulty is circumvented by an algorithm due to Benettin *et al.* [14, 15] and Shimada *et al.* [16], outlined below [17]. The ordered set $\{\lambda_1 \geq \lambda_2 \geq \dots \geq \lambda_L\}$ of exponents is referred to as the Lyapunov spectrum. The Lyapunov exponents geometrically describe the stretching and contraction along linearly-independent phase-space directions of an infinitesimal hypersphere co-moving and co-rotating with the phase-space flow.

2.1 General Properties of Lyapunov Spectra for Particle Systems

The motion is usually described in an extended phase space of dimension $L = 2dN$, and typically has to fulfill a number of constraints, such as the conservation of energy, momentum and/or center of mass. They confine the actual dynamics to a manifold of lower dimension. Since each constraint constitutes a hypersurface in the extended phase space, no perturbation component orthogonal to that surface exists. As a consequence, a number of Lyapunov exponents vanish, one for each dynamical constraint. In addition, one exponent vanishes due to the non-singular expansion of perturbations along the flow direction. As an example we consider the microcanonical dynamics of a two-dimensional particle system in a box with periodic boundaries. There are three constants of the motion, two for the linear momentum and one for the total energy, and they cause three exponents to vanish. Although the system does not conserve the center of mass due to the periodic boundaries, the tangent-space dynamics are not affected by the latter, and two additional exponents accordingly vanish. Including the exponent due to the flow direction, altogether six Lyapunov exponents vanish in this case. The perturbation vectors associated with the vanishing exponents play an important role in the understanding of the Lyapunov modes discussed in Section 3.2. From a numerical point of view, the vanishing exponents are useful to assess the accuracy of the simulation.

All equations of motion in this paper - including the discontinuous-time collisions mentioned above - are time reversible, which means that a reversed movie of the motion obeys the same equations. In such a reversed motion all momenta, $\mathbf{p}_i$, (and momentum-like variables such as the thermostat variables, $\zeta(t)$, of Section 5) change sign. If we rewrite Eq. (8) in terms of a propagator $\mathcal{M}$, $\mathbf{\Gamma}(t) = \mathcal{M} \circ \mathbf{\Gamma}(0)$, the flow is said to be time reversible if it satisfies the operator identity $\mathcal{M} \circ \mathcal{S} \circ \mathcal{M} = \mathcal{S}^{-1}$. Here, $\mathcal{S} = \mathcal{S}^{-1}$ is a time-reversal operator which changes the signs of all velocity components (and of any thermostat variables, if applicable) but leaves the signs of all coordinates unchanged. Two applications of this operator yield the identity $\mathcal{S} \circ \mathcal{S} = \mathcal{I}$. For Hamiltonian systems in thermodynamic equilibrium, and in the absence of magnetic fields, this concept of reversibility[2] has a profound impact on the shape of the Lyapunov spectrum. A perturbation, growing in the time-forward direction, contributes to a positive Lyapunov exponent. If the motion is reversed, the perturbation shrinks and contributes to a negative exponent. Since in equilibrium systems forward and backward motions cannot be distinguished, both exponents must be contained in the Lyapunov spectrum. The spectrum must transform into itself under time reversal. It follows that for each positive exponent there is a conjugate negative exponent such that the pair sum vanishes,

$$\lambda_l + \lambda_{L+1-l} = 0 \, , \quad l = 1, \ldots, L/2 \, . \tag{11}$$

This is the equilibrium formulation of the "conjugate pairing rule"[3], to which we will return in Sect. 7. Thus, the sum of all Lyapunov exponents vanishes in this case, and the volume of a phase extension is invariant under the flow generated by Hamiltonian equations of motion. The latter agrees with the predictions of the classical Liouville equation for Hamiltonian flows.

2.2 Numerical Considerations

From a numerical point of view it is interesting to note that an arbitrary choice of the initial perturbation, $\delta\mathbf{\Gamma}(0)$, (almost) always has a component in the direction of fastest growth, which will eventually take over and lead to the computation of the maximum exponent λ_1 and to a vector $\delta\mathbf{\Gamma}_1(t)$, which points into the direction of fastest growth. This direction varies with time. To reach, and maintain, the proper orientation of any other vector $\delta\mathbf{\Gamma}_l(t)$, $l > 1$, starting from an arbitrary orientation, it is necessary to subtract from $\delta\mathbf{\Gamma}_l(t)$ all vector components parallel to the previously-computed offset vectors $\delta\mathbf{\Gamma}_j(t), 1 \leq j \leq l-1$ corresponding to exponents λ_j larger than λ_l [17]. Starting with an arbitrarily-oriented *orthonormal* vector set $\{\delta\mathbf{\Gamma}_l(0); l = 1, \ldots, L\}$, this can be achieved by periodic (every few time steps)

[2] There are, however, a number of other concepts of reversibility which are not strictly equivalent [18, 19].

[3] To be more precise, the conjugate pairing rule (11) is a consequence of the symplectic nature of the equilibrium equations of motion, which means that the phase flow - viewed as a canonical transformation of the phase space onto itself - leaves the differential two-form $\sum_{i=1}^{L/2} dp_i \wedge dq_i$ invariant [20]. This symmetry also persists for thermostated equilibrium systems [21] as is demonstrated in Fig. 12 of Sect. 7.

re-orthonormalization of the perturbation vectors with a Gram-Schmidt procedure. The Lyapunov exponents, λ_l, are obtained from the time-averaged logarithms of the renormalization factors for $|\delta \mathbf{\Gamma}_l(t)|$, *after* the aforementioned subtraction has been performed. The exponents are automatically ordered according to size. The periodic renormalization also prevents the perturbation vectors from becoming singular due to exponential growths. This algorithm works both for time-continuous flows and for discontinuous collision maps [22] and will be used below. Since the number of equations required for a full Lyapunov spectrum increases with the square of the number of particles, this number is limited to a few thousand given the power of a present-day workstation.

An alternative scheme for a *continuous* orthonormalization of the offset vectors makes use of additional constraint forces in the equations of motion, which force the vectors to remain mutually perpendicular and to evolve at fixed length [23, 24]. These constraints can be imposed by a triangular array of Lagrange multipliers, Λ_{ij}, where $1 \leq j \leq i \leq L$ in an L-dimensional phase space. The usual Lyapunov exponents, λ_j, are the long-time-averaged values of the diagonal Lagrange multipliers, Λ_{jj}: $\lambda_j = \langle \Lambda_{jj} \rangle$, $i\ j = 1, \cdots, L$.

3 Application to Hard-Disk and Hard-Sphere Fluids in Equilibrium

As a first application of this scheme we consider here a two-dimensional system of hard disks in a square box with periodic boundaries. Since the pioneering work of Bernal [25] with steel balls, and the seminal computer simulations of Alder and coworkers [26, 27], hard ball systems are considered good models for the structure of "real" dense fluids and serve as reference systems for highly successful perturbation theories of liquids [28, 29]. The system is assumed to be in thermodynamic equilibrium. We summarize in the following section the relevant formulas obtained from the general scheme presented above [22, 30]. For an extension to nonequilibrium steady states we refer to Sect. 8.

There are N hard disks with diameter σ and mass m, located at $\mathbf{q}_j, j = 1 \ldots N$, which move freely between collisions (streaming) with momentum $\mathbf{p}_j$. The state vector $\mathbf{\Gamma} = \{\mathbf{q}_1, \ldots, \mathbf{q}_N, \mathbf{p}_1, \ldots, \mathbf{p}_N\}$ is a $2dN$-dimensional vector, and so are the tangent vectors $\delta\mathbf{\Gamma} = \{\delta\mathbf{q}_1, \ldots, \delta\mathbf{q}_N, \delta\mathbf{p}_1, \ldots, \delta\mathbf{p}_N\}$. During a streaming period the particles move on straight lines,

$$\mathbf{q}_j\left(t\right) = \mathbf{q}_j\left(0\right) + \mathbf{p}_j\left(0\right) t/m, \quad \mathbf{p}_j\left(t\right) = \mathbf{p}_j\left(0\right); \quad j = 1, \ldots, N. \tag{12}$$

The evolution of the offset vectors is given by

$$\delta\mathbf{q}_j\left(t\right) = \delta\mathbf{q}_j\left(0\right) + \delta\mathbf{p}_j\left(0\right) t/m, \quad \delta\mathbf{p}_j\left(t\right) = \delta\mathbf{p}_j\left(0\right); \quad j = 1, \ldots, N. \tag{13}$$

When two particles, say k and l, collide, their positions do not change, but their momenta change according to the collision map

$$\mathbf{p}_k^f = \mathbf{p}_k^i + (\mathbf{p}\cdot\mathbf{q})\,\mathbf{q}/\sigma^2, \;\; \mathbf{p}_l^f = \mathbf{p}_l - (\mathbf{p}\cdot\mathbf{q})\,\mathbf{q}^i/\sigma^2, \tag{14}$$

where $\mathbf{q} \equiv \mathbf{q}_l^i - \mathbf{q}_k^i$ and $\mathbf{p} \equiv \mathbf{p}_l^i - \mathbf{p}_k^i$ are the relative positions and momenta, respectively, of the colliding particles *immediately before* the collision. As in Eq. (5), the upper indices i and f refer to the initial and final states of the collision. A straightforward application of Eq. (7) yields the linearized collision map [22, 30]. For the non-colliding particles, $j \neq k, l$, one finds,

$$\delta\mathbf{q}_j^f = \delta\mathbf{q}_j^i, \tag{15}$$

$$\delta\mathbf{p}_j^f = \delta\mathbf{p}_j^i, \tag{16}$$

and for the colliding particles k and l,

$$\delta\mathbf{q}_k^f = \delta\mathbf{q}_k^i + (\delta\mathbf{q}\cdot\mathbf{q})\,\mathbf{q}/\sigma^2, \tag{17}$$

$$\delta\mathbf{q}_l^f = \delta\mathbf{q}_l^i - (\delta\mathbf{q}\cdot\mathbf{q})\,\mathbf{q}/\sigma^2, \tag{18}$$

$$\delta\mathbf{p}_k^f = \delta\mathbf{p}_k^i + (\delta\mathbf{p}\cdot\mathbf{q})\,\mathbf{q}/\sigma^2 + \frac{1}{\sigma^2}\left[(\mathbf{p}\cdot\delta\mathbf{q}_c)\,\mathbf{q} + (\mathbf{p}\cdot\mathbf{q})\,\delta\mathbf{q}_c\right], \tag{19}$$

$$\delta\mathbf{p}_l^f = \delta\mathbf{p}_l^i - (\delta\mathbf{p}\cdot\mathbf{q})\,\mathbf{q}/\sigma^2 - \frac{1}{\sigma^2}\left[(\mathbf{p}\cdot\delta\mathbf{q}_c)\,\mathbf{q} + (\mathbf{p}\cdot\mathbf{q})\,\delta\mathbf{q}_c\right]. \tag{20}$$

Again, $\delta\mathbf{q} \equiv \delta\mathbf{q}_l^i - \delta\mathbf{q}_k^i$ and $\delta\mathbf{p} \equiv \delta\mathbf{p}_l^i - \delta\mathbf{p}_k^i$ are the respective relative position and momentum displacements of the colliding particles *before* the collision. The vector

$$\delta\mathbf{q}_c = \delta\mathbf{q} + \mathbf{p}\delta\tau_c\,, \tag{21}$$

with

$$\delta\tau_c = -\frac{(\delta\mathbf{q}\cdot\mathbf{q})}{(\mathbf{p}\cdot\mathbf{q})}\,, \tag{22}$$

denotes the infinitesimal displacement of the collision points of the perturbed trajectory from the reference trajectory. Of course, the components of $\delta\mathbf{\Gamma}$ for particles not partaking in the collision remain unchanged. With these equations, the dynamics in the phase and tangent spaces may be reconstructed.

In our numerical work we use reduced units for which the particle mass, m, the disk diameter, σ, the kinetic energy per particle, K/N, and the Boltzmann constant, k_B, are unity. $K = \sum \mathbf{p}^2/2m$ is the total kinetic energy. With this choice the unit of time is $(m\sigma^2 N/K)^{1/2}$. For the hard-disk system in equilibrium, the temperature is an irrelevant parameter since there is no potential energy, and the Lyapunov exponents strictly scale with $\sqrt{K}$. It is therefore sufficient to consider a single isotherm, which corresponds to a kinetic temperature $T = 2K/d(N-1)k_B$, where the -1 accounts for the center-of-mass velocity not contributing to T. The density, $\rho \equiv N/V$, is the only relevant parameter, where V is the area of the simulation box. $d = 2$ for disks.

The Lyapunov spectra of hard-disk fluids in two dimensions were studied in Refs. [22, 31, 33, 34] and of hard-sphere fluids in three dimensions in

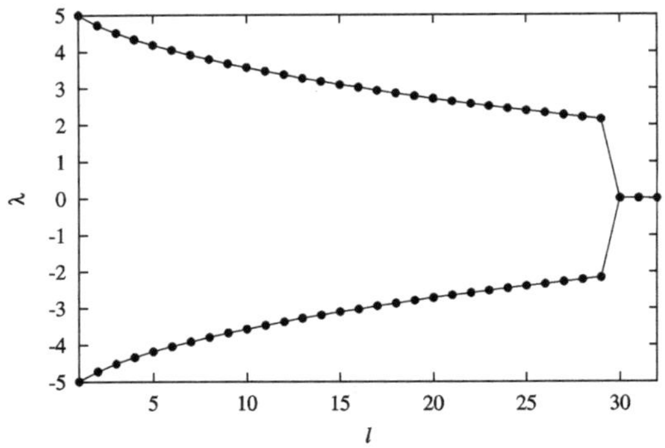

Fig. 2. Lyapunov spectrum of a 16-disk system in a square simulation box with periodic boundaries and a density $\rho = 0.7$

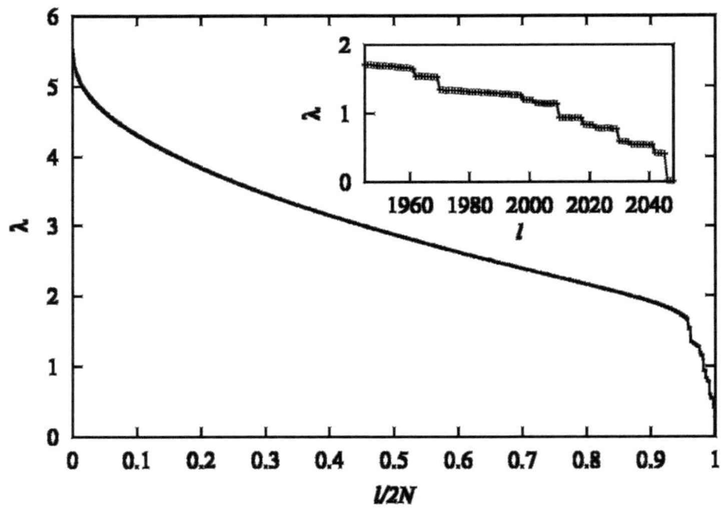

Fig. 3. Lyapunov spectrum of a 1024-disk system in a square simulation box with periodic boundaries. The density $\rho = 0.7$. A reduced index $l/2N$ is used on the abscissa. The steps are magnified in the inset. The non-normalized index, l, is used there. The spectrum is defined only for integer values of l

Refs. [30, 35]. As a first example we show in Fig. 2 the full spectrum of a 16-disk system in a square box with periodic boundaries and a density $\rho = 0.7$, which corresponds to a dense fluid. Conjugate pairs of exponents are labeled by the index l on the abscissa, emphasizing the conjugate pairing symmetry for equilibrium systems. There are three conjugate pairs of vanishing exponents. The spectrum is defined only for integer values of l. In Fig. 3 we show an analogous spectrum for a much larger system containing 1024 particles at the same density, $\rho = 0.7$. In this figure we make explicit use of the pairing symmetry and plot only the positive branch of the spectrum. On the abscissa a normalized index $l/2N$ is used. Although a smooth line is drawn, the spectrum is only defined for integer values of l.

There are various aspects of these spectra that merit a more detailed discussion, to which we turn in the following.

3.1 The Maximum Exponent and the Kolmogorov-Sinai Entropy

The maximum exponent is dominated by the fastest dynamical events taking place in the system. One notices that λ_1 increases slightly from Fig. 2 to Fig. 3 and, hence, seems to increase with N. Does this increase persist for even larger N? We have rather convincing numerical evidence from plots of λ_1 as a function of $1/N$ that it does not and that the thermodynamic limit $\{N, V \to \infty, N/V \text{constant}\}$ for λ_1 and, hence, for the whole spectrum, exists [22, 30, 36]. Theoretical arguments by Sinai [37] also point to the existence of this limit for systems interacting with a pairwise-additive short-range potential. Furthermore, we demonstrate below that the perturbation associated with the maximum exponent is strongly localized in space at any instant of time, which is consistent with the fact that λ_1 is insensitive to N for large N. We conclude that the Lyapunov spectra of hard-disk systems may be confidently extrapolated to the thermodynamic limit. However, Searles *et al.* [38] found a very weak but persistent increase of λ_1 with N for a two-dimensional Lennard-Jones system, which was interpreted as a possible logarithmic singularity. More evidence is needed to settle this question.

Next, we consider the density dependence of λ_1 and of the Kolmogorov-Sinai entropy[4], h_{KS}, which according to Pesin's theorem for closed Hamiltonian systems is equal to the sum of all positive Lyapunov exponents [39, 7]. Both quantities increase monotonically with density, with the exception of a narrow range near the fluid-to-solid phase-transition density, where they display a local maximum [22, 30]. This behavior is fully reflected by the density dependence of the collision frequency, which turns out to be the dominant parameter. The collective motion characteristic of a fluid-to-solid phase transition [40] does not have a noticeable effect on λ_1 and h_{KS}.

So far, all theoretical attempts for the computation of λ_1 and h_{KS} are based on kinetic theory and the Boltzmann equation, and are applicable only to dilute hard-disk and hard-sphere systems. Van Beijeren *et al.* [41] formulated the first successful theory of h_{KS} for low-density gases in two and three dimensions, and extensions of that work in terms of the B.B.G.K.Y. hierarchy have been developed [42]. Van Zon *et al.* [43] were able to compute analytically the maximum exponent of such systems and compared it successfully to simulation results. This work is also summarized in Ref. [44].

[4] The Kolmogorov-Sinai entropy is a measure of dynamical randomness. It is the rate with which information about an initial state is gained by observing the dynamical (time-reversible) system at later times [8].

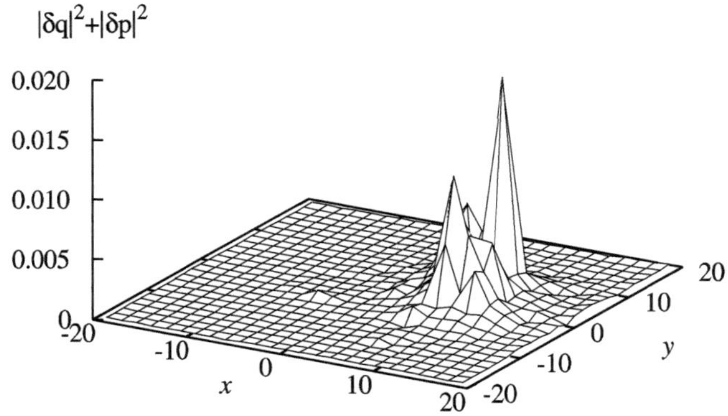

Fig. 4. Spatial localization of the λ_1 perturbation for a 1024-particle hard-disk fluid. The individual particle contributions to the squared perturbation vector belonging to λ_1 are plotted at the positions of the particles in the simulation box. The ensuing surface is linearly interpolated over a periodic grid. The figure is a snapshot for a well-relaxed system

One striking feature is the strong spatial localization of the perturbation associated with the maximum exponent [45, 46, 36, 33]. This is demonstrated by projecting the perturbation vector $\delta\mathbf{\Gamma}_1$ onto the four-dimensional subspaces spanned by the components belonging to the individual particles. The squared norm of this projection, $|\delta\mathbf{q}_i|^2 + |\delta\mathbf{p}_i|^2$, indicates the activity of a particle i in the growth process leading to λ_1 [36, 46]. In Fig. 4 it is plotted for a 1024-disk system at the particle positions (x_i, y_i), where the ensuing surface is interpolated over a periodic grid covering the simulation box. The fastest expansion activity is clearly confined to a few narrow active zones. To understand this localization we note that the offset-vector dynamics are governed by *linear* equations and that a perturbation component has only a fair chance of further growth if its value was already above average before a collision. Each global renormalization indiscriminately reduces all components, large and small, which tends to suppress already-small components even further. This competition for growth favors particles with the largest perturbation components. The active zones move diffusively in space and perform occasional jumps. The spatial localization persists in the thermodynamic limit, and the ratio of particles actively engaged in the growth of $|\delta\mathbf{\Gamma}_1(t)|$ vanishes in that limit [33, 36]. The localization becomes progressively less pronounced for perturbations associated with smaller exponents.

3.2 Lyapunov Modes

In the inset of Fig. 3 an unexpected feature is observed for the small exponents close to the abscissa, which was not present for the much-smaller system in Fig. 2: The Lyapunov exponents become degenerate and the spectrum assumes a step-like structure. The reason for such behavior becomes apparent

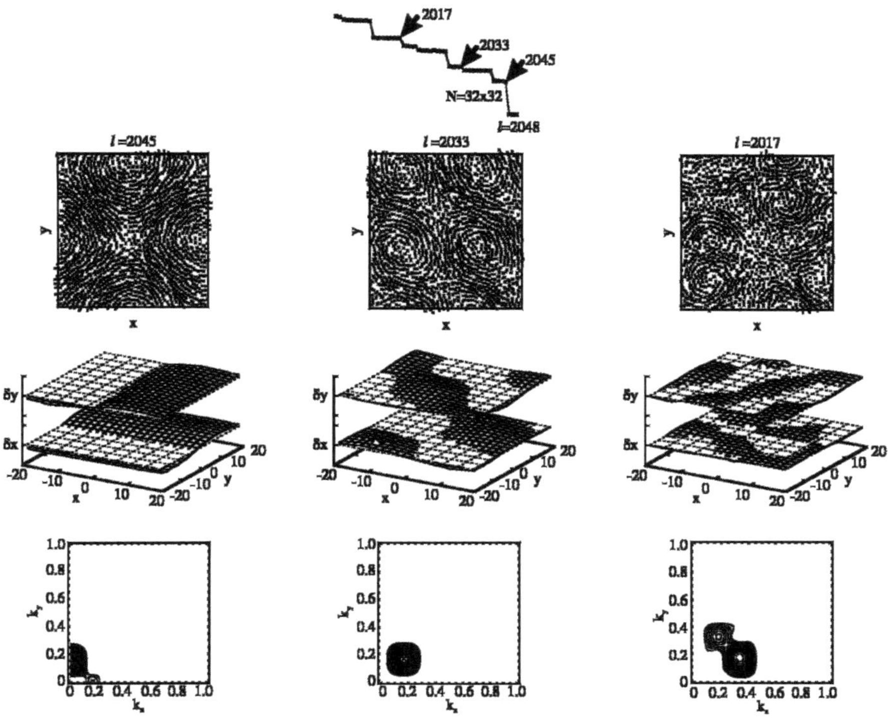

Fig. 5. Top: Mode regime of a Lyapunov spectrum for a 1024-disk system in a square simulation box with periodic boundaries. The density $\rho = 0.7$. *First row*: Snapshot of the vector field $\delta\mathbf{q}(x, y)$ for the position perturbations of the transverse modes indicated by the arrows in the spectrum above. *Second row*: Snapshot of the respective scalar fields $\delta x(x, y)$ and $\delta y(x, y)$ for the same modes. *Third row*: Contour plots of the Fourier transforms of the modes. They are time averages as defined in the main text

if one views the two-dimensional position and momentum perturbations as two-dimensional vector fields $\delta\mathbf{q}(x, y)$ and $\delta\mathbf{p}(x, y)$, respectively, over the domain of the simulation box. In the first row of Fig. 5, snapshots of the fields $\delta\mathbf{q}(x, y)$ are shown for the three exponents indicated by the arrows in the relevant part of the 1024-disk spectrum at the top of this figure. These fields represent coherent patterns in space. This becomes even more apparent if the components $\delta x(x, y)$ and $\delta y(x, y)$ are separately plotted as in the second row of Fig. 5. One observes that these scalar fields are sine- and cosine-like plane waves characterized by wave vectors $\mathbf{k}$ with norm $|\mathbf{k}| \equiv k$. These patterns are reminiscent of the modes of fluctuating hydrodynamics. We refer to them as *Lyapunov modes.* We have found them in hard-disk and hard-sphere systems in two and three dimensions [31, 32, 33], respectively, and in hard-dumbbell systems modeling fluids consisting of linear molecules [46, 36].

The degeneracy of the Lyapunov exponents and, hence, the step-like structure of the spectrum, are a consequence of the symmetry of the simulation box and of the orthonormality of the phase-space perturbation vectors. For a *rectangular* box with sides L_x and L_y and for periodic boundary conditions, the allowed wave numbers, k, of a wave follow from a rectangular reciprocal lattice with a point spacing equal to $2\pi/L_x$ and $2\pi/L_y$ in the k_x and k_y directions, respectively,

$$k = 2\pi\sqrt{\left(\frac{n_x}{L_x}\right)^2 + \left(\frac{n_y}{L_y}\right)^2} . \tag{23}$$

The integers n_x and n_y denote the respective number of modes in the x and y directions. The modes for which $n_x = n_y = 0$ are referred to as zero modes and exist for any translationally-invariant system. They are connected with the vanishing Lyapunov exponents. From the reciprocal lattice construction also follows the degeneracy of the modes as outlined in more detail in Refs. [33, 34]. Degenerate Lyapunov exponents belong to modes with the same k.

Numerically, the wave vectors contributing to a mode are most easily found from the spatial Fourier transforms of the components $\delta x, \delta y, \delta p_x, \delta p_y$, which are denoted by $\delta\tilde{x}, \delta\tilde{y}, \delta\tilde{p}_x, \delta\tilde{p}_y$, respectively. In the bottom line of Fig. 5, contour plots of $\left(\langle \delta\tilde{x}\delta\tilde{x}^* + \delta\tilde{y}\delta\tilde{y}^* + \delta\tilde{p}_x\delta\tilde{p}_x^* + \delta\tilde{p}_y\delta\tilde{p}_y^*\rangle\right)^{1/2}$ are shown for the three modes taken as examples [34]. Here, $\langle\ldots\rangle$ denotes a time average. In all cases the involved wave vectors are readily recognized as reciprocal lattice points. Fourier analysis is particularly useful for more complicated modes with wave vectors not aligned with the simulation box and close to the maximum wave number k_c, beyond which modes cease to exist. k_c is related to the threshold, $l_c/2N \approx 0.96$, for the index l, below which there are no degenerate exponents and modes, as is apparent from Fig. 3.

Next, we consider a whole class of square systems by varying the side L of the simulation box and the particle number N, such that the density $\rho = N/L^2$ is constant. *All* exponents connected with modes may be classified either as *transverse* (T) or *longitudinal* (L), with a unique dependence on k for each set. These so-called "dispersion relations" are shown in Fig. 6 for a hard-disk fluid with a density $\rho = 0.7$. The expressions "longitudinal" and "transverse" refer to the polarisation of the waves with respect to the wave vector. For example, the mode on the left-hand side of Fig. 5 has a perturbation component δx perpendicular to its wave vector and parallel to y, and a component δy with a wave vector parallel to x. It is a transverse mode, as are the other modes depicted in this figure. The transverse modes do not propagate and are *stationary* once formed. The phase of the longitudinal modes *propagates* in space with a unique velocity. This propagation is a consequence of a quasiperiodic rotation of the perturbation vectors in tangent space, effectively switching back and forth between tangent-space directions

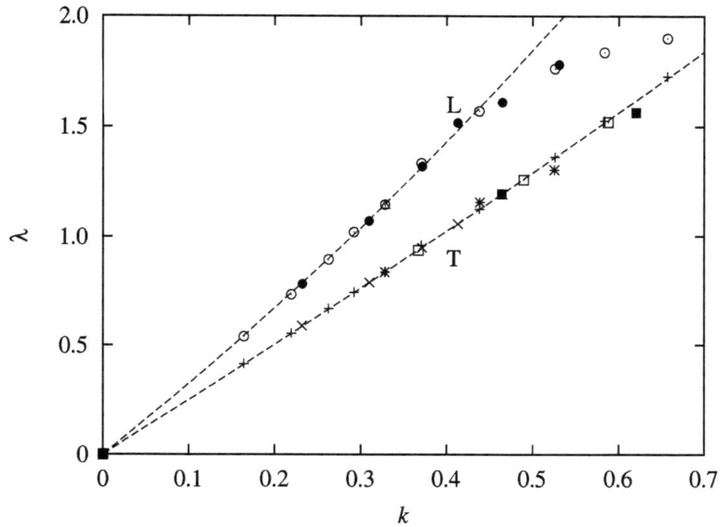

Fig. 6. Dispersion relation for longitudinal (L) and transverse (T) modes. The density of the hard-disk fluid is 0.7

defined by the constraints generating the vanishing exponents. This point is currently being investigated [47, 48].

To second order in k, the dispersion relations in Fig. 6 may be approximated by

$$\lambda_T = 2.48(3)k + 0.23(7)k^2 + \mathcal{O}(k^3)$$
$$\lambda_L = 3.13(7)k + 1.2(2)k^2 + \mathcal{O}(k^3)\ , \tag{24}$$

where the standard deviations affect the last digit of the fit parameters and are given in brackets. These fits are indicated by the dashed lines in Fig. 6. Since the multiplicity of the longitudinal and transverse modes for a given k is determined by the reciprocal lattice, the Eqs. (24) may be used to construct the small wave-number regime (large enough l) for the Lyapunov spectrum of a very large system, otherwise not accessible by computer simulation. The dashed curve in Fig. 7 shows such a construction for a 40,000-disk system at a fluid density of 0.7. k does not exceed 0.3. A power law,

$$\lambda_l = \alpha \left[1 - \frac{l}{2N}\right]^{\beta}\ ,\ 0.99 < \frac{l}{2N} < 1\ , \tag{25}$$

fitted to these points yields $\alpha = 7.81 \pm 0.04$ and $\beta = 0.52 \pm 0.01$ and is shown by the smooth line.[5] The slope of the spectrum near the intersection with

[5] For a simple model of a two-dimensional solid, the number of vibrational modes, dl, between wave numbers k and $k + dk$ is proportional to k. Integrating this relation, and assuming linear dispersion relations for the Lyapunov exponents, we obtain $\lambda \sim l^{1/2}$ [23]. This power is close to the value of β given above.

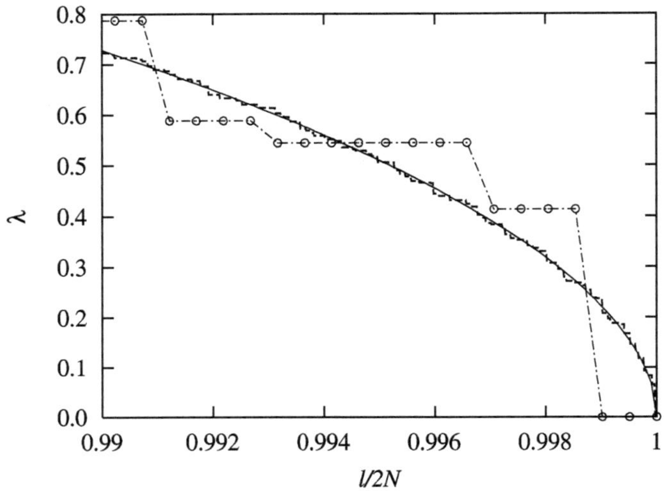

Fig. 7. Constructed Lyapunov spectrum for a fluid of density 0.7, consisting of 40,000 disks (*dashed line*). The *smooth line* is a power-law fit to these points. For comparison, the spectrum for 1024 disks is shown by the *circles*. A normalized index $l/2N$ is used on the abscissa

the abscissa,

$$\lim_{l/2N \to 1} \frac{d\lambda_l}{d(l/2N)} \sim \lim_{l/2N \to 1} \left[1 - \frac{l}{2N}\right]^{-0.48}, \tag{26}$$

diverges for $l/2N \to 1$ and $N \to \infty$. In the thermodynamic limit, no positive lower bound exists for the positive exponents. Due to conjugate pairing, analogous results are true for the negative exponents.

4 Fluids with Soft Particle Interactions

The Lyapunov spectra of soft-particle systems were the first to be studied by computer simulation [23, 49], and their behavior for fluids and solids in two and three dimensions is well known [49]. In Fig. 8 we show the positive branch of a spectrum for a two-dimensional 400-particle system interacting with a purely-repulsive Weeks-Chandler-Anderson (WCA) pair potential,

$$\phi(r) = \begin{cases} 4\epsilon[(\sigma/r)^{12} - (\sigma/r)^6] + \epsilon, & r < 2^{1/6}\sigma \\ 0, & r \geq 2^{1/6}\sigma \,. \end{cases} \tag{27}$$

We use units for which the particle mass, m, the particle diameter, σ, and the energy parameter, ϵ, are unity. The temperature, $T = 0.98$, and the den-

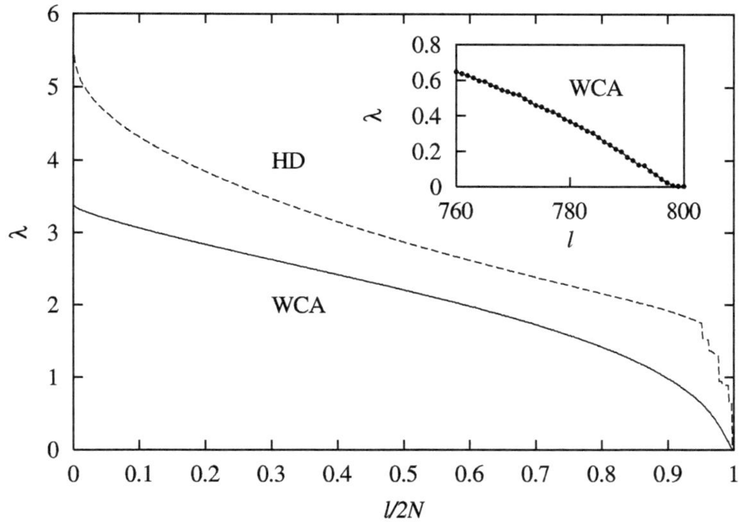

Fig. 8. Equilibrium Lyapunov spectra for 400 soft (WCA) and hard (HD) disks at a density $\rho = 0.7$ and a temperature $T = 0.98$. The simulation box is a square with periodic boundary conditions. A reduced index $l/2N$ is used on the abscissa. In the inset a magnified view of the small-exponent regime for the WCA spectrum is shown. There, l is used without normalization

sity, $N/V = 0.7$, correspond to an undercritical dense gas[6] For comparison, the spectrum of a hard-disk gas at the same density and temperature[7] is also shown. There are surprising differences in the shape of the spectrum and in the absolute value. The latter may be connected to a not-properly-adjusted and temperature-dependent effective hard-disk diameter [50]. Most conspicuous for the WCA fluid, however, is the total absence of degenerate steps and of Lyapunov modes for the perturbations associated with the small Lyapunov exponents. This is clearly demonstrated in the inset of Fig. 8, which shows a magnification of the spectral region of interest. This unexpected behavior has been linked to excessive fluctuations of the respective *time-dependent* exponents [51, 34], which are the instantaneous logarithmic renormalization factors for the Gram-Schmidt re-orthonormalization step mentioned in Sect. 2.2, divided by the renormalization interval. Fourier-transformation methods may be used to demonstrate the existence of Lyapunov modes in this case [52, 53, 54].

[6] Liquids do not exist for a purely repulsive potential. They require that the pair potential becomes attractive for large r, such as the dispersion interaction $\sim -1/r^6$.

[7] For hard-body systems the Lyapunov spectra strictly scale with the square root of the total kinetic energy.

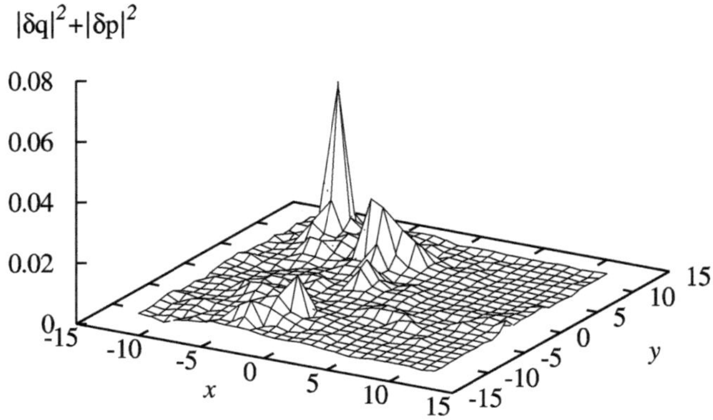

Fig. 9. Spatial localization of the λ_1 perturbation for a planar 400-particle WCA fluid. The individual particle contributions to the squared perturbation vector associated with λ_1 are plotted at the positions of the particles in the simulation box. The ensuing surface is linearly interpolated over a periodic grid. The figure is a snapshot for a well-relaxed system

The spatial localization of the perturbation belonging to the maximum exponent, λ_1, is also observed for the WCA fluid. This is demonstrated in Fig. 9. The localization, although less pronounced than for hard disks in Fig. 4, also persists in the thermodynamic limit [34].

5 Introduction to Nonequilibrium Molecular Dynamics, NEMD

In the following we consider systems driven away from thermal equilibrium by an external perturbation, $X(t)$. For weak-enough forcing, this can been treated by first-order perturbation theory [55, 56, 57]. There it is assumed that all terms of higher-than-first order in X do not contribute. It is easy to see that the irreversibly dissipated heat generated by the external driving is proportional to X^2 and may be safely neglected. The system can be treated adiabatically, and no special provisions, such as thermostats, are required. It is a consequence of this (adiabatic) *linear-response theory* that any transport coefficient can be expressed as a Green-Kubo integral of the equilibrium correlation function of the associated dissipative flux, $J(t)$ [57],

$$L = \beta \int_0^\infty dt \langle J(0)J(t)\rangle_0 . \tag{28}$$

Here, $\beta = 1/k_B T$, k_B is Boltzmann's constant and T is the temperature. The average $\langle \cdots \rangle_0$ is performed with a canonical equilibrium ensemble. However, the highly collective excitations, always present in a dense fluid [58], seriously impede an efficient evaluation of such integrals, since the long-time decay of these correlation functions may be slow. For example, the velocity autocorrelation function follows a power law, $\sim t^{-d/2}$, where d is the space dimension of the system. In two dimensions the respective Green-Kubo integrals diverge logarithmically with system size. In three dimensions the situation is less dramatic since transport coefficients linear in the applied perturbation do exist, but nonlinear coefficients - such as the nonlinear Burnett coefficients for shear flow - still diverge [59].

An alternative method for the evaluation of transport properties was therefore developed which is much closer in spirit to actual experiments carried out in the laboratory than the fluctuation-dissipation approach mentioned above. This method, known as *nonequilibrium molecular dynamics* (NEMD) [4, 6, 5], consists in driving the system away from equilibrium with a *large* external perturbation $X(t)$ for positive times,

$$X(t) = X\Theta(t). \tag{29}$$

Here, Θ is the unit-step function, and $X > 0$ is assumed. The work, continuously performed by X, heats up the system, which will never reach a steady state. Since the rate of internal-energy increase, $-\dot{Q}$, is proportional to X^2, it cannot be ignored in view of the large perturbations anticipated for NEMD. The excess heat must be removed with a thermostat. This is also common laboratory practice if an experiment is to be carried out under nonequilibrium steady-state conditions. In the context of a computer simulation this "thermostat" consists of an extra frictional feedback term $-\zeta(t)\mathbf{p}_i$ added to the forces acting on the particles. Once a stationary flux is generated, the nonequilibrium transport coefficient is defined by

$$L(X) = -\frac{\langle J(t)\rangle}{X}, \tag{30}$$

which is Ohm's law, extended to far-from-equilibrium conditions. For vanishing fields we require that the linear-response result, $L = \lim_{X\to 0} L(X)$, is recovered.

Two types of driving forces are commonly encountered, *mechanical* and *thermal*. A *mechanical* perturbation can always be written as an external field acting on the particle coordinates. It explicitly shows up in the perturbed Hamiltonian. The color field for the conductivity of color-charged particles in Sections 6 and 8 may serve as an example. By contrast, the phenomena associated with the transport of conserved properties, such as linear momentum (viscosity) and energy (thermal conductivity), do not fall into this category. They belong to the second class of *thermally* perturbed systems, for which generally *no* perturbed Hamiltonian can be found. Instead, fictitious "thermal forces" proportional to the particles' momenta, in the case of shear viscosity, and to enthalpy, in the case of heat conduction, must be introduced. Irrespective of their origin, these forces explicitly show up in the equations of motion for the particles,

$$\dot{\mathbf{\Gamma}} = \begin{cases} \dot{\mathbf{q}}_i = \mathbf{p}_i/m + \mathcal{Q}_i(\mathbf{q},\mathbf{p})X(t) \\ \dot{\mathbf{p}}_i = \mathbf{F}_i + \mathcal{P}_i(\mathbf{q},\mathbf{p})X(t) - \zeta\mathbf{p}_i \end{cases} \tag{31}$$

where $\mathbf{q}$ represents all particle positions, and $\mathbf{p}$ all *peculiar* momenta. We have assumed equal mass, m, for all particles. $\mathbf{F}_i = -\partial\Phi/\partial\mathbf{q}_i$ is the intrinsic atomistic force on particle i, and the potential energy $\Phi(\mathbf{q})$ may also include boundary interactions if required. The external driving force, $X(t)$, is coupled to the system by the vector-valued phase functions $\mathcal{Q}(\mathbf{q},\mathbf{p})$ and $\mathcal{P}(\mathbf{q},\mathbf{p})$. For simplicity, X is written as a scalar, such that any vectorial or tensorial character of the perturbation is absorbed in $\mathcal{Q}$ and $\mathcal{P}$. Finally, we recognize the thermostat term, $-\zeta(t)\mathbf{p}_i$, which has the form of a fluctuating frictional feedback added to the force in Eq. (31). In the simplest version, a *single* thermostat variable, $\zeta(t)$, is required. It represents an external heat bath which in the real world consists of many external degrees of freedom. In this sense $\zeta(t)$ can be viewed as a general boundary condition, to which all (*homogeneous* thermostat) or a fraction (*inhomogeneous* thermostat) of the momenta are subjected.

The identification of $\mathbf{p}_i$ in (31) with a *peculiar* momentum implies that $\mathbf{p}_i/m$ is measured relative to a stationary hydrodynamic streaming velocity, $\mathbf{u}(\mathbf{q}_i)$. This restricts the applicability of the method to lower Reynolds number non-turbulent flows [60]. The internal energy is given by

$$H_0 = \sum_i \frac{\mathbf{p}_i^2}{2m} + \Phi(\mathbf{q}), \tag{32}$$

and the temperature is defined by

$$\langle K \rangle = \langle \sum_i^N \frac{\mathbf{p}_i^2}{2m} \rangle = \frac{d}{2}(N-1)k_B T, \tag{33}$$

where d is the dimension. Strictly speaking, T is the thermodynamic temperature only at equilibrium ($X = 0$) and only if the averaging is performed with a canonical ensemble. For general non-equilibrium conditions the distribution of the peculiar momenta is not given by the Maxwell-Boltzmann distribution [61], and (33) becomes the defining equation of a nonequilibrium *kinetic temperature*, which may be determined with a small ideal gas thermometer, and is always defined. For nonequilibrium conditions the definition of a *thermodynamic* temperature, $T \equiv (\partial E/\partial S)_V$, is complicated by the singular properties of Gibbs' nonequilibrium entropy. As will be shown in more detail below, Gibbs' entropy can diverge in nonequilibrium steady states, rendering this definition useless.

The rate of change of the internal energy (32) is given by

$$\dot{H}_0 = \sum_i \left(\frac{\partial H_0}{\partial \mathbf{q}_i} \cdot \dot{\mathbf{q}}_i + \frac{\partial H_0}{\partial \mathbf{p}_i} \cdot \dot{\mathbf{p}}_i\right) - \zeta \sum_i \frac{\mathbf{p}_i^2}{m} \equiv -J(t)X - 2\zeta(t)K(t), \tag{34}$$

where the term proportional to X is the rate with which work is performed on the system. It defines the dissipative flux, $J(t)$,

$$J(t) = \sum \mathcal{Q} \cdot \mathbf{F} - \mathcal{P} \cdot \frac{\mathbf{p}}{m} \,. \tag{35}$$

In a stationary state we require that $\langle \dot{H}_0 \rangle = 0$, where $\langle \ldots \rangle$ is a time average realized by following the stationary nonequilibrium phase trajectory for an infinitely long time.

The general structure of the thermostat term, $-\zeta(t)\mathbf{p}_i$, follows from Gauss' principle of least constraint [62, 63]. It requires that any force constraining the dynamics becomes minimal in the least-squared sense. This allows us to treat nonholonomic constraints, i.e. constraints depending not only on the particle positions but also on the velocities, such as the kinetic or total energies. The variable $\zeta(t)$ is the Lagrange multiplier for the minimization procedure. For its explicit determination two methods are most commonly used, which differ in the ensemble they generate under equilibrium conditions ($X = 0$):

Gauss Thermostat: The peculiar kinetic energy, $K(\mathbf{p}) = \sum_i \mathbf{p}_i^2/2m$, is a constant of the motion, K_0. Differentiation with respect to time and insertion of Eq. (31) yields for the Lagrange multiplier $\zeta(t)$:

$$\zeta(t) = \frac{\sum (\mathbf{p}/m) \cdot [\mathbf{F}(\mathbf{q}) + \mathcal{P}(\mathbf{q}, \mathbf{p}) X]}{\sum (\mathbf{p} \cdot \mathbf{p})/m} . \tag{36}$$

It acts as a fluctuating thermostat variable. For ergodic systems in equilibrium ($X = 0$), the phase-space probability distribution,

$$f_0^{(G)} \sim \exp\{-\beta \Phi(\mathbf{q})\} \delta(K(\mathbf{p}) - K_0), \tag{37}$$

is canonical with respect to the potential energy, Φ, and microcanonical with respect to the kinetic energy, K. It is referred to as an *isokinetic ensemble* or as a *Gauss thermostat.* K_0 is the kinetic energy prescribed by the initial conditions. If the method of Gauss is used to fix the total internal energy, H_0, and not only its kinetic-energy contribution, the friction variable for the ensuing *isoenergetic ensemble* becomes

$$\zeta(t) = \frac{J(t) X}{\sum (\mathbf{p} \cdot \mathbf{p})/m} . \tag{38}$$

This is also referred to as an *ergostat.* In equilibrium ($X = 0$) this ensemble becomes microcanonical, and the ergostat does not affect the equations of motion at all. All schemes based on Gauss' principle have in common that $\zeta(\mathbf{q}(\mathbf{t}), \mathbf{p}(\mathbf{t}))$ is a dynamical variable and that the dimension of phase space is solely determined by the number of particles in the system.

Nosé-Hoover Thermostat: For $X = 0$ this method generates a trajectory with a probability distribution conforming to the *canonical ensemble.* It was invented by Nosé [64, 65] and put into a more practical form by Hoover [66, 4]. Assuming equations of motion with a thermostating force, $-\zeta \mathbf{p}_i$, as in (31), the thermostat variable, ζ, is considered an additional independent variable, thus increasing the number of state variables by one. The time evolution of the phase-space density, $f(\mathbf{q}, \mathbf{p}, \zeta)$, in the extended $(2dN + 1)$-dimensional phase space is given by the continuity equation

$$\frac{\partial f}{\partial t} + \sum \frac{\partial}{\partial \mathbf{q}} \cdot (f\dot{\mathbf{q}}) + \sum \frac{\partial}{\partial \mathbf{p}} \cdot (f\dot{\mathbf{p}}) + \frac{\partial}{\partial \zeta}(f\dot{\zeta}) = 0, \tag{39}$$

which is a generalized version of the Liouville equation. If we require that at equilibrium $(X = 0)$ the stationary solution $(\partial f/\partial t = 0)$ of (39) is canonical with respect to the internal energy and Gaussian for the momentum-like thermostat variable,

$$f(\mathbf{q}, \mathbf{p}, \zeta) \sim \exp(-\beta H_0) \exp(-\zeta^2 \tau^2/2) \ , \tag{40}$$

then the time evolution of ζ follows from the feedback equation

$$\dot{\zeta} = \frac{1}{\tau^2} \left(\sum_i^N \frac{\mathbf{p}_i^2}{dmk_B T} - 1 \right) . \tag{41}$$

(31) and (41) together constitute the Nosé-Hoover equations of motion for a $(2Nd + 1)$-dimensional phase point $\mathbf{\Gamma} = (\mathbf{q}, \mathbf{p}, \zeta)$. T is the kinetic temperature the thermostat is expected to generate by the time average of K. The arbitrary parameter τ determines the response time of the thermostat and should be chosen to be on the order of the duration of the fastest dynamical event. For particular cases, the Nosé-Hoover thermostat may not be mixing enough to prevent undesired regular features from surviving, such as in the phase space of the thermostated one-dimensional harmonic oscillator [67]. Remedies have been found which use at least two thermostating variables rather than a single friction coefficient, ζ. For a thorough investigation we refer to Ref. [68].

All thermostated systems mentioned above have in common that the equations of motion are *time reversible* and that the fluctuating thermostat variable behaves like a momentum variable under time reversal: it changes sign. $\langle \zeta \rangle$ vanishes for equilibrium systems and becomes positive for nonequilibrium stationary states: The thermostat removes the excessive heat from the system.

The Gibbs entropy of a nonequilibrium system is defined by

$$S(t) = -k_B \int d^L \Gamma f(\mathbf{\Gamma}, t) \ln f(\mathbf{\Gamma}, t) \equiv -k_B \langle \ln f(\mathbf{\Gamma}, t) \rangle, \tag{42}$$

where L is the phase-space dimension, and where the state vector $\mathbf{\Gamma}$ may or may not include ζ as an independent variable, depending on the thermostat. The notation in Eq. (42) emphasizes that the probability for a fixed volume element centered at $\mathbf{\Gamma}$ in phase space varies with time according to the generalized Liouville equation [9, 71],

$$\frac{\partial f}{\partial t} + \frac{\partial}{\partial \mathbf{\Gamma}} \cdot (\dot{\mathbf{\Gamma}} f) = 0. \tag{43}$$

This is the continuity equation for the conservation of phase-space probability. For the rate of change of S one finds the following important chain of identities:[8]

$$\frac{\dot{S}}{k_B} = \left\langle \frac{\partial}{\partial \mathbf{\Gamma}} \cdot \dot{\mathbf{\Gamma}} \right\rangle = -Nd\langle \zeta \rangle = \left\langle \frac{d \ln \delta V}{dt} \right\rangle = \sum_{l=1}^{L} \lambda_l \leq 0 \ . \tag{44}$$

Eq. (43) and integration by parts are used to relate $\dot{S}$ in the first equality to the ensemble-averaged phase-space divergence $\langle (\partial/\partial \mathbf{\Gamma}) \cdot \dot{\mathbf{\Gamma}} \rangle$ [72]. Insertion of the equations of motion for $\dot{\mathbf{\Gamma}}$ yields the second equality, where $\langle \zeta \rangle$ is the time-averaged thermostat variable. The third equality is also a consequence of the Liouville equation and relates $\dot{S}/k_B$ to the rate of change of an arbitrary differentially small volume element $\delta V(\mathbf{\Gamma}(t))$, co-moving with - and time averaged over - an ergodic trajectory visiting all of phase space. Finally, the fourth equality follows from the definition of the Lyapunov exponents in Section 2, according to which the rate constant for the exponential time evolution of a volume element δV is given by the sum of all Lyapunov exponents. The quantity $-Nd\langle \zeta \rangle$ is equal to $\dot{Q}/k_B T$, where Q is interpreted as the outgoing heat extracted by the frictional forces characterized by a kinetic temperature, T. In equilibrium, $X = 0$, all these terms vanish; in nonequilibrium stationary states they are negative as is indicated by the rightmost inequality in Eq. (44). This follows from $\langle \zeta \rangle > 0$ as mentioned above.

The consequences of these simple, but general, relations are far-reaching [69, 4, 9]. From $\langle \dot{H}_0 \rangle = 0$ in Eq. (34) it follows for the Nosé-Hoover thermostat (and similarly for the Gauss case) that

$$\langle J(t) \rangle = -2 \langle \zeta(t) \rangle \langle K(t) \rangle / X \ , \tag{45}$$

where we have used that $\langle \zeta K \rangle = \langle \zeta \rangle \langle K \rangle$, since $\langle \zeta^2 \rangle$ is bounded and constant. In combination with the defining equation (30) we find that the transport coefficient,

$$L(X) = -\frac{k_B T \sum_{l=1}^{2dN+1} \lambda_l}{X^2} = \frac{dN k_B T \langle \zeta \rangle}{X^2} > 0, \tag{46}$$

is always positive. This is in agreement with the Second Law of Thermodynamics. Furthermore, Eq. (44) indicates that the Gibbs entropy diverges to

[8] Since we assume that the system is ergodic, we do not distinguish between ensemble averages and time averages in our notation.

$-\infty$ for long times. To better understand this strange result, it is instructive to consider in the next section a low-dimensional model, for which a visualisation of the phase space is possible.

In Eq. (46) the transport coefficient is also expressed in terms of the sum of all Lyapunov exponents. This provides another motivation to compute Lyapunov spectra for thermostated nonequilibrium systems. For a smooth interaction potential such a computation poses no further difficulties and proceeds exactly as outlined in Sect. 2. Examples are presented in Sect. 6 and Sect. 7. Hard-particle systems need a little more care and are treated in Sect. 8.

6 One-Dimensional Conductivity

In order to visualize what happens in phase space we study here a very simple and low-dimensional model [69, 70]. It consists of a charged particle of mass m moving along the q-axis under the influence of a periodic potential and a constant external field, X. This is an example of *mechanical* driving, for which the perturbed Hamiltonian exists,

$$H(q,p) = H_0 - cXq = \frac{p^2}{2m} + \epsilon(1 - \cos(q/q_0)) - cXq. \tag{47}$$

In the following we use reduced units, for which the energy, ϵ, the charge, c, the mass, m, the length, q_0, and Boltzmann's constant, k_B, are all unity. Using a Nosé-Hoover thermostat to maintain an average kinetic temperature $T = 1$, the equations of motion follow:

$$\begin{aligned} \dot{q} &= p \\ \dot{p} &= -\sin(q) + X - \zeta p \\ \dot{\zeta} &= \frac{1}{\tau^2}(p^2 - 1) \ . \end{aligned} \tag{48}$$

Apart from the thermostat term, $-\zeta p$, the model corresponds to a one-dimensional version of the solid-state Frenkel-Kontorova model for a single particle [70]. The phase space is periodic in the q direction with period 2π. For $\tau^2 = 10$ and a field $X = 0.3$, the system is chaotic, and a time-averaged current $\langle J(t)\rangle = -c\langle p\rangle/m$ develops according to Eq. (35). The Lyapunov spectrum, $\{0.0398, 0, -0.0857\}$, has a negative sum as expected. This means that any phase extension shrinks continuously. In Fig. 10 three Poincaré maps of the phase flow for the three planes defining the first quadrant of the phase space are shown. Surprisingly, the probability density is not a smooth and differentiable function but is a multifractal object with an information dimension [13] $D_1 = 2.42$, which is smaller than the phase-space dimension $L = 3$. $D_1 < L$ means that in the limit $t \to \infty$ the natural measure shrinks onto

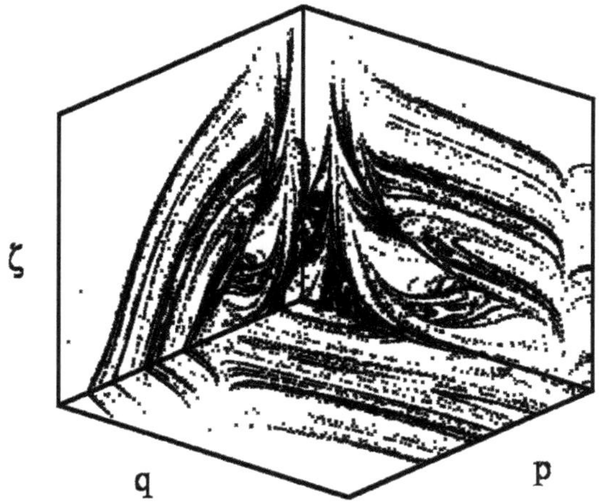

Fig. 10. Poincaré maps for three orthogonal planes defining the first quadrant of the phase space for the driven one-dimensional conductivity model. The external field $X = 0.3$, and $\tau^2 = 10$

a multifractal attractor which occupies a vanishing phase-space volume, and which is mapped onto itself by the evolution governed by the Liouville equation (43). The existence of this fractal attractor is a geometrical manifestation of the Second Law.

Other low-dimensional examples for which the fractal attractors may be visualized include the periodic Lorentz gas [74, 22, 3], for which the existence of the fractal attractor has theoretically been proven [75], and a doubly-thermostated harmonic oscillator [76, 77] with a space-dependent Nosé-Hoover thermostat.

7 Shear Flow in Two Dimensions

In the low-dimensional model of the previous section, D_1 can be easily computed by a box-counting method. However, for systems with $L > 10$ this becomes increasingly impractical [2]. It has been argued by Kaplan and Yorke [73, 13], that for most practical purposes the information dimension can be obtained from the Lyapunov spectrum according to

$$D_1 \approx J + \frac{\mu(J)}{|\lambda_{J+1}|}, \tag{49}$$

where the cumulative sum of the first J exponents, $\mu(J) \equiv \sum_{i=1}^{J} \lambda_i$ (with the exponents ordered according to $\lambda_i \geq \lambda_{i+1}$), changes sign between J and $J + 1$. This conjecture allows us to compute the information dimension also for many-particle systems and underlines the usefulness of the Lyapunov spectrum for statistical mechanics.

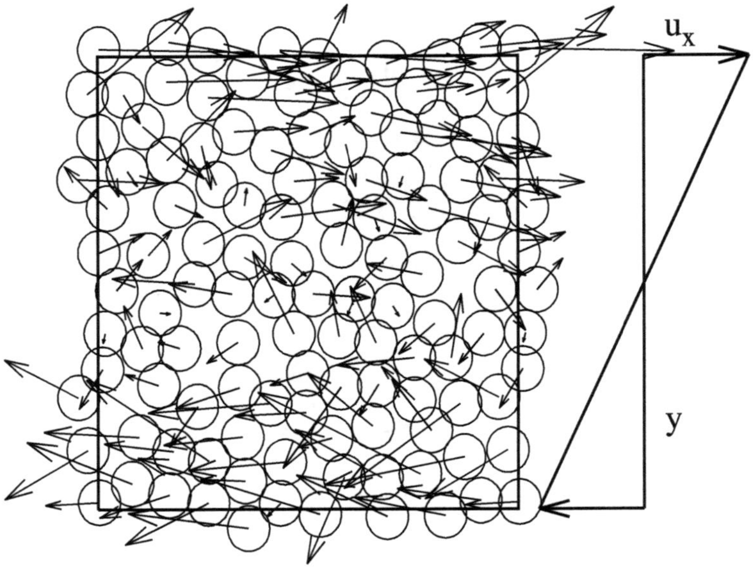

Fig. 11. Snapshot of a typical nonequilibrium steady-state configuration of a homogeneously sheared 100-particle SLLOD simulation at a density $N/V = 1$ and a strain rate $\dot{\epsilon} = 0.5$. The time-averaged velocity profile is shown on the right

We apply this method to a two-dimensional shear-flow problem [79, 78], which involves N particles in a square box of area V. The particles interact with a smooth pair potential,

$$\phi_{ij}(r) = 100\epsilon[1 - (r/\sigma)^2]^4 \quad r/\sigma < 1. \tag{50}$$

The box is aligned with the axes (see Fig. 11). The fluid is homogeneously sheared parallel to x by a *thermal* perturbation, X, which is identified with the strain rate $\dot{\epsilon} \equiv du_x/dy$. Here, $\mathbf{u}(x,y) = (u_x, u_y) \equiv (\dot{\epsilon}y, 0)$ is the hydrodynamic velocity field, which has a *linear* velocity profile relative to which the peculiar momenta $\mathbf{p}_i$ of the particles are defined. The so-called SLLOD equations of motion [6] for a particle become

$$\begin{aligned} \dot{x} &= p_x/m + \dot{\epsilon}y \\ \dot{y} &= p_y/m \\ \dot{p}_x &= F_x - \dot{\epsilon}p_y - \zeta p_x \\ \dot{p}_y &= F_y - \zeta p_y, \end{aligned}$$

and are of the form given by Eq. (31). $\mathbf{F} = (F_x, F_y)$ is the total force exerted on the particle by all the others. The dissipative flux belonging to the "force" $\dot{\epsilon}$ is given by [4, 6]

$$J(t) = VP_{yx} \equiv \sum_i \sum_{j>i} y_{ij} F_{x,ij} + \sum_i p_{x,i} p_{y,i}/m \, , \tag{51}$$

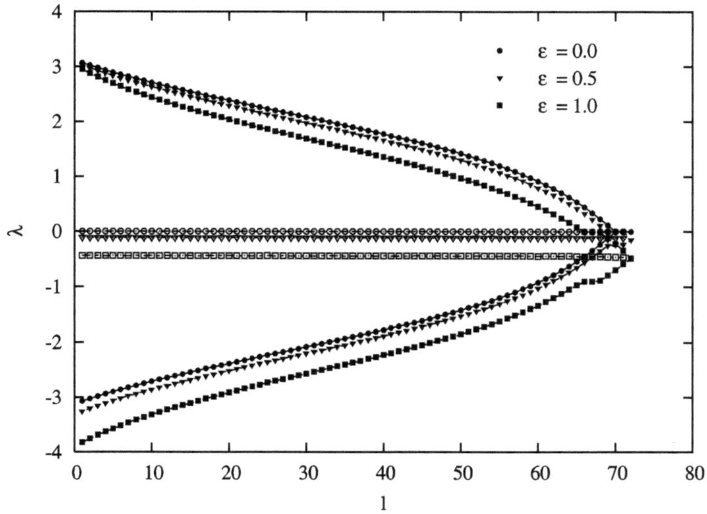

Fig. 12. Lyapunov spectra for two-dimensional shear-flow with strain rates $\dot{\epsilon}$ as indicated: $\dot{\epsilon} = 0.0$, *full circles*; 0.5, *full triangles*; 1.0, *full squares*. The respective open symbols denote the arithmetic means of the conjugate pairs of exponents. All 36 particles are ergostated

where P_{yx} is an off-diagonal element of the pressure tensor [80], and V is the box volume. Here, $y_{ij} = y_i - y_j$ and $F_{x,ij} = F_{x,i} - F_{x,j}$. For Gaussian isoenergetic control the ergostat variable ζ follows from Eq. (38):

$$\zeta = -\dot{\epsilon} V P_{yx}/2K .$$

The SLLOD equations must be complemented by proper periodic boundaries [80, 6] suggested by Lees and Edward [81]. Finally, the strain-rate dependent shear viscosity is given by $\eta(\dot{\epsilon}) = -\langle P_{xy} \rangle / \dot{\epsilon}$, and it reduces to its linear-response value for $\dot{\epsilon} \to 0$.

As an example we show in Fig. 12 Lyapunov spectra for various rates for 36 particles with a density $N/V = 0.2$. Reduced units are used for which m and the potential parameters in Eq. 50, σ and ϵ, are unity. We observe that in the nonequilibrium case the whole spectrum is shifted to more negative values and that the sum over all exponents becomes negative, as expected. This indicates the existence of a strange attractor in phase space and confirms our general results.

In Sect. 2.1 we have mentioned the conjugate-pairing rule for symplectic systems in thermodynamic equilibrium. An extension of this rule to nonequilibrium states [21] has been strictly proven for *isokinetically-thermostated* systems [84, 9]. According to this theorem, conjugate Lyapunov exponents sum to minus the time average of the isokinetic friction,

$$\lambda_l + \lambda_{L+1-l} = -\langle \zeta \rangle \, , \tag{52}$$

if the thermostat acts on all the particles in the system (homogeneous thermostat). Although no proof exists for *ergostated* systems, the spectra in Fig. 12 seem to obey this rule. This is demonstrated by the open symbols which represent the arithmetic means for conjugate exponents. The constant value obtained for all pairs of a spectrum agrees with the negative time-averaged friction.

A careful analysis [79, 78] of spectra such as those depicted in Fig. 12 has far-reaching consequences, which are generally true for the dynamically-thermostated many-body systems believed to be ergodic:

1) The information dimension of the fractal attractors may be computed from the Kaplan-Yorke formula (49). The dimensionality reduction $L - D_1$ for homogeneously-thermostated systems is proportional to X^2, the square of the external driving force.
2) $L - D_1$ is an extensive quantity and, hence, increases with system size [82]. Recently, very convincing arguments have been given which show that this extensivity persists also for nonhomogeneous systems far from equilibrium [24].
3) The systems remain ergodic for not-too-strong driving[9]. The support of the fractal measure is the full equilibrium phase space (where all constants of the motion are explicitly taken into account). Thus, the full equilibrium phase space is required to support the measure in the nonequilibrium steady state. The Hausdorff dimension, D_0, which is the dimension of the support of the measure [83, 13], agrees with the dimension of this space and is not fractal. One finds $D_0 > D_1$, and the attractor turns out to be a multifractal. This has been mathematically proven for the periodic Lorentz gas [75] mentioned above.[10]
4) The equations of motion are time-reversal invariant. A time-reversal transformation maps the Lyapunov-stable attractor states, for which $\sum_{l=1}^{L} \lambda_l < 0$, into Lyapunov-unstable repellor states, for which $\sum_{l=1}^{L} \lambda_0 > 0$. The repellor states, occupied in the distant past, act as a source of the phase flow in the same way as the attractor states act as a sink and are occupied in the distant future. This breaks the time-reversal symmetry for macroscopic processes which require an infinitely long ergodic trajectory that traces out all phase space. The flow always leads from the repellor to the attractor regardless of the direction of time [3, 4].

These general results hold regardless of the driving and the particular thermostating mechanism, as long as the equations of motion are time reversible and the system is ergodic. The system need not be homogeneous and may be

[9] For extremely strong external perturbations any system trajectory traces out a global limit cycle.

[10] For the one-dimensional conductivity model of Sec. 6 even the Hausdorff dimension is fractal and smaller than the phase-space dimension. $L > D_0 > D_1$ holds in this case, and this system is not ergodic.

controlled by more than one thermostat. We shall argue in Sect. 9 that even non-reversible stochastic thermostats can be treated to fit into this scheme.

8 Hard-Disk Systems in Nonequilibrium Steady States

Another popular transport process is color conductivity. The system consists of N particles of mass m, which carry positive and negative color charges $c_i = \pm 1$. Charge neutrality is assumed: $\sum_{i=1}^{N} c_i = 0$. The charges interact with a homogeneous and constant external field, $\mathbf{E}$, with an interaction energy $-\sum_{i=1}^{N} c_i(\mathbf{E}\cdot\mathbf{q}_i)$. Half of the particles are accelerated parallel, the other half antiparallel, to the field. The interaction between the particles is given by a pair potential but does not include charge-charge interactions. By this simplifying assumption, complications from long-range interactions are avoided. The driving force is purely mechanical and a Hamiltonian for the perturbed system can be written down (which, of course, does not include the thermostat). For smooth interaction potentials any kind of reversible thermostat may be used, and Lyapunov exponents have been computed [23, 79] for systems as large as 10^5 particles [45]. All the general conclusions of the previous section are confirmed by these results.

In Sect. 3 we have treated in some detail the computation of the Lyapunov spectra for hard-disk systems in two dimensions, and for hard spheres in three. For completeness, we indicate here what complications arise for the color-conductivity simulation of such particles [22] if a Gaussian thermostat is used.[11] We use the same notation as in Sect. 3, where the upper indices i and f always refer to states immediately before and after the collision, respectively, and where the lower indices j, k, l point to particles.

During the streaming period between the collisions the equations of motion become

$$\begin{aligned}\dot{\mathbf{q}}_j &= \mathbf{p}_j/m \\ \dot{\mathbf{p}}_j &= c_j\mathbf{E} - \zeta\mathbf{p}_j,\end{aligned} \tag{53}$$

where the friction

$$\zeta = \frac{\sum_{j=1}^{N}\left(c_j\mathbf{E}\cdot\mathbf{p}_j\right)}{\sum_{j=1}^{N}\mathbf{p}_j^2}. \tag{54}$$

For simplicity we do not thermostat the system with respect to the local streaming velocity of the two particle species [85], but keep the total kinetic

[11] For hard particles there is no difference between a thermostat and an ergostat, since hard particles do not have a potential energy and the kinetic energy agrees with the total energy.

energy in the laboratory frame constant. Thus, $\mathbf{p}_j$ is the total momentum of j and includes the systematic drift velocity parallel to $c_j\mathbf{E}$. This has no qualitative influence on the results [23]. Due to the thermostat the particles do not move on straight lines any more, and Eqs. (12) must be replaced by the numerical solutions of Eqs. (53). Similarly, the motion of the perturbation vectors in Eq. (13) follows from the solution of the linearized equations

$$\delta\dot{\mathbf{q}}_j = \delta\mathbf{p}_j/m, \tag{55}$$

$$\delta\dot{\mathbf{p}}_j = -\zeta\delta\mathbf{p}_j - \left(\frac{\sum_{k=1}^{N}(c_k\mathbf{E}-2\mathbf{p}_k\zeta)\cdot\delta\mathbf{p}_k}{\sum_{k=1}^{N}(\mathbf{p}_k\cdot\mathbf{p}_k)}\right)\mathbf{p}_j. \tag{56}$$

The collision map (14) remains unchanged. However, due to the friction and field terms in the motion equations (53) for streaming, rewritten as Eq. (1), additional terms appear in the linearized collision map (6). Let us assume that particles k and l collide. For the non-colliding particles, $j \neq k, l$, one finds

$$\delta\mathbf{q}_j^f = \delta\mathbf{q}_j^i\ , \tag{57}$$

$$\delta\mathbf{p}_j^f = \delta\mathbf{p}_j^i + \Delta\zeta\ \mathbf{p}_j^i\delta\tau_c\ , \tag{58}$$

where

$$\begin{aligned}\Delta\zeta &\equiv \zeta(\{\mathbf{p}_n^f, n=1,\ldots,N\}) - \zeta(\{\mathbf{p}_n^i, n=1,\ldots,N\} \\ &= -c(\mathbf{p}\cdot\mathbf{q})(\mathbf{E}\cdot\mathbf{q})/(\sigma^2\sum_{n=1}^{N}\mathbf{p}_n\cdot\mathbf{p}_n)\end{aligned} \tag{59}$$

is the change of ζ due to the collision of particles k and l. Since $c \equiv (c_l - c_k)$ denotes the charge difference of the colliding pair, $\Delta\zeta$ vanishes for collisions between equally charged particles. As before, the delay time, $\delta\tau_c$, is obtained from Eq. (22). For the colliding particles k and l the linearized map becomes

$$\delta\mathbf{q}_k^f = \delta\mathbf{q}_k^i + (\delta\mathbf{q}\cdot\mathbf{q})\,\mathbf{q}/\sigma^2, \tag{60}$$

$$\delta\mathbf{q}_l^f = \delta\mathbf{q}_l^i - (\delta\mathbf{q}\cdot\mathbf{q})\,\mathbf{q}/\sigma^2, \tag{61}$$

$$\begin{aligned}\delta\mathbf{p}_k^f &= \delta\mathbf{p}_k^i + (\delta\mathbf{p}\cdot\mathbf{q})\,\mathbf{q}/\sigma^2 + \frac{1}{\sigma^2}\left[(\mathbf{p}\cdot\delta\mathbf{q}_c)\,\mathbf{q} + (\mathbf{p}\cdot\mathbf{q})\,\delta\mathbf{q}_c\right] \\ &\quad + \left(\Delta\zeta\ \mathbf{p}_k^i + \frac{c}{\sigma^2}(\mathbf{E}\cdot\mathbf{q})\,\mathbf{q}\right)\delta\tau_c,\end{aligned} \tag{62}$$

$$\begin{aligned}\delta\mathbf{p}_l^f &= \delta\mathbf{p}_l^i - (\delta\mathbf{p}\cdot\mathbf{q})\,\mathbf{q}/\sigma^2 - \frac{1}{\sigma^2}\left[(\mathbf{p}\cdot\delta\mathbf{q}_c)\,\mathbf{q} + (\mathbf{p}\cdot\mathbf{q})\,\delta\mathbf{q}_c\right] \\ &\quad + \left(\Delta\zeta\ \mathbf{p}_l^i - \frac{c}{\sigma^2}(\mathbf{E}\cdot\mathbf{q})\,\mathbf{q}\right)\delta\tau_c.\end{aligned} \tag{63}$$

As in Sect. 3, the quantities $\mathbf{q} \equiv \mathbf{q}_l^i - \mathbf{q}_k^i$, $\mathbf{p} \equiv \mathbf{p}_l^i - \mathbf{p}_k^i$, $\delta\mathbf{q} \equiv \delta\mathbf{q}_l^i - \delta\mathbf{q}_k^i$, and $\delta\mathbf{p} \equiv \delta\mathbf{p}_l^i - \delta\mathbf{p}_k^i$ for the colliding particles are evaluated immediately before the collision.

Lyapunov spectra computed with these expressions [22] obey the conjugate pairing rule, as expected for isokinetic thermostats. It allows us, in principle although not advisable in practice, to determine transport coefficients from the knowledge of a single conjugate pair of exponents, usually the most positive and the most negative [21].

9 Beyond Time-Reversible Dynamical Thermostats

One important ingredient for the previous discussion is the use of dynamical thermostats such as the Gauss or Nosé-Hoover thermostats introduced in Sect. 5. They have very desirable properties, such as simplicity and time reversibility, and are well suited for stationary-state simulations far from thermal equilibrium. In equilibrium they trace out states belonging to particular ensembles such as Gibbs' microcanonical or canonical ensembles or one in between, and generalizations to the grand canonical ensemble are easily formulated. In this sense different dynamical thermostats are equivalent, as are the classical ensembles of equilibrium statistical mechanics. In the limit of weak external driving, the transport properties computed with the help of such thermostats changing "their help" to "the help of such thermostats"? converge to the linear-response result.

However, as elegant as these thermostats might be, there is one flaw: they cannot be built in the laboratory. Therefore they are considered by many physicists as an admittedly very practical tool for the study of nonequilibrium transport but with limited significance for laboratory physics. According to them, thermostats constitute an interesting class of models, but with weak connection to reality. Numbers such as transport coefficients are accepted, but some theoretical concepts such as the fractality of the phase-space probability are thought to have little consequence in general.

In the following we try to counter this view. What happens if the dynamical thermostats of Sect. 5 are replaced by *stochastic* thermostats representing a heat bath with infinitely many degrees of freedom? Näively one would expect that Lyapunov spectra cannot be defined at all in this case. However, if the heat flow through a system is controlled by a stochastic map, it is reasonable to assume that the reference trajectory and the perturbed trajectories are affected by the same sequence of random numbers. In addition, we *assume* that the rate of heat transfer from the system to the bath is still connected to a shrinkage of infinitesimal phase-volume elements co-moving with the flow. This allows us to construct a Lyapunov spectrum from stochastic equations of motion that has all the desirable properties, such as a fractal phase-space distribution, with connections to the Second Law.

This idea has been applied to heat conduction along a one-dimensional chain of particles with stochastic thermostats at each end [86], and, even more convincingly, to the color conductivity (see Fig. 13) of N heavy Brownian particles immersed in a heat bath and accelerated by an external color field

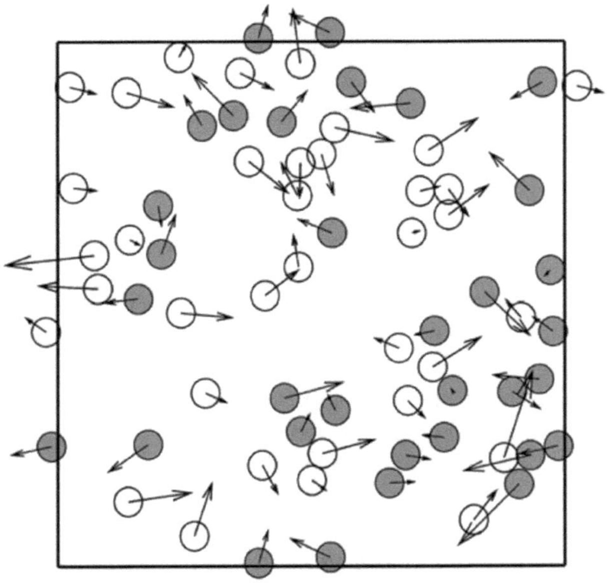

Fig. 13. Color conductivity of 64 Brownian particles: snapshot of a typical nonequilibrium steady-state configuration for a density $N/V = 0.2$ and a friction constant $\zeta = 0.5$. The color field $\mathbf{E} = (1, 0)$ points into the x direction

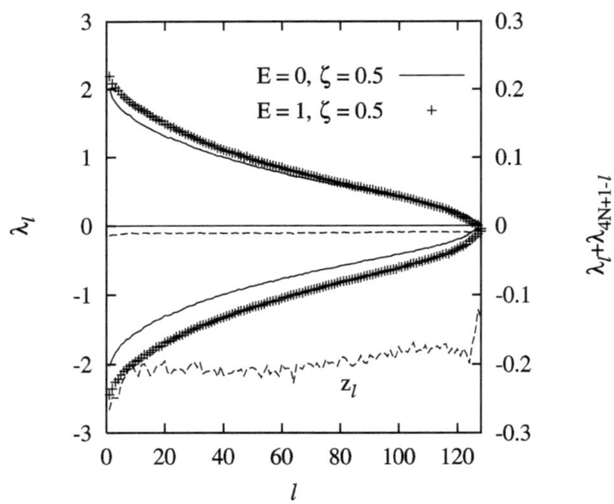

Fig. 14. Lyapunov spectra for 64 Brownian disks of unit mass m and particle density $N/V = 0.2$ immersed in a bath with friction constant $\zeta = 0.5$. E is the strength of the applied field. The arithmetic mean of conjugate Lyapunov exponents for $E = 1$ is shown by the dashed line parallel to the abscissa. Twice that value, $z_l \equiv \lambda_l + \lambda_{4N-l+1}$, is also shown on an expanded vertical scale on the right-hand side

analogous to Sect. 8. In this model the heavy Brownian particles of mass m interact with each other through a pair potential given by Eq. (50). They carry color charges, $c_i = \pm 1$, which couple to a constant external field $\mathbf{E}$. Their motion is assumed to follow a single-particle Langevin equation,

$$\begin{aligned} \dot{\mathbf{q}}_i &= \mathbf{p}_i/m, \\ \dot{\mathbf{p}}_i &= -\mathbf{F}_i(\{\mathbf{q}\}) + c_i\mathbf{E} - \zeta\mathbf{p}_i + \mathbf{A}_i(t), \end{aligned} \tag{64}$$

where $\mathbf{F}_i$ is the total force exerted on i from all the other Brownian particles. Total charge neutrality is assumed, $\sum_i c_i = 0$. The stochastic forces $\mathbf{A}_i(t)$ are δ-correlated Gaussian random variables and are connected to a *constant* friction ζ by the detailed-balance condition

$$\langle \mathbf{A}_i(0) \cdot \mathbf{A}_i(t) \rangle = 2d\zeta k T_0 \delta(t) \ . \tag{65}$$

Here, ζ is an input parameter. The rate of heat extracted by the bath turns out to be [87]

$$\dot{Q} = 2K\zeta - \sum_i \frac{\mathbf{p}_i}{m} \cdot \bar{\mathbf{A}}_i, \tag{66}$$

where $K = \sum_i \mathbf{p}_i^2/2m$ is the kinetic energy of the Brownian particles, and where $\bar{\mathbf{A}}_i \equiv \frac{1}{\Delta t}\int_t^{t+\Delta t} \mathbf{A}_i(\tau)d\tau$ is an average of the stochastic force over an interval Δt identified with a time step of the simulation. Its components are well-defined Gaussian random variables. If we *assume*, in analogy to Eq. (44), that $\langle d\ln \delta V/dt \rangle = -\langle \dot{Q} \rangle / k_B T = \sum_{l=1}^{L} \lambda_l$, it is possible to derive from Eq. (64) the linearized equations of motion for the offset vectors.

Using reduced units for which the mass, m, and the potential parameters, σ and ϵ, are unity, we show in Fig. 14 Lyapunov spectra for equilibrium ($E = 0$, smooth line) and for a field $E = 1$ (crosses). The particle density $N/V = 0.2$, and the friction constant $\zeta = 0.5$. The figure demonstrates that it is possible to construct a Lyapunov spectrum for a stochastically thermostated system which has the desired properties of Eq. (44). This lends support to the hope that the important results obtained for time-reversibly thermostated systems may be extended to models governed by nonreversible stochastic equations of motion. It is interesting to note that the conjugate-pairing rule is far from being obeyed in Fig. 14.

Acknowledgements

We are grateful to many colleagues and friends for the illuminating discussions and fruitful collaborations we have enjoyed over the years, and, in particular, to Professor William G. Hoover, co-founder of nonequilibrium molecular dynamics, with whom many of the topics that are addressed here have been investigated. This work has been assisted by financial support from the Austrian "Fonds zur Förderung der wissenschaftlichen Forschung" (FWF), most recently by the grants P11428 and P15348.

References

1. D. Ruelle, "Irreversibility Revisited", in *Highlights of Mathematical Physics*, edited by A. Fokas, J. Halliwell, T. Kibble, and B. Zegarlinski (American Mathematical Society, 2002), p.233.
2. H.A. Posch, W.G. Hoover, "Nonequilibrium molecular dynamics of classical fluids", in *Molecular Liquids: new perspectives in Physics and Chemistry*, edited by José J.C.Teixeira-Dias, NATO-ASI Series C: Mathematical & Physical Sciences (Kluwer Academic Publishers, Dordrecht, 1992), p.527.
3. H.A. Posch, Ch. Dellago, W.G. Hoover,O. Kum, "Microscopic Time-Reversibility and Macroscopic Irreversibility - Still a Paradox?", in *Pioneering Ideas for the Physical and Chemical Sciences: Josef Loschmidt's Contributions and Modern Developments in Structural Organic Chemistry, Atomistics, and Statistical Mechanics* W. Fleischhacker, T. Sch"onfeld, eds., p. 233 - 248, Plenum, New York, 1997.
4. W.G. Hoover, *Time Reversibility, Computer Simulation, and Chaos*, World Scientific, Singapore, 1999.
5. W.G. Hoover, *Computational Statistical Mechanics*, Elsevier, Amsterdam, 1991.
6. D.J. Evans, G.P. Morriss, *Statistical Mechanics of Nonequilibrium Liquids* (Academic Press, London, 1990).
7. P. Gaspard, *Chaos, Scattering and Statistical Mechanics*, Cambridge University Press, Cambridge, 1998.
8. J.R. Dorfman, *An Introduction to Chaos in Nonequilibrium Statistical Mechanics*, Cambridge University Press, Cambridge 1999.
9. D. Ruelle, "Smooth Dynamics and New Theoretical Ideas in Nonequilibrium Statistical Mechanics", J. Stat. Phys. **95**, 393-468 (1999).
10. D. Szasz, editor, *Hard Ball Systems and the Lorenz Gas*, Encyclopedia of the mathematical sciences **101**, Springer Verlag, Berlin (2000).
11. V.I. Oseledec,"A multiplicative ergodic theorem. Lyapunov characteristic numbers for dynamical systems", Trans. Moscow Math. Soc. **19**, 197 (1968).
12. J.-P. Eckmann, D. Ruelle,"Ergodic theory of chaos and strange attractors", Rev. of Modern Phys. **57**, 617 (1985).
13. E. Ott, *Chaos in dynamical systems*, Cambridge University Press, Cambridge, 1993.
14. G. Benettin, L. Galgani, A. Giorgilli, J.M. Strelcyn, "Théorie ergodic-Tous les nombres caractéristiques de Lyapunov sant effectivement caculables", C. R. Acad. Sci., Ser. A **286**, 431 (1978).
15. G. Benettin, L. Galgani, A. Giorgilli, J.M. Strelcyn, "Lyapunov Characteristic Exponents for Smooth Dynamical, Systems and for Hamiltonian Systems, A Method for Computing all of them", Meccanica **15**, 9 (1980).
16. I. Shimada, T. Nagashima, "A numerical approach to ergodic problem of dissipative dynamical systems", Progr. Theor. Phys. **61**, 1605 (1979).
17. A. Wolf, J.B. Swift, H.L. Swinney, J.A. Vastano, " Determining Lyapunov exponents from a time series", Physica **16** D, 285 (1985).
18. R. Illner, H. Neunzert, "The concept of irreversibility and statistical physics", Transport Theory and Statistical Physics **16**(1), 89 (1987).
19. J.A.G. Roberts, G.R.W. Quispel, "Chaos and time-reversal symmetry. Order and chaos in reversible dynamical systems", Physics Reports **216**, 63 (1992).

20. V.I. Arnold, *Mathematical methods of classical mechanics* (Springer, Berlin, 1989).
21. D.J. Evans, E.G.D. Cohen, G.P. Morriss, "Viscosity of a fluid from its maximal Lyapunov exponents", Phys. Rev. A **42**, 5990 (1990).
22. C. Dellago, H.A. Posch, W.G.Hoover, "Lyapunov instability of hard disks in equilibrium and nonequilibrium steady states", Phys. Rev. E **53**, 1485 (1996).
23. H.A. Posch, W.G. Hoover, "Lyapunov instability of dense Lennard-Jones fluids", Phys. Rev. A **38**, 473 (1988).
24. H.A. Posch, W.G. Hoover, "Large-system phase-space dimensionality loss in stationary heat flows", Physica D **187**, 281 (2004).
25. J.D. Bernal, S.V. King, in *Physics of Simple Liquids*, H. N. V. Temperley, J.S. Rowlinson, G.S. Rushbrooke eds., page 231, (North-Holland, Amsterdam, 1968).
26. B.J. Alder, T.E. Wainwright, "Molecules in Motion", Sci. Am. **201**(4), 113 (1959).
27. T. Einwohner, B.J. Alder, "Molecular Dynamics VI. Free-Path Distributions and Collision Rates for Hard-Sphere and Square-Well Molecules", J. Chem. Phys. **49**, 1458 (1968).
28. J.-P. Hansen, I.R. McDonald, *Theory of simple liquids*, (Academic Press, London, 1991).
29. T.M. Reed, K.E. Gubbins, *Applied Statistical Mechanics*, (McGraw Hill, Tokyo, 1973).
30. C. Dellago, H.A. Posch, "Kolmogorov-Sinai entropy and Lyapunov spectra of a hard sphere gas", Physica A, **240**, 68 (1997).
31. H.A. Posch, R. Hirschl, "Simulation of Billiards and of Hard-Body Fluids", pages 269 - 310 in *Hard Ball Systems and the Lorenz Gas*, edited by D. Szasz, Encyclopedia of the mathematical sciences **101**, Springer Verlag, Berlin (2000).
32. W.G. Hoover, H.A. Posch, C. Forster, C. Dellago, M. Zhou, "Lyapunov instability of two-dimensional many-body systems; soft disks, hard disks, and rotors", J. Stat. Phys. **109**, Nos. 3/4, 765 (2002).
33. C. Forster, R. Hirschl, H.A. Posch, W.G. Hoover, "Perturbed phase-space dynamics of hard-disk fluids", Physica D **187**, 294 (2004).
34. C. Forster, R. Hirschl, H.A. Posch, "Analysis of Lyapunov modes for hard-disk fluids", Proceedings for the Math.-Physics conference, Lissabon, July 2003.
35. C. Dellago, H.A. Posch, "Mixing, Lyapunov instability, and the approach to equilibrium in a hard-sphere gas", Phys. Rev. E **55**, R9 (1997).
36. Lj. Milanović, H.A. Posch, "Localized and delocalized modes in the tangent space dynamics of planar dumbbell fluids", J. Molec. Liquids, **96-97**, 221 - 244 (2002).
37. Y.G. Sinai, "A remark concerning the thermodynamical limit of the Lyapunov spectrum", Int. J. of Bifurcation and Chaos Appl. Sci. Eng. **6**, 1137 (1996).
38. D.J. Searles, D.J. Evans, D.J. Isbister, "The number dependence of the maximum Lyapunov exponent", Physica **240A**, 96 (1997).
39. Y.B. Pesin, "Lyapunov characteristic exponents and ergodic properties of smooth dynamical systems with an invariant measure", Sov. Math. Dokl. **17**, 196 (1976).
40. C. Dellago, H.A. Posch, "Lyapunov instability, local curvature and the fluid-solid phase transition in two dimensions", Physica A, **230**, 364 (1996).
41. H. van Beijeren, J.R. Dorfman, H.A. Posch, C. Dellago, "The Kolmogorov-Sinai entropy for dilute gases in equilibrium", Phys. Rev. E **56**, 5272 (1997).

42. J.R. Dorfman, A. Latz, H. van Beijeren, "B.B.G.K.Y. Hierarchy Methods for Sums of Lyapunov Exponents for Dilute Gases", preprint.
43. R. van Zon, H. van Beijeren, C. Dellago, "Largest Lyapunov Exponent for Many Particle Systems at Low Densities", Phys. Rev. Lett. **80**, 2035 (1998).
44. R. van Zon, H. van Beijeren, J.R. Dorfman, "Kinetic theory estimates for the Kolmogorov-Sinai entropy, and the largest Lyapunov exponents for dilute, hard ball gases and for dilute, random Lorentz gases", in *Hard Ball Systems and the Lorenz Gas*, edited by D. Szasz, Encyclopedia of the mathematical sciences **101**, Springer Verlag, Berlin (2000).
45. W.G. Hoover, K. Boercker, H.A. Posch, "Large-system hydrodynamic limit for color conductivity in two dimensions", Phys. Rev. E **57**, 3911 (1998).
46. L. Milanović, H.A. Posch, W.G. Hoover, "What is 'Liquid'? Understanding the States of Matter", Molec. Phys., **95**, 281 (1998).
47. J.-P. Eckmann, C. Forster, H.A. Posch,E. Zabey, "Lyapunov modes in hard-disk systems", J. Stat. Phys., submitted (2004); arXiv: nlin.CD/0404007.
48. T. Taniguchi, G. Morriss, "Time-oscillating Lyapunov modes and autocorrelation functions for quasi-one-dimensional systems", preprint; arXiv: nlin.CD/0404052.
49. H.A. Posch, W.G. Hoover, "Equilibrium and Nonequilibrium Lyapunov Spectra for Dense Fluids and Solids", Phys. Rev. A **39**, 2175 - 2188, (1989).
50. E. Wilhelm, "Thermodynamic properties of a hard disk fluid with temperature dependent effective hard sphere diameter", J. Chem. Phys. *60*, 3896 (1974).
51. C. Dellago, W.G. Hoover, H.A. Posch, "Fluctuations, convergence times, correlation functions, and power laws from many-body Lyapunov spectra for soft and hard disks and spheres", Phys. Ref. E **65**, 056216 (2002).
52. C. Forster, H.A. Posch, "Lyapunov modes in soft-disk systems", preprint.
53. G. Radons, H. Yang, "Static and dynamic correlations in many-particle Pyapunov vectors", preprint; arXiv: nlin.CD/0404028.
54. H. Yang, G. Radons, "Lyapunov instabilities of Lennard-Jones fluids", preprint; arXiv: nlin.CD/0404027.
55. M.S. Green, " Markoff Random Processes and the Statistical Mechanics of Time-Dependent Phenomena", J. Chem. Phys. **20**, 1281 (1952); "Markoff Random Processes and the Statistical Mechanics of Time-Dependent Phenomena. II Irreversible Processes in Fluids", **22**, 398 (1954).
56. R. Kubo, "Statistical-Mechanical Theory of Irreversible Processes I. General Theory and Simple Applications to Magnetic and Conduction Problems", J. Phys. Soc. (Jpn.) **12**, 570 (1957).
57. R. Zwanzig, "Time-Correlation Functions and Transport Coefficients in Statistical Mechanics", Ann. Rev. Phys. Chem. **16**, 67 (1965).
58. B.J. Alder, T.E. Wainwright, "Decay of the Velocity Autocorrelation Function", Phys. Rev. A **1**, 18 (1970).
59. B.J. Alder, E.E. Alley, in *Perspectives in Statistical Physics*, edited by H. J. Raveche (North-Holland, Amsterdam, 1981), p. 3.
60. W. Loose, G. Ciccotti, "Temperature and temperature control in nonequilibrium-molecular-dynamics simulations of the shear flow of dense liquids", Phys. Rev. A **45**, 3859 (1992).
61. W. Loose, S. Hess, "Anisotropy in velocity space induced by transport processes", Physica A **174**, 47 (1991).
62. K.F. Gauss, "Über ein neues allgemeines Grundgesetz der Mechanik", Crelles Journ. f. Math. 4 (1829); gesammelte Werke **5**, p. 23.

63. A. Sommerfeld, *Vorlesungen über Theoretische Physik, Band 1: Mechanik*, Akademische Verlagsgesellschaft, Leipzig, 1962.
64. S. Nosé, "A molecular dynamics method for simulations in the canonical ensemble", Molec. Phys. **52**, 255 (1984).
65. S. Nosé, "A unified formulation of the constant temperature molecular dynamics methods", J. Chem. Phys. **81**, 511 (1984).
66. W.G. Hoover, "Canonical dynamics: Equilibrium phase-space distributions", Phys. Rev. A **31**, 1695 (1985).
67. H.A. Posch, W.G. Hoover, F.J. Vesely, "Canonical dynamics of Nosé oscillator: Stability, order, and chaos", Phys. Rev. A **33**, 4253 (1986).
68. B.L. Holian, "The character of the nonequilibrium steady state: Beautiful formalism meets ugly reality," in*Monte Carlo and Molecular Dynamics of Condensed Matter Systems*, K. Binder and G. Ciccotti, eds., Vol. 49 (Proceedings of the Euroconference on Computer Simulation in Condensed Matter Physics and Chemistry, Italian Physical Society, Bologna, 1996), p.791.
69. B.L. Holian, W.G. Hoover, H.A. Posch, "Resolution of Loschmidt's paradox: The origin of irreversible behavior in reversible atomistic dynamics", Phys. Rev. Lett. **59**, 10 (1987).
70. W.G. Hoover, H.A. Posch, B.L. Holian, M.J. Gillan, M. Mareschal, C. Massobrio, "Dissipative irreversibility from Nosé's reversible mechanics", Molec. Simulation **1**, 79 (1997).
71. J. Ramshaw, "Remarks on Non-Hamiltonian Statistical Mechanics", Europhys. Lett. **59**, 319-323 (2002).
72. B.L. Holian, "Entropy evolution as a guide for replacing the Liouville equation", Phys. Rev. A **34**, 4238 (1986).
73. J. Kaplan, J. Yorke, in *Functional Differential Equations and the Approximation of Fixed Points*, Vol. 730 of Lecture Notes in Mathematics, eds.: H.O. Peitgen, H.O. Walther (Springer-Verlag, Berlin, 1980).
74. B. Moran, W.G. Hoover, "Diffusion in a Periodic Lorentz Gas", J. Stat. Phys. **48**, 709 (1987).
75. N.I. Chernov, G.L. Eyink, J.L. Lebowitz, Y.G. Sinai, "Derivation of Ohm‘s Law in a Deterministic Mechanical Model", Phys. Rev. Lett. **70**, 2209 (1993).
76. H.A. Posch, unpublished.
77. W.G. Hoover, H.A. Posch, C.G. Hoover, "Fluctuations and Asymmetry via Local Lyapunov Instability in the Time-Reversible Doubly-Thermostated Harmonic Oscillator", J. Chem. Phys. **115**, 5744 - 5750 (2001).
78. W.G. Hoover, H.A. Posch, "Shear Viscosity *via* Global Control of Spatio-Temporal Chaos in Two-Dimensional Isoenergetic Dense Fluids, Phys. Rev. E **51**, 273 - 279 (1995).
79. W.G. Hoover, H.A. Posch, "Second-Law Irreversibility, and Phase-Space Dimensionality Loss, from Time-Reversible Nonequilibrium Steady-State Lyapunov Spectra", Phys. Rev. E, **49**, 1913 - 1920 (1994).
80. M.P. Allen, D.J. Tildesley, *Computer Simulations of Liquids* (Clarendon Press, Oxford, 1987).
81. A.W. Lees, S.F. Edwards, "The computer study of transport processes under extreme conditions", J. Phys. **C5**, 1921 (1972).
82. K. Aoki, D. Kusnezov, "Lyapunov Exponents, Transport, and the Extensivity of Dimensionality Loss", nlin.CD/0204015.
83. J.D. Farmer, E. Ott, J.A. Yorke, "The Dimension of Chaotic Attractors", Physica **7D**, 153-180 (1983).

84. C.P. Dettmann, G.P. Morriss, "Proof of Lyapunov pairing for systems at constant kinetic energy", Phys. Rev. E **53**, R5541 (1996).
85. S. Sarman, D.J. Evans, G.P. Morriss, "Conjugate-pairing rule and thermal transport coefficient", Phys. Rev. A **45**, 2233 (1992).
86. H.A. Posch, W.G. Hoover, "Heat conduction in one-dimensional chains and nonequilibrium Lyapunov spectrum", Phys. Rev. E, **58**, 4344 - 4350 (1998).
87. H.A. Posch, R. Hirschl, W.G. Hoover, "Multifractal phase-space distributions for stationary nonequilibrium systems", in *Dynamics: Models and Kinetic Methods for Nonequilibrium Many-Body Systems*, J. Karkheck, ed., NATO ASI Series E: Applied Sciences - Vol. 371, p. 169 - 189 (Kluwer, Dordrecht, 2000).

On Symbolic Dynamics of Space–Time Chaotic Models

Wolfram Just

School of Mathematical Sciences, Queen Mary/University of London,
Mile End Road, London E1 4NS, UK
w.just@qmul.ac.uk

1 Introduction and Overview

The foundations of Statistical Mechanics rest to some extent on dynamical theories, as far as questions like relaxation and the approach to thermodynamic equilibrium as well as the justification of canonical distributions are concerned. Thus it is obvious that strong links between Statistical Physics and Nonlinear Dynamics exist. While from the traditional perspective the Hamiltonian character of the microscopic equations of motion and the thermodynamic limit play a crucial role we will sketch in this tutorial that the links between both fields extend to more general dynamical systems, e.g. even systems with dissipation or with few degrees of freedom.

During the last decades Nonlinear Dynamics and chaotic motion has attracted much interest from the principle and the experimental point of view and a huge amount of textbooks addressing different aspects are available (e.g. [1, 2, 3]). Contrary to the integrable case an individual trajectory is not a relevant object for chaotic motion since sensitivity with respect to the initial condition prevents in reality the reproduction of a particular trajectory. Thus it is obvious that statistical approaches become relevant. In particular, concepts developed in Statistical Physics for the treatment of many particle systems can be applied fruitfully to understand the motion in general dynamical systems. It is the scope of the present chapter to demonstrate such a link between Nonlinear Dynamics and (equilibrium) Statistical Physics with very elementary tools. The tutorial presented here just requires material which is usually contained in undergraduate physics or mathematics courses. Of course, much more general and sometimes quite elaborate presentations can be found in the literature (e.g. [4, 5, 6]) and the reader may find such references useful for further studies.

For our purpose we consider the simplest type of dynamical system, i.e. time discrete maps. Such a choice is quite common in Nonlinear Dynamics as long as principle and universal features are concerned in order to avoid all the technicalities which are related with real physical models. As a particular benefit we will be able to perform all our calculations explicitly by analytical means without imposing any approximation[1].

All concepts which establish a link between dynamics and statistics rely on a coarse grained description of the motion. Such approaches which are summarised in the term symbolic dynamics use suitable partitions of the full phase space. The trajectories are labelled by symbol sequences which tell in which order the different parts of the phase space are visited. It is one of the key observations that such a coarse graining does not reduce the amount of information provided one chooses a suitable partition. We will illustrate this fact in detail in Sect. 2. The power of such a method goes in fact far beyond the simple examples we are addressing here (cf. e.g. [7, 8]) and one may use such concepts to investigate the topological complexity of the orbits in quite different settings.

An important characteristic of a dynamical system is given by the relative probability how often a "typical" trajectory visits certain parts of the phase space. Knowing such probabilities one may express time averages by phase space averages. In a more specific sense these questions are addressed by ergodic theory (cf. [9, 10, 6]). For the examples studied here we are able to answer such a question, i.e. closed formulas for the probabilities can be written down. Thus one can associate a Markov model with the deterministic equations of motion which describe the full motion from the statistical point of view. In some respect such a result provides the simplest explanation how one can relate deterministic dynamics with stochastic equations of motion and how irreversibility can be derived from time reversible dynamics [11]. In addition, the probabilistic approach tells us how frequently a finite symbol string appears in the symbolic dynamics mentioned above. Such a probability can be formally written as a Boltzmann weight of a spin chain and thus all dynamical mean values are rewritten in terms of a canonical equilibrium model. We will develop the details of such a statistical mechanics of dynamical systems in Sect. 3. Up to that stage the correspondence between dynamical systems and Statistical Physics is more a formal trick and gives no essential new insight apart from the fact that the dependence of mean values on system parameters can be understood on a quite general level.

Qualitatively new phenomena show up when we consider dynamical models with many degrees of freedom, e.g. spatially extended dynamical systems. It is already a severe problem to carry over the notion of temporal chaos to the spatio–temporal case and to establish the concept of spatio–temporal chaos on some rigorous basis. Lacking a satisfactory definition which takes the spatial character of the degrees of freedom appropriately into account one may understand spatio–temporal chaos as states where spatio–temporal correlations decay exponentially [12, 13, 14]. There are alternative concepts available in the literature for characterising spatio–temporal chaos [15, 16, 17, 18], but

[1] Our presentation is formally exact but not rigorous, since we skip all the mathematically delicate questions.

most of them depend on some details of the underlying system. We will illustrate in Sect. 4 how the above mentioned statistical description will shed some light on such a situation and how one can understand that systems of infinite extent behave differently compared to systems with a finite number of degrees of freedom. On a phenomenological level such a difference is already visible by considering the transient dynamics of dynamical models. Frequently one observes transient times which increase exponentially with the system size [19, 20] so that the model with a large number of degrees of freedom may never reach its stationary state. Our considerations will show that such a behaviour may be understood as a proper equilibrium phase transition appearing on the level of symbolic dynamics in the corresponding equilibrium spin system. Thus such a type of instability is intimately related to the limit of infinite system size and may be termed a phase transition in the proper sense (cf. [21]).

2 Piecewise Linear Markov Maps

We will illustrate the power of symbolic dynamics with a simple example, a time discrete dynamical system. In order to keep our discussion elementary and entirely analytic we restrict ourselves to one–dimensional maps and investigate the motion of the map depicted in Fig. 1. We do not need the corresponding analytical expression but of course it can be easily written down as

$$x_{n+1} = T(x_n) = \begin{cases} 2x + 1/3 & \text{if } 0 \le x \le 1/3 \\ -3/2(x-1) & \text{if } 1/3 < x \le 1 \end{cases} \quad . \tag{1}$$

The domain of the map splits in a natural way in two parts U_σ, $\sigma \in \{0, 1\}$, where $U_0 = [0, 1/3]$ and $U_1 = [1/3, 1]$ denote (closed) intervals. Roughly speaking we may associate to any initial condition x_0 an infinite sequence of symbols $(\sigma_0\sigma_1\sigma_2\ldots)$ where σ_n tells us the interval which contains the trajectory at time n, $x_n \in U_{\sigma_n}$. Such a construction can be performed regardless of the particular partition $\{U_\sigma\}$ provided we discard for a moment the endpoints of the intervals where the symbols are not uniquely assigned[2]. The representation of x_0 in terms of its symbol sequence $(\sigma_0\sigma_1\sigma_2\ldots)$ has a nice property with respect to the dynamics. Since $(\sigma_n\sigma_{n+1}\ldots)$ and $(\sigma_{n+1}\sigma_{n+2}\ldots)$ are the sequences corresponding to the phase space points x_n and x_{n+1} respectively, the dynamics induced by the map, $x_{n+1} = T(x_n)$ just boils down to a simple shift in symbol sequences. Without further precaution it is of course completely unclear whether the prescription $x \mapsto (\sigma_0\sigma_1\ldots)$ is one to one and whether we loose some information when we describe the motion in terms of symbol sequences.

[2] Such a lack of uniqueness is also well known from the binary representation of real numbers.

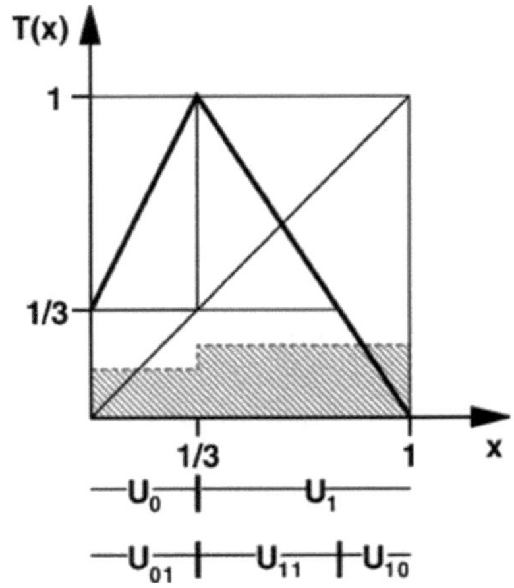

Fig. 1. Piecewise linear Markov map $T : [0,1] \to [0,1]$ (thick line). The Markov partition $\{U_0, U_1\}$ and the first generation of cylinder sets $U_{\sigma\tilde{\sigma}}$ is indicated (cf. eq. (4)). *Dashed line* and *hatched area* represent the invariant density (grey) (cf. eq. (8))

To obtain the uniqueness of the labelling one needs special properties of the partition $\{U_\sigma\}$. From figure 1 it is obvious that T maps U_σ onto (a union of) $U_{\tilde{\sigma}}$ in a monotonic way. We can summarise such a behaviour by

$$T: \quad \begin{array}{l} U_0 \to U_1 \\ U_1 \to U_0 \cup U_1 \end{array} \quad . \tag{2}$$

Thus (when we disregard endpoints) transitions from U_0 to U_1 and transitions from U_1 to U_0 and U_1 are permitted whereas the transition from U_0 to U_0 is prohibited. We may condense these relations in a transition matrix $A_{\sigma\tilde{\sigma}}$ such that the matrix element is one when the transition $U_\sigma \to U_{\tilde{\sigma}}$ is permitted and zero otherwise. For our particular example the transition matrix reads

$$\underline{\underline{A}} = \begin{pmatrix} 0 & 1 \\ 1 & 1 \end{pmatrix} \quad . \tag{3}$$

If we consider symbol sequences $(\sigma_0\sigma_1\ldots)$ which contain only permitted nearest neighbour transitions $\sigma_n \to \sigma_{n+1}$ then we are indeed able to show that our construction $x \mapsto (\sigma_0\sigma_1\ldots)$ can be inverted, i.e. there is only one initial condition x_0 that generates the symbol sequence $(\sigma_0\sigma_1\ldots)$. To achieve such a goal we introduce the so called cylinder sets. Consider a finite symbol string $\sigma_0\sigma_1\ldots\sigma_{n-1}$ of n symbols which contains no forbidden transition. Then the set

$$U_{\sigma_0\ldots\sigma_{n-1}} := \{x \,|\, T^k(x) \in U_{\sigma_k},\, 0 \le k \le n-1\} \tag{4}$$

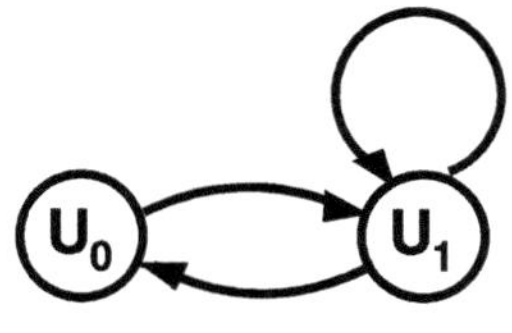

Fig. 2. Directed graph which generates the topological Markov chain of the dynamical system depicted in figure 1 (cf. eq. (2))

contains all initial conditions of trajectories which travel through the intervals U_{σ_k} during the finite time span $0 \leq k \leq n-1$. It is easy to check that eq. (4) yields a finite interval [3]. Obviously the cylinder sets obey the dynamical rule

$$T(U_{\sigma_0 \ldots \sigma_{n-1}}) = U_{\sigma_1 \ldots \sigma_{n-1}} \quad . \tag{5}$$

Since the map T is linear and in particular expansive on U_σ, i.e. has slope of modulus larger than one, the intervals (4) decrease exponentially in size if we consider longer symbol strings. In addition, $U_{\sigma_0 \ldots \sigma_n} \subseteq U_{\sigma_0 \ldots \sigma_{n-1}}$ holds. Thus eq. (4) yields a sequence of nested intervals which single out one phase space point x in the limit $n \to \infty$. That particular initial condition generates by construction the infinite symbol sequence $(\sigma_0\sigma_1 \ldots)$. Hence we can invert our construction $x \mapsto (\sigma_0\sigma_1 \ldots)$ as long as we confine ourself to admissible symbol sequences[4]. In particular, no information is lost when we change from the phase space to the coarse grained description in terms of symbol sequences.

We have developed a complete topological description of the dynamics of eq. (1) in terms of symbol sequences and the shift operation. The symbolic dynamics is completely determined by the transition matrix (3), i.e. by nearest neighbour transitions. Thus the selection rule for admissible symbol sequences, the so called grammar, is quite simple. Since just a nearest neighbour memory is involved we have obtained a (topological) Markov chain or a subshift of finite type. There is a simple way to represent such a process. Admissible symbol sequences may be generated by following the paths in the finite directed graph depicted in Fig. 2. Each path on that graph generates a symbol sequence according to the sites which are visited by the path. Such paths represent the symbol chain as well as the dynamics of the underlying deterministic map.

The nice description of the dynamics in terms of symbol sequences rests on the proper choice of the partition. As the previous example illustrates there are two essential constraints which the partition $\{U_\sigma\}$ has to fulfil. The map

---

[3] For forbidden symbol strings the set is either empty or consists just of isolated points.

[4] $(\sigma_0\sigma_1 \ldots) \mapsto x$ is not injective. That property reflects the lack of uniqueness for phase space points which are mapped on the boundaries of U_σ. Thus, one has to match certain symbol sequences like for the binary representation of real numbers.

T must be expansive on each set U_σ and it must map boundary points of U_σ on boundary points. Expansiveness guarantees that the description in terms of symbol sequences becomes one to one. The condition on the boundary points ensures that we end up with the simple Markov property. Therefore one calls partitions which obey both constraints Markov partitions. If one of these constraints is relaxed the situation may become much more elaborate. If the constraint on the boundary points is dropped one still might have a generating partition [9] but the rules for the admissible symbol sequences may contain a long memory. If the expansiveness is violated then the uniqueness between state phase and symbol space may be lost but it is still possible to obtain useful information about the dynamics [7]. Here we just stick to the setting with the Markov property.

So far we have discussed the topological properties of the dynamics, i.e. how to characterise trajectories of the system. We have not addressed the problem "how frequently" a particular symbol sequence occurs. Such a topic is part of ergodic theory (cf. [9]). We do not give a full account here but just describe some basic features in elementary terms. Consider a "typical" orbit (i.e. an orbit emerging from initial conditions of a suitable set of finite Lebesgue measure). Fix a finite symbol string $\sigma_0 \dots \sigma_{n-1}$ and compute the relative frequency how often the string occurs in the symbol sequence which is generated by the orbit. The relative frequency coincides with the probability that the orbit visits the cylinder set (4). If we denote this probability by $\mu(U_{\sigma_0 \dots \sigma_{n-1}})$, observe the dynamical law (5), and require the stationarity of the probabilities then the balance condition of the probabilities in each time step tells us that

$$\mu(U_{\sigma_1 \dots \sigma_{n-1}}) = {\sum_{\sigma_0}}' \mu(U_{\sigma_0 \dots \sigma_{n-1}}) = \sum_{\sigma_0} A_{\sigma_0 \sigma_1} \mu(U_{\sigma_0 \dots \sigma_{n-1}}) \quad . \tag{6}$$

Here $\sum'$ denotes the sum with respect to all admissible symbol sequences. Of course it can be rewritten using the transition matrix (3). There is a second trivial constraint the probabilities have to obey. Recalling the definition of cylinder sets (4) it is obvious that skipping the last symbol we obtain cylinder sets of the previous generation, i.e. $\cup'_{\sigma_{n-1}} U_{\sigma_0 \dots \sigma_{n-1}} = U_{\sigma_0 \dots \sigma_{n-2}}$. Thus the sum rule

$$\mu(U_{\sigma_0 \dots \sigma_{n-2}}) = {\sum_{\sigma_{n-1}}}' \mu(U_{\sigma_0 \dots \sigma_{n-1}}) = \sum_{\sigma_{n-1}} \mu(U_{\sigma_0 \dots \sigma_{n-1}}) A_{\sigma_{n-2} \sigma_{n-1}} \tag{7}$$

has to be fulfilled.

Equation (6) is nothing else but the famous Frobenius–Perron equation (cf. e.g. [10]). It admits many different solutions. We are here interested in typical orbits and in that case the probabilities scale with the local expansion rates which the orbit points visit through its itinerary. If $\gamma_\sigma := |T'(x)|$, $x \in U_\sigma$

denotes the modulus of the slope of the map on U_σ we make the ansatz[5]

$$\mu(U_{\sigma_0\ldots\sigma_{n-1}}) = h_{\sigma_0}\frac{1}{\gamma_{\sigma_0}} \times \ldots \times \frac{1}{\gamma_{\sigma_{n-2}}}\nu_{\sigma_{n-1}} \quad . \tag{8}$$

The reason for such an ansatz is quite simple (although the mathematical proof has been a challenge from the rigorous point of view). Since the map expands on each time step by a factor γ_{σ_k} the geometric length of the cylinder set (4) scales as $(\gamma_{\sigma_0}\ldots\gamma_{\sigma_{n-1}})^{-1}$. Thus the probabilities according to eq. (8) yield a continuous probability density[6]. In the first and the last symbol the expression differs from local expansion rates in order to take care of the two conditions (6) and (7). In fact, plugging the ansatz (8) into eq. (7) we just get

$$\nu_\sigma = \sum_{\tilde{\sigma}} \frac{A_{\sigma\tilde{\sigma}}}{\gamma_\sigma}\nu_{\tilde{\sigma}} \tag{9}$$

and the matrix appearing on the right hand side is usually called the transfer matrix of the system. The other boundary contribution is obtained from eq. (6) as

$$h_\sigma = \sum_{\tilde{\sigma}} h_{\tilde{\sigma}}\frac{A_{\tilde{\sigma}\sigma}}{\gamma_{\tilde{\sigma}}} \tag{10}$$

which is just the adjoint equation of the expression (9). Finally normalisation of the probabilities results in

$$1 = \sum_\sigma \mu(U_\sigma) = \sum_\sigma h_\sigma \nu_\sigma \quad . \tag{11}$$

Since the probabilities obey by construction the condition of stationarity (cf. eq. (6)) the corresponding probability density is usually called the invariant measure (actually the SRB measure). For our particular example eq. (1) we obviously have $\gamma_0 = 2$, $\gamma_1 = 3/2$, $(\nu_0, \nu_1) = (1/4, 1/2)$, and $(h_0, h_1) = (1, 3/2)$. Thus $\mu(U_0) = 1/4$, $\mu(U_1) = 3/4$ and the corresponding probability density (cf. figure 1) is constant on U_σ.

With the probabilities (8) we are able to compute average values of phase space functions. According to ergodic theory these averages may also be understood as long time averages. For analytical computations evaluation of phase space averages using eq. (8) is of course simpler. In order to avoid some technicalities we restrict to phase space functions $h(x)$ which are constant on cylinder sets (4). The reader may consult e.g. [23] for the treatment

[5] Other measures, which can be obtained by different scaling factors, correspond to non typical orbits. These measures are useful for the multifractal analysis of dynamical systems [4, 22].

[6] Ergodic theory then tells us that these probabilities are generated by (Lebesgue) almost all initial conditions.

of more general cases, which however do not pose essential problems. Assume that for some fixed value N the phase space function $h(x)$ is constant on $U_{\sigma_0 \ldots \sigma_{N-1}}$ and let $h_{\sigma_0 \ldots \sigma_{N-1}}$ denote the corresponding value of $h(x)$. Then the average value is given by

$$\langle h(x) \rangle = \sum_{\sigma_0, \ldots, \sigma_{N-1}} h_{\sigma_0 \ldots \sigma_{N-1}} \mu(U_{\sigma_0 \ldots \sigma_{N-1}}) = \lim_{n \to \infty} \frac{1}{n} \sum_{k=0}^{n-1} h(x_k) \quad . \tag{12}$$

Taking the relation between the temporal dynamics and the symbol shift into account (cf. eq. (5)) correlation functions may be evaluated too, e.g.

$$\langle h(T^k(x)) h(x) \rangle = \sum_{\sigma_0, \ldots, \sigma_{N+k-1}} h_{\sigma_k \ldots \sigma_{N+k-1}} h_{\sigma_0 \ldots \sigma_{N-1}} \mu(U_{\sigma_0 \ldots \sigma_{N+k-1}}) \quad . \tag{13}$$

Thus for piecewise linear Markov maps we are able to compute all dynamical features by analytical methods. In that sense linear Markov maps are a suitable testing ground for problems in Nonlinear Dynamics (cf. e.g. [24]). The scheme that has been sketched in the present section is of course not restricted to the simple example eq. (1). It works for general Markov maps. The notion of a Markov map is quite flexible and allows the modelling of quite complex situations, e.g. intermittent motion [25, 26, 27]. In such cases the complexity of the motion is represented by a richer alphabet of symbols from which the symbolic dynamics is derived and sometimes by a more complicated grammar.

3 Statistical Mechanics of Simple Maps

The symbolic description developed in section 2 permits to link dynamical properties with equilibrium statistical mechanics on a formal level. If we inspect the dynamical averages (12) or (13) then these quantities can be rewritten as canonical mean values of an observable $h_{\sigma_0 \ldots \sigma_{N-1}}$ in spin chain provided we identify the probability with a Boltzmann weight and define the spin Hamiltonian according to

$$\exp(-H_{\sigma_0 \ldots \sigma_{n-1}}) := \mu(U_{\sigma_0 \ldots \sigma_{n-1}}) \quad . \tag{14}$$

At the moment such a formulation is just a formal trick. Nevertheless we see that dynamical averages are equivalent to the thermostatics of a spin chain $\sigma_0 \ldots \sigma_{n-1}$ with Hamiltonian being given by eq. (14). The lattice translation in the spin chain corresponds to the temporal evolution of the dynamical system (cf. eq. (13) or eq. (5)) and the thermodynamic limit $n \to \infty$ yields a finer resolution by cylinder sets (4), i.e. the long time limit. As for the structure of the Hamiltonian eq. (8) tells us that the quantity for piecewise linear Markov maps reads

$$H_{\sigma_0 \ldots \sigma_{n-1}} = \sum_{k=0}^{n-1} \ln \gamma_{\sigma_k} - \ln h_{\sigma_0} - \ln(\nu_{\sigma_{n-1}}/\gamma_{\sigma_{n-1}}) \quad . \tag{15}$$

The Hamiltonian describes a model consisting of two level systems, where the single site quantity $\sigma_k = 0, 1$ indicates which level at site k is occupied. Using the notation of lattice gases the index σ_k indicates whether the lattice site is occupied by a particle, whereas if we identify the index σ_k with the quantum number of a spin then the Hamiltonian (15) describes a spin chain. Here we adopt the latter interpretation. Thus apart from boundary contributions at $k = 0$ and $k = n - 1$ the Hamiltonian describes a spin system without interaction but subjected to a magnetic field of strength $\ln(\gamma_1/\gamma_0)$[7]. Of course one has to take into account that only spin states according to the transition matrix (cf. eq. (3)) are permitted. Such a constraint can be viewed as an infinitely strong hard core repulsion between neighbouring lattice sites. Nevertheless common wisdom of equilibrium statistical mechanics tells us that no phase transition occurs, i.e. that the mean values (12) depend analytically on the system parameters, that is on the local slopes of the map. Furthermore correlations (13) decay exponentially.

One should note that in the just mentioned formulation there does not exist any temperature. Since the statistical approach just reproduces dynamical averages, there is nothing like the zero law of thermodynamics available. Nevertheless one may introduce a temperature as a formal parameter.Then different kinds of invariant measures are singled out which can be fruitfully used to analyse multifractal properties and the statistics of fluctuations of finite time averages. For such advanced topics we refer the interested reader to the literature [28, 29, 30, 4].

So far we have restricted the discussion to piecewise linear maps. In order to study the influence of a finite curvature we analyse a simple example. Let us consider the one–dimensional map shown in Fig. 3. A Markov partition consisting of two sets $U_\pm$ is indicated. On these sets the map has a simple curvature, since the slope takes two different values. It will simplify the subsequent analysis considerably that on a finer partition the map is again a piecewise linear Markov map. For our particular example not all initial conditions stay in the domain of the map. On iteration a Cantor set is singled out which contains all points that remain in the domain forever. That is easily seen if we compute successive cylinder sets. As indicated in Fig. 3 the cylinder sets yield the usual construction of a simple multifractal[8].

---

[7] Using $\ln \gamma_\sigma = \sigma \ln \gamma_1 + (1-\sigma) \ln \gamma_0$ the Hamiltonian may be cast in the standard form for a spin system.

[8] The invariant set studied here is a chaotic saddle with nontrivial fractal dimension in contrast to attractors which have been studied in the previous section.

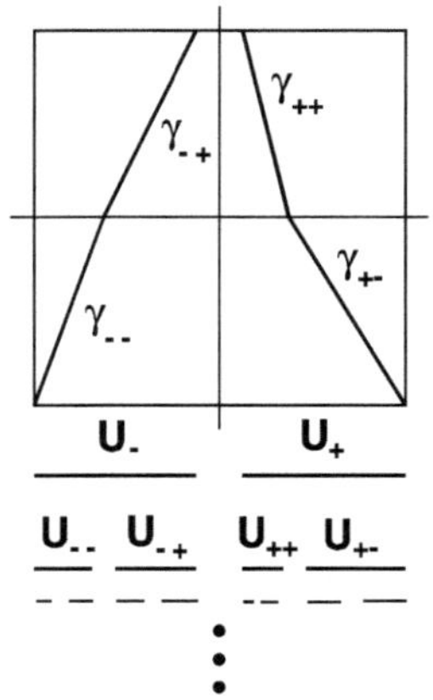

Fig. 3. One–dimensional Ising map. A Markov partition $\{U_+, U_-\}$ and the first generations of cylinder sets are indicated. Moduli of the slopes of the map are denoted by $\gamma_{\sigma\tilde{\sigma}}$

If we restrict the analysis to the invariant set then most considerations of the previous section can be applied. For our map all transitions $U_\sigma \to U_{\tilde{\sigma}}$ are possible. Thus no nontrivial transition matrix appears and all symbol sequences are admissible. For the probabilities a slight modification of eq. (8) applies since the slopes of the map are not constant on U_σ. But by construction the slopes are constant on the cylinder sets $U_{\sigma\tilde{\sigma}}$ of the first generation. Therefore, we apply essentially the same idea that lead us to eq. (8) for obtaining the invariant measure by using the partition given by $U_{\sigma\tilde{\sigma}}$. Thus the invariant measure is again determined by the local slopes and it reads

$$\mu(U_{\sigma_0 \ldots \sigma_{n-1}}) = h_{\sigma_0\sigma_1} \frac{\alpha^{-1}}{\gamma_{\sigma_0\sigma_1}} \times \ldots \times \frac{\alpha^{-1}}{\gamma_{\sigma_{n-3}\sigma_{n-2}}} \nu_{\sigma_{n-2}\sigma_{n-1}} \quad . \tag{16}$$

Here the factor $\alpha < 1$ takes into account that we are dealing with a repelling set, i.e. on each iteration step a fraction $\alpha < 1$ of phase space points will leave the domain. As a consequence the escape factor α appears in eq. (16) (cf. e.g. [22] for a more extensive discussion of chaotic saddles and [31] for the relation between chaotic saddles, fractals, iterated function systems, and concepts of modern ergodic theory). Thus plugging eq. (16) into the constraint eq. (7) and observing that the transitions $U_{\sigma_0\sigma_1} \to U_{\sigma_1\sigma_2}$ between cylinder sets of the first generation are permitted we obtain the eigenvalue equation (cf. eq. (9) for the case of attracting sets)

$$\alpha \nu_{\sigma_0\sigma_1} = \sum_{\sigma_2} \frac{1}{\gamma_{\sigma_0\sigma_1}} \nu_{\sigma_1\sigma_2} \quad . \tag{17}$$

Accordingly the balance condition (6) yields the adjoint problem

$$\alpha h_{\sigma_1\sigma_2} = \sum_{\sigma_0} h_{\sigma_0\sigma_1} \frac{1}{\gamma_{\sigma_0\sigma_1}} \quad . \tag{18}$$

These eigenvalue equations determine the invariant probability as well as the escape factor[9].

Our considerations show that the dynamical properties of the map with curvature are again equivalent to a spin chain with Hamiltonian being given by

$$H_{\sigma_0\ldots\sigma_{n-1}} \simeq \sum_{k=0}^{n-2} \ln \gamma_{\sigma_k\sigma_{k+1}} + n \ln \alpha \tag{19}$$

where the boundary contributions have not been written down explicitly. Thus we obtain an Ising model with nearest neighbour interaction. Written in conventional notation[10] the exchange constant is given by $\ln(\gamma_{--}/\gamma_{-+}) + \ln(\gamma_{++}/\gamma_{+-})$. Thus, it is just the curvature which causes a short range interaction in the corresponding spin Hamiltonian. In addition an asymmetry of the map is reflected in eq. (19) by an external field, which can be read off as $\ln(\gamma_{++}/\gamma_{--})$. Finally the escape factor enters the ground state energy of the Hamiltonian (19).

The features that have been deduced for the special example are quite typical. In fact, since expansiveness causes an exponential decrease of the length of cylinder sets a map with curvature will always result in a spin Hamiltonian with exponentially decreasing interactions, although such interactions may involve many spins. Thus no phase transition, i.e. a qualitative change of the dynamics, can be caused by curvatures of expanding maps. Such a property is relevant in different contexts too, i.e. for nice convergence properties of expansions in terms of periodic orbits [32].

We have illustrated that general expansive Markov maps result in one–dimensional spin Hamiltonians with short range interactions. Such systems do not display phase transitions, i.e. averages depend analytically on the model parameters and correlations decay exponentially. Roughly speaking temporal chaos corresponds to the paramagnetic or high temperature phase of the corresponding equilibrium model.

Qualitative changes in the dynamics (e.g. bifurcations) may occur if the dynamical system looses expansivity, i.e. when the dynamics becomes non-hyperbolic. Then long range interactions may be induced in the spin Hamiltonian, as was explicitly shown for intermittent dynamics [26]. The emerging phase transition displays the qualitatively different types of motion which are related with the particular bifurcation (cf. e.g. [27]). For the purpose of our tutorial we will stick to the simple case of expansive motion. We will

[9] For the case of attracting sets the escape factor turns out to be $\alpha = 1$ by Perron's theorem.

[10] $\ln \gamma_{\sigma\tilde{\sigma}} = [(1+\sigma)(1+\tilde{\sigma}) \ln \gamma_{++} + (1+\sigma)(1-\tilde{\sigma}) \ln \gamma_{+-} + (1-\sigma)(1+\tilde{\sigma}) \ln \gamma_{-+} + (1-\sigma)(1-\tilde{\sigma}) \ln \gamma_{--}]/4$.

demonstrate that in spatially extended systems a different mechanism which is intimately related to the limit of large system size may induce qualitative changes of the motion. Such a transition may be termed a phase transition in the proper sense (cf. [21]).

4 Spatially Extended Chaotic Models

To investigate the qualitatively new features which are related with a spatial degree of freedom we again resort to the analysis of a simple class of model systems. It has turned out that the study of coupled map lattices provides useful information [33] at least if one wants to understand dynamics from a qualitative and phenomenological point of view. Consider a spatially one–dimensional lattice of length L and assume that each lattice site ν, $0 \leq \nu \leq L-1$ is associated with a variable $x^{(\nu)}$. The full state of the system is described by $\underline{x} = (x^{(0)}, \ldots, x^{(L-1)})$. For the dynamics we suppose that on each lattice site there acts a map $f(x)$ and neighbouring lattice sites are coupled through a coupling function $\Phi : [-1,1]^L \rightarrow [-1,1]^L$. Then the whole time discrete motion is determined by $\underline{x}_{n+1} = T(\underline{x}_n)$ where[11]

$$\underline{x}_{n+1} = T(\underline{x}_n) = \Phi\left[f(x_n^{(0)}), f(x_n^{(1)}), \ldots, f(x_n^{(L-1)})\right] . \tag{20}$$

Boundary conditions will play no crucial role for our purpose, thus one may consider periodic boundary conditions. Although no microscopic derivation of such dynamical models has been performed from first principles, models of the type eq. (20) had turned out quite useful for numerical studies [34]. Particular numerical investigations where the system size becomes important concern the exponential increase of transient motion [19], the study of nonequilibrium phase transitions, and the related critical behaviour [35, 36]. These investigations already indicate that in the limit of large L spatially extended models may display qualitative new features. Hence we will focus on the influence of the system size on the motion of the system.

In order to keep our discussion as simple as possible we will focus on a particular coupled map lattice [37]. For the single site map f we consider the system already introduced in section 3 (cf. Fig. 3). To get an idea how the dynamics will behave let us first consider the trivial case of uncoupled maps, where Φ is the identity map. At each lattice site just the map f acts on $x_n^{(\nu)}$ and induces its symbol sequence $(\sigma_0^{(\nu)}\sigma_1^{(\nu)}\ldots)$. Thus the motion of the whole state vector $\underline{x}_n$ is described by a two–dimensional symbol lattice

[11] A very popular choice for the coupling consists in diffusive coupling where

$$x_{n+1}^{(\nu)} = f(x_n^{(\nu)}) + \varepsilon\left[f(x_n^{(\nu+1)}) + f(x_n^{(\nu-1)}) - 2f(x_n^{(\nu)})\right]$$

i.e. $(\Phi[\underline{x}])^{(\nu)} = x^{(\nu)} + \varepsilon(x^{(\nu+1)} + x^{(\nu-1)} - 2x^{(\nu)})$. But we do not restrict ourselves to such a particular case.

$$\begin{matrix} \sigma_0^{(0)} & \sigma_1^{(0)} & \sigma_2^{(0)} & \dots \\ \sigma_0^{(1)} & \sigma_1^{(1)} & \sigma_2^{(1)} & \dots \\ \vdots & \vdots & \vdots & \\ \sigma_0^{(L-1)} & \sigma_1^{(L-1)} & \sigma_2^{(L-1)} & \dots \end{matrix} =: (\underline{\sigma}_0\underline{\sigma}_1\underline{\sigma}_2\dots) \quad . \tag{21}$$

$\sigma_n^{(\nu)}$ tells us which interval $U_\pm$ is occupied by (the corresponding component of) the phase space point at time n on lattice site ν. The first dimension of such a spin lattice corresponds to the spatial extension of the dynamical system whereas the second dimension reflects the temporal motion of the original system. The shift operation in the lattice yields either the spatial translation or the temporal evolution, as in the case of nonextended systems (cf. section 2). Obviously the uncoupled map lattice is a piecewise linear Markov map, with phase space being given by $[-1,1]^L$.

For the introduction of probabilities we again consider the associated cylinder sets, which are related with the time evolution over a finite period $0 \le k \le n-1$ (cf. eq. (4)). Different cylinder sets are labelled by finite two–dimensional spin lattices $(\underline{\sigma}_0 \dots \underline{\sigma}_{n-1})$. Since we consider the uncoupled case the probabilities are just the products of the quantities of the individual maps (cf. eq. (16)) and we thus have

$$H_{\underline{\sigma}_0\cdots\underline{\sigma}_{n-1}} := -\ln\mu(U_{\underline{\sigma}_0\cdots\underline{\sigma}_{n-1}}) \simeq \sum_{\nu=0}^{L-1}\sum_{k=0}^{n-2} \ln\gamma_{\sigma_k^{(\nu)}\sigma_{k+1}^{(\nu)}} + nL\ln\alpha \quad . \tag{22}$$

For simplicity of notation the boundary contributions have not been written down explicitly. Thus the corresponding spin Hamiltonian describing our dynamical system consists of a two–dimensional lattice where only a nearest neighbour intra–chain interaction occurs. That is of course not surprising since we started from a dynamical system without spatial interaction. One already expects that on turning on the spatial interaction we will obtain an associated spin system with real two–dimensional interaction. Such a scenario opens the possibility for a phase transition in the thermodynamic limit $L \to \infty$.

To prepare the formal steps that we need to investigate the coupled case let us first have a look at the phase space structure of the uncoupled map lattice. For the purpose of visualisation we concentrate on the case of two coupled maps $L = 2$, but the considerations go along the same lines for general L. The Markov partition of the uncoupled map lattice consists of sets which are just the direct product of the partition of the single site maps, $U_{\underline{\sigma}} = U_{\sigma^{(0)}} \otimes \dots \otimes U_{\sigma^{(L-1)}}$. The full phase space $[-1,1]^L$ is divided by that Markov partition $\{U_{\underline{\sigma}}\}$ in 2^L parts (cf. figure 4). Cylinder sets of the first generation $U_{\underline{\sigma}\underline{\tilde{\sigma}}}$ divide the space in 2^{2L} parts and so on. Cylinder sets of the generation $n-1$ are labelled by finite symbol lattices of size $L \times n$. The lack of spatial coupling in the map lattice is reflected by the fact that the cylinder

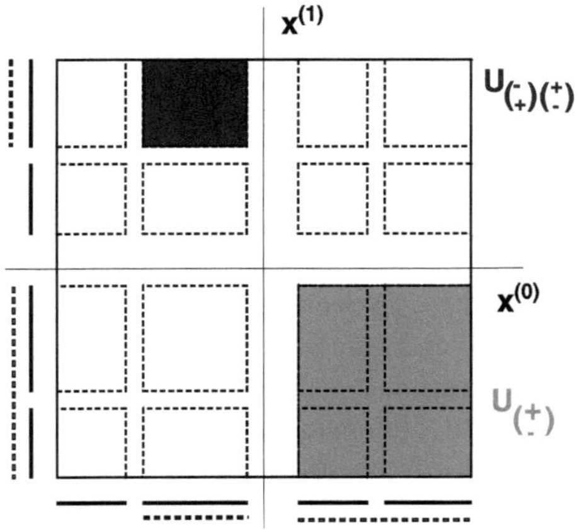

Fig. 4. Phase space structure of the uncoupled map lattice T_0 for $L = 2$. Cylinder sets of the first generation $U_{\underline{\sigma\tilde{\sigma}}}$ (*dashed rectangles*), and cylinder sets of the single site maps (*full lines*, cf. Fig. 3). Shaded areas show a cylinder set (dark) and its image (light), broken lines indicate the sets and its images for the single site maps

sets admit a product structure (i.e. the extension of neighbouring sets in the direction of the coordinate axis is equal). The uncoupled map lattice itself, which we denote by T_0, maps successive cylinder sets in a linear way (cf. eq. (5)), i.e.

$$T_0 : U_{\underline{\sigma\tilde{\sigma}}} \to U_{\underline{\tilde{\sigma}}} \tag{23}$$

is a linear map with Jacobian $\prod_{\nu=0}^{L-1} \gamma_{\sigma^{(\nu)}\tilde{\sigma}^{(\nu)}}$ when restricted to the set $U_{\underline{\sigma\tilde{\sigma}}}$.

The introduction of a coupling Φ will remove the just mentioned product structure of the cylinder sets. For an arbitrary choice of the coupling the Markov partition may become quite complicated. In order to keep the analysis as simple as possible we will define the coupling function in such a way that a simple pattern for the cylinder sets will remain so that no tedious analytical computations become necessary. Consider the cylinder sets $U_{\underline{\sigma\tilde{\sigma}}}$ depicted in Fig. 4. Now move the "inner vertex" of each set by some amount as shown in Fig. 5. If we require that the new cylinder sets, which we will denote by $V_{\underline{\sigma\tilde{\sigma}}}$, are still mapped in a linear fashion like in the uncoupled case (cf. eq. (23))

$$T : V_{\underline{\sigma\tilde{\sigma}}} \to U_{\underline{\tilde{\sigma}}} \tag{24}$$

we obtain a new dynamical system T which is now a coupled map lattice. Since this new model is still a piecewise linear Markov map we are able to solve the dynamics along the lines of Sect. 2.

In what follows we do not need the analytical formula of the just defined map lattice T (cf. the relevance of eq. (1) for the considerations in Sect. 2). Nevertheless we will supply the expression for completeness. The coupling Φ is constructed in such a way that it maps the cylinder set $V_{\underline{\sigma\tilde{\sigma}}}$ of the full system

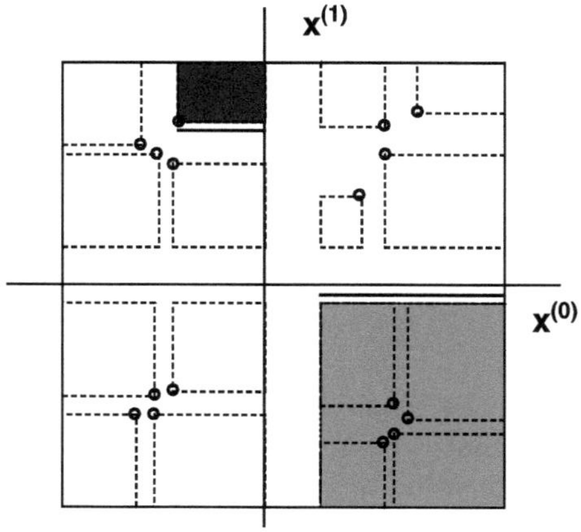

Fig. 5. Phase space structure of the coupled map lattice T for $L = 2$. *Dashed rectangles*: cylinder sets of the first generation $V_{\underline{\sigma}\tilde{\underline{\sigma}}}$, *open circles*: inner vertexes of the cylinder sets. Shaded areas depict a cylinder set (dark) and its image (light) (cf. Fig. 4), full lines indicate an edge of a cylinder set and its image

(cf. Fig. 5) to the cylinder set $U_{\underline{\sigma}\tilde{\underline{\sigma}}}$ of the map lattice without interaction (cf. Fig. 4). Then the dynamics is given by

$$x_{n+1}^{(\nu)} = f\left((\Phi[\underline{x}])^{(\nu)}\right) \quad . \tag{25}$$

In order to write down the analytical expression for Φ consider a cylinder set $V_{\underline{\sigma}\tilde{\underline{\sigma}}}$ and denote its outer vertex, i.e. the common vertex of $V_{\underline{\sigma}\tilde{\underline{\sigma}}}$ and $U_{\underline{\sigma}\tilde{\underline{\sigma}}}$ by $\underline{a}_{\underline{\sigma}\tilde{\underline{\sigma}}}$. Consider the edges of $V_{\underline{\sigma}\tilde{\underline{\sigma}}}$ and of $U_{\underline{\sigma}\tilde{\underline{\sigma}}}$ which are parallel to the ν–axis and denote the ratio of the length of these edges by $r_{\underline{\sigma}\tilde{\underline{\sigma}}}^{(\nu)}$. Then the coupling function reads

$$(\Phi[\underline{x}])^{(\nu)} = r_{\underline{\sigma}\tilde{\underline{\sigma}}}^{(\nu)}(x^{(\nu)} - a_{\underline{\sigma}\tilde{\underline{\sigma}}}^{(\nu)}) + a_{\underline{\sigma}\tilde{\underline{\sigma}}}^{(\nu)}, \qquad \underline{x} \in V_{\underline{\sigma}\tilde{\underline{\sigma}}} \quad . \tag{26}$$

Spatial coupling is in particular mediated by the factor $r_{\underline{\sigma}\tilde{\underline{\sigma}}}^{(\nu)}$ which may depend on symbols of neighbouring lattice sites. Thus the spatial coupling removes the product structure visible in Fig. 4. Finally, in order to convert our map lattice (25) to the form (20) one just introduces new coordinates $\underline{z} = \Phi(\underline{x})$ and applies the coupling Φ to both sides of eq. (25).

For the following analysis we will just need the Jacobian of the map lattice T. That quantity is of purely geometric origin, since it is determined by the volume ratio of $U_{\tilde{\underline{\sigma}}}$ and $V_{\underline{\sigma}\tilde{\underline{\sigma}}}$ (cf. eq. (24) and Fig. 5). Let $t_{\underline{\sigma}\tilde{\underline{\sigma}}}^{(\nu)}$ denote the ratio of the edges of $U_{\tilde{\underline{\sigma}}}$ and $V_{\underline{\sigma}\tilde{\underline{\sigma}}}$ which are parallel to the ν–axis (cf. Fig. 5)[12]. Then, the Jacobian of the map T reads.

[12] $t_{\underline{\sigma}\tilde{\underline{\sigma}}}^{(\nu)} = \gamma_{\sigma^{(\nu)}\tilde{\sigma}^{(\nu)}} r_{\underline{\sigma}\tilde{\underline{\sigma}}}^{(\nu)}$.

$$|\det DT(\underline{x})| = \prod_{\nu=0}^{L-1} t^{(\nu)}_{\underline{\sigma}\underline{\tilde{\sigma}}} \qquad (\text{if } \underline{x} \in V_{\underline{\sigma}\underline{\tilde{\sigma}}}) \quad . \tag{27}$$

According to our general considerations of sections 2 and 3 these expansion rates determine the invariant measure and we end up with (cf. eq. (19))

$$H_{\underline{\sigma}_0\cdots\underline{\sigma}_{n-1}} = -\ln \mu(V_{\underline{\sigma}_0\cdots\underline{\sigma}_{n-1}}) \simeq \sum_{k=0}^{n-2}\sum_{\nu=0}^{L-1} \ln t^{(\nu)}_{\underline{\sigma}_k\underline{\sigma}_{k+1}} + nL\ln\alpha \tag{28}$$

where we have dropped the boundary terms, and α denotes the escape factor of our map lattice. Since α just determines the overall normalisation of the probabilities (i.e. the value of the ground state energy) we refrain from writing down the corresponding eigenvalue problem explicitly (cf. eq. (17) or (18)). The quantities $t^{(\nu)}_{\underline{\sigma}_k\underline{\sigma}_{k+1}}$ are the important parameters which enter the coupled map lattice and determine the coupling (cf. eq. (26)) as well as the invariant measure. In dynamical terms these parameters determine the local slope which the phase space point experiences at lattice site ν.

Depending on the particular choice of our coupling parameters $t^{(\nu)}_{\underline{\sigma}\underline{\tilde{\sigma}}}$ the map lattice T is equivalent to the thermostatics of a Hamiltonian on a two–dimensional lattice with two–dimensional interaction. If we choose for instance

$$t^{(\nu)}_{\underline{\sigma}\underline{\tilde{\sigma}}} = t^{(\nu)}_{\underline{\sigma}\underline{\tilde{\sigma}}}(h,J) := \exp[-h\sigma^{(\nu)} - J\sigma^{(\nu)}(\tilde{\sigma}^{(\nu)} + \sigma^{(\nu+1)}) + e_0] \tag{29}$$

then eq. (28) yields the famous nearest neighbour coupled Ising Hamiltonian. It is well known that the thermodynamic equilibrium state develops a phase transition at $h = 0$, $J = J_c = \operatorname{Artanh}(\sqrt{2} - 1)$ in the thermodynamic limit. In the high temperature phase $0 \leq J < J_c$ correlations decay exponentially whereas for $J > J_c$ two pure phases develop. These standard results now transfer to the dynamical system. In the regime $J < J_c$, i.e. when the quantities (29) do not depend strongly on neighbouring lattice sites, spatio–temporal correlations decay exponentially. Such a behaviour, sometimes called space–time chaos, is expected in the regime of weak spatial coupling. When the spatial coupling exceeds a critical value, then the invariant measure decomposes into two different ergodic components, which are not mixed by the dynamics and which correspond to the pure phases in the associated equilibrium system. If one adds a small asymmetry, i.e. a small field h, then the two limits $h \uparrow 0$ and $h \downarrow 0$ result in two different dynamical states where the magnetisation $\sum_{n,\nu} \sigma^{(\nu)}_n$ attains a finite value in the thermodynamic limit. In dynamical terms the magnetisation is identical to the dynamical mean value of $S_n = \sum_\nu \sigma^{(\nu)}_n$ (cf. eq. (12)). Thus the majority of single site maps stays in their left or right part of the domain although the single site dynamics looks like the graph in Fig. 3. To achieve such a phase transition a modulation of the local slopes by a factor of size ~ 4 is already sufficient, according to eq. (29). For finite system size L no phase transition occurs at all, long time averages depend analytically on system parameters, all correlations decay exponentially, and the system is mixing. Thus the behaviour described here is indeed a transition which is caused just by the limit of large system size.

The behaviour illustrated by our special model reflects typical features which can be found in quite general coupled map lattices. But the analysis is often very delicate. It has been shown rigorously that the weak coupling regime of a coupled map lattice can be mapped onto the high–temperature phase of a spin–lattice so that the dynamics displays a spatio–temporal mixing state, i.e. the motion generates exponentially decaying spatio–temporal correlations (cf. e.g. [12, 13, 14]) provided certain technical constraints on the single site dynamics and the coupling are fulfilled. Such a behaviour may be used for the definition of space–time chaos. But one has to admit that no fully satisfactory definition is available in the literature (cf. e.g. [15, 16, 38] for alternative concepts). The corresponding spin Hamiltonian is in general far more complicated as in our simple example since complicated anisotropic and many particle interactions appear. Thus the Statistical Mechanics of a typical coupled map lattice seems to resemble the description of spin glasses instead of simple ferromagnetic phase transitions. For larger coupling strength it is conjectured that the breakdown of the spatio–temporal chaotic state takes place through some phase transition, as in our simple example, although the nature of the transition in the typical case is still unclear (cf. e.g. [39, 40] for some explicit examples and [21] for some general considerations). There are of course alternatives for the onset of coherent pattern formation e.g. through dynamical bifurcations (cf. e.g. [18]) which are well known from the theory of low–dimensional dynamical systems. As such a mechanism is not related to the limit of large system size one should not call such a transition a phase transition in the proper sense.

So far we have emphasised how the invariant measure (i.e. the probabilities) are modified by the spatial coupling in the limit of infinite systems size, keeping the topological properties, i.e. the symbolic dynamics, rather simple. It is in fact less known about, how the coupling may influence topological properties, and whether these effects are relevant in the thermodynamic limit. Case studies in simple situations (e.g. two–dimensional phase spaces) show already [41] that a quite complicated grammar may appear. In fact such questions have to be addressed in order to get a full overview of the dynamical behaviour in terms of associated spin systems.

Within our tutorial we have confined ourselves to the construction of symbolic dynamics through Markov partitions. There are of course alternatives for obtaining coarse grained descriptions of dynamical models (cf. e.g. [42] for the relation between extended dynamical systems and kinetic Ising models). Although there are a plenty of unanswered problems the elementary introduction presented here indicates, that the link between nonlinear dynamical systems and Statistical Mechanics is very fruitful for understanding principal features. It gives a new and unexpected perspective on the field of nonequilibrium and coherent pattern formation.

References

1. J. Guckenheimer, P. Holmes, *Nonlinear Oscillations, Dynamical Systems, and Bifurcations of Vector Fields* (Springer, New York, 1986).
2. H.G. Schuster, *Deterministic chaos* (VCH, Weinheim, 1995).
3. C. Robinson, *Dynamical systems: stability, symbolic dynamics, and chaos* (CRC Press, Boca Raton, 1995).
4. C. Beck, F. Schlögl, *Thermodynamics of chaotic systems* (Cambridge University Press, Cambridge, 1995).
5. D. Ruelle, *Thermodynamic formalism* (Addison-Wesley, Reading, 1978).
6. A. Katok, B. Hasselblatt, *Introduction to the modern theory of dynamical systems* (Cambridge University Press, Cambridge, 1996).
7. P. Collet, J.P. Eckmann, *Iterated maps on the interval as dynamical systems* (Birkhäuser, Basel, 1980).
8. B.L. Hao, *Elementary symbolic dynamics and chaos in dissipative systems* (World Scientific, Singapore, 1989).
9. P. Walters, *An introduction to ergodic theory* (Springer, New York, 1982).
10. A. Lasota, M.C. Mackey, *Chaos, Fractals, and Noise: Stochastic Aspects of Dynamics* (Springer, New York, 1994).
11. H.H. Hasegawa, W.C. Saphir, *Unitarity and irreversibility in chaotic systems*, Phys. Rev. A **46**, 7401 (1992).
12. L.A. Bunimovich, Y.G. Sinai, *Spacetime chaos in coupled map lattices*, Nonlin. **1**, 491 (1988).
13. J. Bricmont, A. Kupiainen, *Coupled analytic maps*, Nonlin. **8**, 379 (1995).
14. T. Fischer, H.H. Rugh, *Transfer operators for coupled analytic maps*, Erg. Theor. Dyn. Syst. **20**, 109 (2000).
15. J.P. Eckmann, D. Ruelle, *Ergodic theory of chaos and strange attractors*, Rev. Mod. Phys. **57**, 1115 (1985).
16. R.J. Deissler, K. Kaneko, *Velocity-dependent Lyapunov exponents as a measure of chaos for open-flow systems*, Phys. Lett. A **119**, 397 (1987).
17. A. Politi, A. Torcini, *Towards a Statistical Mechanics of Spatiotemporal Chaos*, Phys. Rev. Lett. **69**, 3421 (1992).
18. M.C. Cross, P.C. Hohenberg, *Pattern-formation outside of equilibrium*, Rev. Mod. Phys. **65**, 851 (1993).
19. K. Kaneko, *Supertransients, spatiotemporal intermittency and stability of fully developed spatiotemporal chaos*, Phys. Lett. A **149**, 105 (1990).
20. W. Just, *Globally Coupled Maps: Phase Transitions and Synchronization*, Physica D **81**, 317 (1995).
21. M. Blank, L. Bunimovich, *Multicomponent dynamical systems: SRB measures and phase transitions*, Nonlin. **16**, 387 (2003).
22. P. Gaspard, *Chaos, Scattering and Statistical Mechanics* (Cambridge University Press, Cambridge, 1998).
23. B.C. So, N. Yoshitake, H. Okamoto, H. Mori, *Correlations and Spectra of an Intermittent Chaos near Its Onset Point*, J. Stat. Phys. **36**, 367 (1984).

24. H. Mori, B.S. So, T. Ose, *Time–correlation functions of one–dimensional transformations*, Prog. Theor. Phys. **66**, 1266 (1981).
25. K. Shobu, T. Ose, H. Mori, *Shapes of the Power Spectrum of Intermittent Turbulence near Its Onset Point*, Prog. Theor. Phys. **71**, 458 (1984).
26. X.J. Wang, *Statistical physics of temporal intermittency*, Phys. Rev. A **40**, 6647 (1989).
27. W. Just, H. Fujisaka, *Gibbs Measures and Power Spectra for Type I Intermittent Maps*, Physica D **64**, 98 (1993).
28. H. Fujisaka, M. Inoue, *Statistical–physical theory of multivariate temporal fluctuations: Global characterization and temporal correlation*, Phys. Rev. A **41**, 5302 (1990).
29. H. Mori, H. Hata, T. Horita, T. Kobayashi, *Statistical Mechanics of Dynamical Systems*, Prog. Theor. Phys. Suppl. **99**, 1 (1989).
30. P. Grassberger, R. Badii, A. Politi, *Scaling laws for invariant measures on hyperbolic and nonhyperbolic attractors*, J. Stat. Phys. **51**, 135 (1988).
31. K. Falconer, *Techniques in fractal geometry* (Wiley, Chicester, 1997).
32. R. Artuso, E. Aurell, P. Cvitanovich, *Recycling of strange sets*, Nonlin. **3**, 325 (1990).
33. K. Kaneko, *Theory and applications of coupled map lattices* (Wiley, Chichester, 1993).
34. K. Kaneko, *Pattern dynamics in spatiotemporal chaos*, Physica D **34**, 1 (1989).
35. J. Miller, D.A. Huse, *Macroscopic equilibrium from microscopic irreversibility in a chaotic coupled map lattice*, Phys. Rev. E **48**, 2528 (1993).
36. P. Marcq, H. Chaté, P. Manneville, *Universality in Ising–like phase transitions of coupled chaotic maps*, Phys. Rev. E **55**, 2606 (1997).
37. W. Just, *Equilibrium Phase Transitions in Coupled Map Lattices: A Pedestrian Approach,*, J. Stat. Phys. **105**, 133 (2001).
38. A. Torcini, P. Grassberger, A. Politi, *Error propagation in extended chaotic systems*, J. Phys. A **27**, 4533 (1995).
39. G. Gielis, R.S. MacKay, *Coupled map lattices with phase transition*, Nonlin. **13**, 867 (2000).
40. W. Just, *Analytical Approach for Piecewise Linear Coupled Map Lattices*, J. Stat. Phys. **90**, 727 (1998).
41. P. Cvitanovich, G.H. Guaratne, I. Procaccia, *Topological and metric properties of Henon–type strange attractors*, Phys. Rev. A **38**, 1503 (1988).
42. F. Schmüser, W. Just, H. Kantz, *On the relation between coupled map lattices and kinetic Ising models*, Phys. Rev. E **61**, 3675 (2000).

Complex Adaptive Systems

Heinz Georg Schuster

Institut für Theoretische Physik, Universität Kiel,
Olshausenstrasse 40, 24098 Kiel, Germany
schuster@theo-physik.uni-kiel.de

1 Introduction

Examples for complex adaptive systems (CAS) are the brain, whose neurons are connected by synapses to a complex network, the network of chemical reactions in prebiotic evolution, the control network of our immune system, and social or economical networks whose dynamics could be described as an evolutionary game.

All of these systems have a common structure: They consist of a network of elements (e.g. neurons, genes, agents in a game,...), interacting nonlinearly with themselves and with their environment. They receive information from their environment (that could also consist of other species), and transform this knowledge into actions yielding certain advantages (reward, more descendants). These transformations, using also information from memory, are in fact a very general form of a computation.

The main element which distinguishes complex adaptive systems from pure logical machines is their ability to adapt or to change their computations and programs in dependence of the reward, where the goal is to achieve an optimal advantage. In other words, the systems considered are adaptive, while the respective systems determining the reward associate a meaning to the computation, a fact which is missing when pure logic machines are concerned. Being coupled nonlinear systems, their behaviour can be described using methods of nonlinear dynamics.

My lectures at the Heraeus Seminar in Chemnitz were mostly based on sections of my new introductory book: Complex Adaptive Systems [1]. In the following we discuss the behavior of two complex adaptive systems which were also presented in the lectures but which are not described in this book.

The first example provides the arguably simplest model of a system that can learn something about itself. We will show how a system can find its status in a hierarchy of agents by using the reflexes of other agents as a mirror.

The second example is the so called minority game. There we will show how a group of adaptive agents can predict the optimal occupancy of a restaurant or the least crowded, but still promising stock on a market.

2 Learning the Status in a Hierarchy

It has only been recently that the study of the problem how an organism can become aware if its surroundings and its own self has become a subject of scientific inquiry [2]. One of the simplest manifestations of self-recognition is the fact that already animals can learn to know their status in a hierarchy [3]. Here we investigate the question how an animal can detect its standing in a hierarchy via the rewards obtained by interacting with others, while trying to keep its losses in this process at a minimum.

We consider a game with N players, where each has a status $s_i = i$, $(i = 1..N)$. The status of a player (think of his size) is known to all others but not to himself. After an encounter of player i with player j, the former, i receives a reward r_{ij} which is larger than zero if the own standing s_i is larger than that of the opponent s_j and $r_{ij} \leq 0$ for $s_i \leq s_j$, such that $sign(r_{ij}) = sign(s_i - s_j)$. In each round of the game all players are randomly paired with each other and receive rewards according to their status. The reactions of the other players provide a mirror in which each player can detect his own standing.

The simplest model for a player is a lookup table in which he records after each encounter his reward together with the status of the opponent. His own status is then defined by the border between negative and positive rewards. This shows that the information provided by the rewards received is enough to infer the own a status. However such a model of the player requires a lot of memory, i.e. neural hardware. Instead here we will consider a simpler model in which each player is characterized only by a single variable E_i which represents an estimate of his own status. E_i provides also an estimate of the predicted reward $\hat{r}_{ij}$ whose sign changes according to $sign(\hat{r}_{ij}) = sign(E_i - s_j)$. One can think of E_i either as a hormone level which changes according to the rewards received - and determines whether an animal attacks the other or not, depending on the sign of $\hat{r}_{ij}$ - or as the threshold of a single neuron with input $s_j$and output $sign(E_i - s_j)$. The variable E_i represents the only memory of the player which increases if the player makes a correct prediction about the sign of the expected reward when he encounters s_j and decreases otherwise, i.e the learning rule which describes the change of the status E_i^t at time t after one encounter becomes:

$$E_i^{t+1} - E_i^t = \alpha \left[(sign(r_{ij}^t) - sign(E_i^t - s_j^t)\right] \ . \tag{1}$$

Here the status s_j^t of the opponent j encountered at times t acts as a random input variable which drives the learning process and α is a proportionality constant which measures the learning rate. By scaling E_i^t and s_k^t appropriately we can put $\alpha = 1$.

The first question is whether such a simple model converges to the correct standings, i.e. whether $\lim_{t\to\infty} E_i^t = s_i$. Since the s_j^t are chosen randomly from a uniform distribution, the Frobenius Perron equation [4] for the probability $P^t(E_i)$ for E_i at time reads

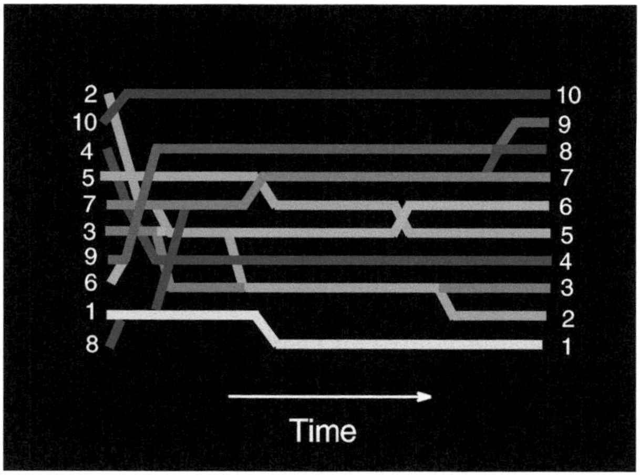

Fig. 1. Time evolution of the standings of ten species that were initially randomly distributed across the steps of the hierarchy

$$P^{t+1}(E_i) = \sum_{E_i'} K\left(E_i \mid E_i'\right) P^t(E_i') \tag{2}$$

with a kernel

$$K\left(E_i \mid E_i'\right) = \frac{1}{N} \sum_{s_j} \Delta\left[E_i - E_i' - sign(r_{ij}) + sign(E_i' - s_j^t)\right] \tag{3}$$

where Δ denotes the Kronecker Delta. It can be verified by direct computation that the largest eigenvalue of $K\left(E_i \mid E_i'\right)$ is one and belongs to the eigenfunction $P^{\infty}(E_i) = \Delta(E_i - s_i)$. The second largest eigenvalue which determines the rate by which an initial distribution $P^0(E_i)$ converges to $P^{\infty}(E_i)$ has the value $1 - \frac{1}{N}$.This demonstrates that the estimated levels E_i^t of our simple model converge to the correct standings s_i as shown for an example of 10 species in Fig. 1.

Since each encounter of a player with an opponent bears the risk of getting hurt, we arrive at the question how to select opponents which help to find out the own status with the smallest number of encounters in the shortest time. We have found that the simple rule "win-stay, lose-shift" in which the systems "looks at its own internal status E_i^t " and only selects a new value for s_j^{t+1} if $E_i^{t+1} - E_i^t = 0$, works very well. During the "learning state" in which the internal state is still changing i.e. for $E_i^{t+1} - E_i^t \neq 0$ the value of s_j^t is kept constant and new costly encounters are avoided. Fig. 2 shows that the feedback from the inner status to the selection of new opponents drastically reduces the number of encounters that are needed for finding out one's own standing.We note that such a feedback loop in which the measured change of the own status leads to a reconsideration of the inputs provides the probably simplest realization of a "second order map" suggested by Damasio [5].

In order to get an understanding for the power of the rule "win-stay, lose-shift" we consider a simple optimization problem which can be solved by nonlinear programming and which generates just this rule.

The update rule

$$e^{t+1} = f(e^t, k^t) = e^t - 1 + k^t \tag{4}$$

describes the linear decay of a quantity e^t which is externally driven by a positive input k^t which has only two possible values $k^t = 0$ or $k^t = K > 0$. This models for example a learning process driven by k^t where e^t measures the information which becomes linearly absorbed by the system. The task is to optimize the number of time steps for which $e_i^t > 0$, i.e. for which enough information for learning remains available while keeping the number of time steps with positive input $k^t = K$ minimal. The cost function for such a process is

$$L = \sum_{t=0}^{T} \hat{\theta}(e^t) \left[1 - \hat{\theta}(k^t)\right] \tag{5}$$

where $\hat{\theta}(e^t) = 1$ for $e^t > 0$ and zero otherwise. Equations (4) and (5) describe an optimization problem that can be solved by dynamical programming [1]. We can find values k^t which maximize the sum by starting the optimization process stepwise from above:

$$\max_{k^T} \left\{\hat{\theta}(e^T) \left[1 - \hat{\theta}(k^T)\right]\right\} \to k^T = k^T\left(e^T\right) \tag{6}$$

$$\max_{k^{T-1}} \left\{\hat{\theta}(e^{T-1}) \left[1 - \hat{\theta}(k^{T-1})\right] + \hat{\theta}(e^T) \left[1 - \hat{\theta}(k^T)\right]\right\} \to k^{T-1} = k^{T-1}\left(e^{T-1}\right) \tag{7}$$

because we obtain from equations (4) and (5)

$$\hat{\theta}(e^T) \left[1 - \hat{\theta}(k^T)\right] = \hat{\theta}(f(e^{T-1}, k^{T-1})) \left[1 - \hat{\theta}(k^T(f(e^{T-1}, k^{T-1}))\right] . \tag{8}$$

This generates a sequence $k^t = k^t\left(e^t\right)$ and the corresponding e^t values follow by forward iteration $e^{t+1} = f(e^t, k^t\left(e^t\right))$. Our example yields (if we start initially with $e^0 = 0$) the solutions $k^{t+1} = 0$ for $e^t > 0$ and $k^{t+1} = K$ for $e^t = 0$ which is just the rule "win-stay, lose-shift".

To summarize: We have presented a simple model in which players endowed only with a single variable which acts as a memory can find their rank in a hierarchy by observing rewards obtained by interacting with the other members. The number of encounters needed for learning, each of which could

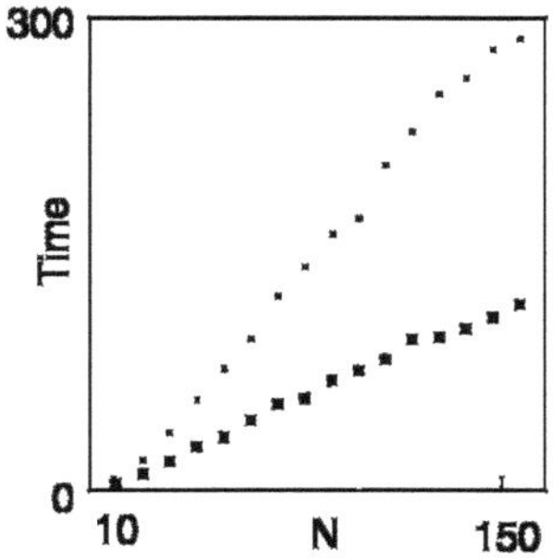

Fig. 2. The times needed for converge of species $i = 15$ to its status as a function of the total number of players N, averaged over 100 runs. *Upper curve*: convergence times without feedback. *Lower curve*: convergence times with feedback via "win-stay, lose-shift"

lead to a damage of the player, can be reduced by a feedback loop which works according to the rule "win-stay, lose-shift" and allows for an intake of new information only if the internal learning process halts. We have shown that this rule optimizes rigorously the information transfer in a simple learning process. These results show, on an elementary level, how learning of the own standing can be achieved by using interactions with an environment as a mirror and self inspection via a feedback loop provides the probably simplest realization of Damasio's idea of a "second order map". It has been shown by Sigmund and Nowak [6] that cooperation between members of a group based on mutual reciprocity depends crucially on the ability of the player to estimate their own status and that of the opponent. Therefore the recognition of the own rank which can even change in a game and be inferred from watching others is an important ingredient for the development of cooperative structures.

3 Adaptive Booleans Solve the Minority Problem

There has been much interest in the dynamical behavior of interacting agents which can be used to model systems that arise naturally in biology [7, 8] and in the social sciences, especially in economy [9]. Here we will focus on the dynamical behavior of agents in the minority game which has been devised as a fundamental model for a market of competitive agents. The minority game was originally introduced by Arthur [10] in terms of "El Farol's bar problem". If too many people attend the bar, the patrons will not enjoy the evening. Therefore each of the N possible visitors i must decide whether to go ($\sigma_i = 1$) or stay home ($\sigma_i = -1$). Only those agents who are in the minority, i.e. who went, when the bar was less then half full, or stood home when the bar was overcrowded, win. Decisions are made independently, i.e. the agents cannot communicate with each other directly. The only information available to them is the decision of the minority in the last L time-steps which determines the reward which they receive, after having made their decision.

There exist many publications on this problem in which the agents used either a small number of randomly chosen decision tables (Boolean functions) that prescribed their actions based on the previous history [10, 11, 12, 13], or adaptive neural networks such as perceptrons which learned to predict the next successful action in the game [14] or stochastic strategies [15].

Here we want to make the agents as clever as possible. Therefore we model each agent by an adaptive probabilistic Boolean function [1] with L inputs. The weights, which determine the structure of the Boolean, are adapted according as to how much the agent deviated from the optimal solution [13]. We will show that the corresponding learning algorithm, which makes only use of global information that is available to all agents, leads to a complete solution of the problem. In our formulation the minority problem is mapped to the problem of finding the energy minima of an all-to-all coupled Ising antiferromagnet. The learning algorithm self-adapts the local temperatures of the spins in such a way that they adjust with a time schedule which is optimal for simulated annealing [16]. This implies that the minima will indeed be found and then the external stimuli (see below) generate a hopping between different minima, which corresponds to solutions where different groups of agents visit the bar in such a way that it has always optimal occupancy.

Our model can be described as follows: There is an odd number of N players, each of which receives a binary input string $\boldsymbol{\mu} = (x_1, ..., x_L)$, where $x_\alpha = \pm 1$. Each player i reacts to this input with a binary output σ_i which becomes $\sigma_i = \pm 1$ with a probability

$$P_{\boldsymbol{\mu}}(\sigma_i) = \frac{1}{2}\left[1 + \sigma_i \tanh(\omega^i_{\boldsymbol{\mu}})\right]. \tag{9}$$

The 2^L adaptive weights $\omega^i_{\boldsymbol{\mu}}$ determine which random Boolean models player i. If player i who is stimulated at time t with $\boldsymbol{\mu}$ reacts with an output σ_i and receives a reward $r(\sigma_i, \boldsymbol{\mu})$ his weight $\omega^i_{\boldsymbol{\mu}}$ changes according to

$$\Delta\omega^i_{\boldsymbol{\mu}} = \eta r(\sigma_i, \boldsymbol{\mu})\left[\sigma_i - \langle\sigma_i\rangle\right] \tag{10}$$

where η determines the learning rate. This means that the weight $\omega^i_{\boldsymbol{\mu}}$ increases (decreases) if he received a positive (negative) reward and it stops changing if σ_i equals the average value $\langle\sigma_i\rangle = \tanh(\omega^i_{\boldsymbol{\mu}})$ [17].

In the minority game each player receives at time t the same input $\boldsymbol{\mu}$ and should adjust its weights in such a way that the sign of his output σ_i is opposite to the sign of majority $\sum_{i=1}^N \sigma_i$. In the language of the original El Farol bar problem this means, he should have gone to the bar ($\sigma_i = 1$) or stood home ($\sigma_i = -1$) only if it was less (more) than half full, i.e. if $sign\left(\sum_{i=1}^N \sigma_i\right) = -1$ or $(+1)$. Accordingly his reward r becomes, for a specific input $\boldsymbol{\mu}$ which specifies σ_i according to (9)

$$r(\sigma_i, \boldsymbol{\mu}) = -\sigma_i \sum_{j=1}^N \sigma_j\,. \tag{11}$$

Here we used a reward introduced in [13] which measures not only $sign\left(\sum_{i=1}^{N}\sigma_i\right)$ but the strength of the deviation from the minority via $\sum_{j=1}^{N}\sigma_j$. Although in the original game [10, 11] the input $\boldsymbol{\mu}$ was taken to be the correct output during the last L time steps, i.e. the task of the agent was to predict the next correct output, given L previous ones, it has been shown in [18] that the result of the game is the same when the inputs $\boldsymbol{\mu}$ are drawn from a uniform random distribution. The goal of the game is to maximize the overall averaged reward

$$\frac{1}{2^L N}\sum_{i,\boldsymbol{\mu}}\langle r(\sigma_i,\boldsymbol{\mu})\rangle \tag{12}$$

or to minimize the cost function

$$C=-\frac{1}{2^L N}\sum_{i,\boldsymbol{\mu}}\langle r(\sigma_i,\boldsymbol{\mu})\rangle=\frac{1}{2^L}\sum_{\boldsymbol{\mu}}\left\langle\frac{1}{N}\left(\sum_{j=1}^{N}\sigma_j\right)^2\right\rangle . \tag{13}$$

The term $\frac{1}{N}\left(\sum_{j=1}^{N}\sigma_j\right)^2=\frac{1}{N}\sum_{i,j=1}^{N}\sigma_i\sigma_j$ describes the energy of an all-to-all antiferromagnetically coupled Ising model with coupling $J=1$ [19]. This means the task of the agents is to adjust their Booleans in such a way that the averaged energy of this antiferromagnet will be minimized. By performing in (13) the average $\langle ...\rangle$ with $\prod_{i,\mu}P_{\boldsymbol{\mu}}(\sigma_i)$ the cost function becomes

$$C=1-\frac{1}{2^L N}\sum_{\boldsymbol{\mu}}\left\{\sum_i\left[\tanh(\omega^i_{\boldsymbol{\mu}})\right]^2-\left[\sum_i\tanh(\omega^i_{\boldsymbol{\mu}})\right]^2\right\} . \tag{14}$$

If we assume that in the learning equation the change of weights occurs slowly as compared to the spin flips, we can take in (10) the average as well and obtain

$$\Delta\omega^i_{\boldsymbol{\mu}}=\eta\,\langle r(\sigma_i,\boldsymbol{\mu})\,[\sigma_i-\langle\sigma_i\rangle]\rangle \tag{15}$$

$$=-\eta\left[1-\left(\tanh(\omega^i_{\boldsymbol{\mu}})\right)^2\right]\sum_{j\neq i}\tanh(\omega^j_{\boldsymbol{\mu}}) . \tag{16}$$

The right hand side of (15) is proportional to $\partial C/\partial\omega^i_{\boldsymbol{\mu}}$, i.e. the cost function C is a Lyapunov function for the averaged learning dynamics of the weights and its fixed points are the minima of C.

The antiferromagnetically all-to-all coupled Ising model with N spins has $M=N!/\left[((N-1)/2)!\,((N+1)/2)!\right]$ different groundstates. The dynamics for different inputs $\boldsymbol{\mu}$ is uncoupled and the equations of motion (10) could at most detect 2^L different minima. For the case $L = 0$ where each player has only one weight ω^i, the groundstates correspond to strategies where at most $N/2 + 1$ players will attend the bar at the same time i.e. the cost function reduces to $C_{\min} = 1/N$. But the strategies are fixed by $\{\omega^i\}$, i.e. always the same people will visit the bar. This changes already for $L = 1$. In this case two goundstates will be found by the dynamical equations. Random initial conditions $\{\omega^i_\mu(t=0)\}$ generate in the long time limit different solutions $\{\omega^i_{\mu=1}\}$ and $\{\omega^i_{\mu=-1}\}$ and random inputs $\mu = \pm 1$ will generate via (9) jumps of $\boldsymbol{\sigma}(t) = (\sigma_i, ..., \sigma_N)$ between these two different groundstates. This means that two different optimal populations will visit the bar randomly in time. Generally $2^L \leq M$ minima will be visited, triggered by randomly drawn input patterns $\boldsymbol{\mu}$.

These considerations are supported by numerical simulations shown in Fig. 3. There we updated the spins according to (9) and we used for $\Delta\omega^i_{\boldsymbol{\mu}} = \omega^i_{\boldsymbol{\mu}}(t+1) - \omega^i_{\boldsymbol{\mu}}\ (t)$ the equation of motion (10) where we replaced $\langle\sigma_i\rangle$ by its time average $\langle\sigma_i\rangle(t) = \frac{1}{t+1}\sum_{\tau=0}^{t}\sigma_i(t)$. In this way we made sure that each player i received only the allowed information $\sigma_i(t)\sum_{j=1}^{N}\sigma_j(t)$ from its neighbors (and not directly the weights $\{\omega^j_{\boldsymbol{\mu}}\ ,\ j \neq i\}$).

Our method derived above finds the minima of a spin Hamiltonian $H\{\sigma_i\}$ by averaging it in such a way that different spin values become (watching $(\sigma_i)^2 = 1$) replaced by their thermal averages $\tan(\omega_i)$ and the weights ω_i are updated according to $\dot{\omega}_i \propto -\frac{\partial}{\partial\omega_i}H\{\tan(\omega_j)\}$, where the dot denotes the time derivative. The transformation $\langle H\{\sigma_i\}\rangle = H\{\tan(\omega_j)\}$ generates a smooth energy landscape in the variables $\{\omega_j\}$ whose absolute values could be interpreted as site dependent inverse local temperatures. The equation of motion of these weights can also be written as

$$\dot{\omega}_i \propto - - \frac{1}{\cosh^2(\omega_i)}\frac{\partial}{\partial[\tan(\omega_i)]}H\{\tan(\omega_j)\}\,. \tag{17}$$

In the long time limit the weights move to $|\omega_i| \gg 1$. Then the time-schedule of the adaptation process is essentially determined by the prefactor $1/\cosh^2(\omega_i)$ which leads to $\dot{\omega}_i \propto -e^{-|\omega_i|}$ such that $|\omega_i|^{-1} \simeq 1/\log(t)$ for $t \gg 1$. This means that in the long time limit the time dependence of the effective local temperature $|\omega_i|^{-1}$ self-adapts to the optimal schedule for simulated annealing [16].

To summarize: By modeling different agents by different random adaptive Booleans we have found a solution to the minority problem which reduces the fluctuations to their minimal value $C = 1/N$. Our approach maps the solutions of the minority problem to different energy minima of an all-to- all coupled antiferromagnet and different random inputs to the Booleans generate fluctuations between these minima. The weights of the agents adapt dynamically in such a way that the averaged equations of motions describe a gradient descend in a smooth energy landscape. The absolute values of the weights can be interpreted as inverse local temperatures and the self-adapted

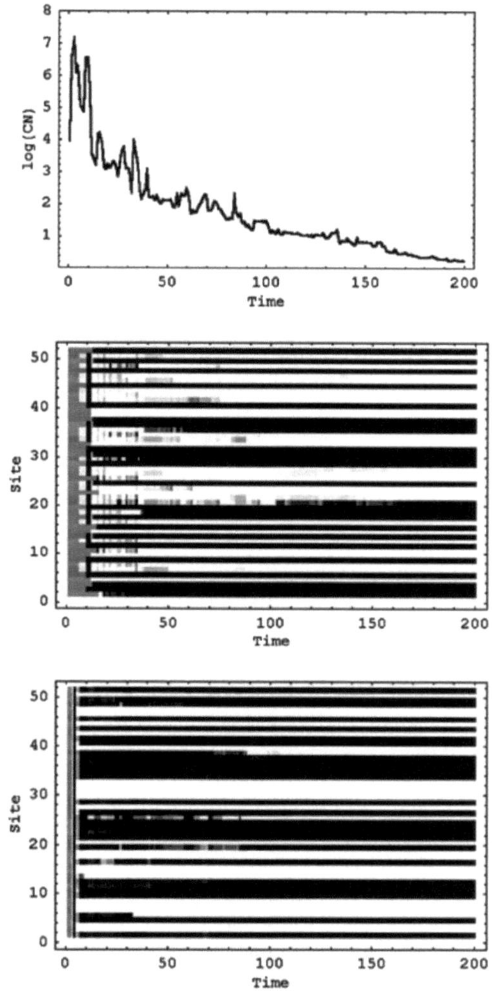

Fig. 3. Typical time evolution of a population of $N = 51$ agents with a learning parameter $\eta = 0.5$. The *upper panel* shows the time dependence of the logarithm of the cost function C (multiplied by N) which converges to zero in the long time limit. The two *lower panels* show the time evolution of the "inner structures" $\tanh(\omega^i_{\mu=+1})$ and $\tanh(\omega^i_{\mu=-1})$ of the agents at "sites" $i = 1...N$ in a linear gray-scale which runs from *white* $= -1$ to *black* $= +1$. It follows that the populations decompose in the long time limit into two almost equally sized groups of agents which stay home (*white*) and which attend the bar (*black*), but the structure of the population is different for $\mu = 1$ (*middle panel*) and $\mu = -1$ (*lowest panel*)

time dependence of the effective local temperatures corresponds to the optimal time schedule which has been found for simulated annealing. Generally the smoothening of spin Hamiltonians $H\{\sigma_i\} \to H\{\tan(\omega_j)\}$ with a subsequent self-regulated gradient descend, introduced here, provides an effective tool for finding minima in spin systems. Our procedure of modeling interacting agents by adaptive random Booleans can also be extended to many other situations where for example competing agents at a financial market are sending signals to each other [20], or to the prisoner's dilemma game where cooperation could emerge between egoistic agents [21]. In these cases we expect that the dynamical behavior will become more complex [22] because there is no common random input $\boldsymbol{\mu}$, but $\boldsymbol{\mu}(t)$ will be related to the dynamical outputs of different agents.

It is a pleasure to thank C. Adami and R. Grigoriev for stimulating discussions, C. Koch for the hospitality extended to me at Caltech and the Volkswagen Foundation for financial support.

References

1. H.G. Schuster: *Complex Adaptive Systems* (Scator Publisher, 2001, see also http://www.theo-physik.uni-kiel.de).
2. F. Crick, C. Koch, Nature, **391**,245(1998); P.J.Fitzgerald, S.S.Hsiao, K.O.Johnson, E.Niebur, Nature, **404**, 187 (2000).
3. Allman: *Evolving Brains* (W.H.Freeman and Company, New York, 1999).
4. H.G. Schuster: *Deterministic Chaos* (Wiley-VCH,Weinheim, 1994).
5. A.R. Damasio: *The Feeling of What Happens p.178*, (Harcourt Brace and Company,New York,1999).
6. M. Nowak, K. Sigmund, Nature, **393**, 573 (1998).
7. S. Kauffman, *The Origins of Order* (Oxford University Press, New York, 1993).
8. P. Bak, K. Sneppen, Phys. Rev. Lett. 71, 4083 (1993); M. Paczuski, S. Maslov, P. Bak, Phys. Rev. **E 53**, 4141 (1996).
9. P.W. Anderson, K. Arrow, D. Pines (eds.), *The Economy as an Evolving Complex System*, (Addison Wesley, Redwood City, CA, (1988).
10. W.B. Arthur, Am. Econ. Assoc. Papers Proc. **84**, 406 (1994); Science **284**, 107 (1999).
11. D. Challet, Y.-C. Zhang, Physica (Amsterdam) **246 A**, 407 (1997).
12. D. Challet, M. Marsili, R. Zecchina, cond-mat/9904392.
13. A. Cavagna, J.P. Garrahan, I. Giardini, D. Sherrington, Phys. Rev. Lett. **83**, 4429 (1999).
14. R. Metzler, W. Kinzel, I. Kanter, Phys. Rev. **E 62**, 2555 (2000).
15. G. Reents, R. Metzler, W. Kinzel, cond-mat/0007351.
16. S. Geman, D. Geman, IEEE Trans. on Pattern Analysis and Machine Intelligence **6**, 721 (1984); S. Kirckpatrick, C.D. Gellat, M.P. Vecchi, Science **220**, 671 (1983).
17. A.G. Barto, P. Anadan, IEEE Trans. on Systems Man and Cybernetics, **15**, 630 (1985).
18. A. Cavagna, Phys.Rev. **E 59**, R 3783 (1999).
19. M. Mezard, G. Parisi, M.A. Virasoro, *Spin Glass Theory and Beyond* (World Scientific, 1987).
20. T. Lux, M. Marchesi, Nature **397**, 498 (1999).

21. J. Hofbauer, K. Sigmund, *Evolutionary Games and Population Dynamics* (Cambridge University Press, Cambridge, England, 1998).
22. M. Paczuski, K.E. Bassler, A. Corral, Phys. Rev. Lett. **84**, 3185 (2000).

Index

Printing: Strauss GmbH, Mörlenbach
Binding: Schäffer, Grünstadt

MIX
Papier aus verantwortungsvollen Quellen
Paper from responsible sources
FSC® C105338

If you have any concerns about our products,
you can contact us on
ProductSafety@springernature.com

In case Publisher is established outside the EU,
the EU authorized representative is:
Springer Nature Customer Service Center GmbH
Europaplatz 3, 69115 Heidelberg, Germany

Printed by Libri Plureos GmbH
in Hamburg, Germany